Beginning Deep Learning with TensorFlow

Work with Keras, MNIST Data Sets, and Advanced Neural Networks

Liangqu Long
Xiangming Zeng

Apress®

Beginning Deep Learning with TensorFlow: Work with Keras, MNIST Data Sets, and Advanced Neural Networks

Liangqu Long
Shenzhen, Guangdong, China

Xiangming Zeng
State College, PA, USA

ISBN-13 (pbk): 978-1-4842-7914-4
https://doi.org/10.1007/978-1-4842-7915-1

ISBN-13 (electronic): 978-1-4842-7915-1

Managing Director, Apress Media LLC: Welmoed Spahr
Acquisitions Editor: Aaron Black
Development Editor: James Markham
Coordinating Editor: Jessica Vakili

Distributed to the book trade worldwide by Springer Science+Business Media New York, 233 Spring Street, 6th Floor, New York, NY 10013. Phone 1-800-SPRINGER, fax (201) 348-4505, e-mail orders-ny@springer-sbm.com, or visit www.springeronline.com. Apress Media, LLC is a California LLC and the sole member (owner) is Springer Science + Business Media Finance Inc (SSBM Finance Inc). SSBM Finance Inc is a **Delaware** corporation.

For information on translations, please e-mail booktranslations@springernature.com; for reprint, paperback, or audio rights, please e-mail bookpermissions@springernature.com.

Apress titles may be purchased in bulk for academic, corporate, or promotional use. eBook versions and licenses are also available for most titles. For more information, reference our Print and eBook Bulk Sales web page at http://www.apress.com/bulk-sales.

Any source code or other supplementary material referenced by the author in this book is available to readers on the Github repository: https://github.com/Apress/Beginning-Deep-Learning-with-TensorFlow. For more detailed information, please visit http://www.apress.com/source-code.

Printed on acid-free paper

Table of Contents

About the Authors

Liangqu Long is a well-known deep learning educator and engineer in China. He is a successfully published author in the topic area with years of experience in teaching machine learning concepts. His two online video tutorial courses, "Deep Learning with PyTorch" and "Deep Learning with TensorFlow 2," have received massive positive comments and allowed him to refine his deep learning teaching methods.

Xiangming Zeng is an experienced data scientist and machine learning practitioner. He has over ten years of experience in using machine learning and deep learning models to solve real-world problems both in academia and industry. Xiangming is familiar with deep learning fundamentals and mainstream machine learning libraries such as TensorFlow and scikit-learn.

About the Technical Reviewer

Vishwesh Ravi Shrimali graduated in 2018 from BITS Pilani, where he studied mechanical engineering. Since then, he has worked with Big Vision LLC on deep learning and computer vision and was involved in creating official OpenCV AI courses. Currently, he is working at Mercedes-Benz Research and Development India Pvt. Ltd. He has a keen interest in programming and artificial intelligence (AI) and has applied that interest in mechanical engineering projects. He has also written multiple blogs on OpenCV and deep learning on LearnOpenCV, a leading blog on computer vision. He has also coauthored *Machine Learning for OpenCV 4* (second edition) by Packt. When he is not writing blogs or working on projects, he likes to go on long walks or play his acoustic guitar.

Acknowledgments

It's been a long journey writing this book. This is definitely a team
effort, and we would like to thank everyone who is part of this process,
especially our families for their support and understanding, the reviewers
of this book for providing valuable feedback, and of course the Apress
crew – especially Aaron and Jessica for working with us and making this
book possible! We are also grateful for the open source and machine
learning communities who shared and continue sharing their knowledge
and great work!

CHAPTER 1

Introduction to Artificial Intelligence

What we want is a machine that can learn from experience.

—Alan Turing

1.1 Artificial Intelligence in Action

Information technology is the third industrial revolution in human history. The popularity of computers, the Internet, and smart home technology has greatly facilitated people's daily lives. Through programming, humans can hand over the interaction logic designed in advance to the machine to execute repeatedly and quickly, thereby freeing humans from simple and tedious repetitive labor. However, for tasks that require a high level of intelligence, such as face recognition, chat robots, and autonomous driving, it is difficult to design clear logic rules. Therefore, traditional programming methods are powerless to those kinds of tasks, whereas artificial intelligence (AI), as the key technology to solve this kind of problem, is very promising.

   With the rise of deep learning algorithms, AI has achieved or even surpassed humanlike intelligence on some tasks. For example, the AlphaGo program has defeated Ke Jie, one of the strongest human Go

© Liangqu Long and Xiangming Zeng 2022
L. Long and X. Zeng, *Beginning Deep Learning with TensorFlow*,
https://doi.org/10.1007/978-1-4842-7915-1_1

players, and OpenAI Five has beaten the champion team OG on the Dota 2 game. In the meantime, practical technologies such as face recognition, intelligent speech, and machine translation have entered people's daily lives. Now our lives are actually surrounded by AI. Although the current level of intelligence that can be reached is still a long way from artificial general intelligence (AGI), we still firmly believe that the era of AI has arrived.

Next, we will introduce the concepts of AI, machine learning, and deep learning, as well as the connections and differences between them.

1.1.1 Artificial Intelligence Explained

AI is a technology that allows machines to acquire intelligent and inferential mechanisms like humans. This concept first appeared at the Dartmouth Conference in 1956. This is a very challenging task. At present, human beings cannot yet have a comprehensive and scientific understanding of the working mechanism of the human brain. It is undoubtedly more difficult to make intelligent machines that can reach the level of the human brain. With that being said, machines that archive similar to or even surpass human intelligence in some way have been proven to be feasible.

How to realize AI is a very broad question. The development of AI has mainly gone through three stages, and each stage represents the exploration footprint of the human trying to realize AI from different angles. In the early stage, people tried to develop intelligent systems by summarizing and generalizing some logical rules and implementing them in the form of computer programs. But such explicit rules are often too simple and are difficult to be used to express complex and abstract concepts and rules. This stage is called the inference period.

In the 1970s, scientists tried to implement AI through knowledge database and reasoning. They built a large and complex expert system to simulate the intelligence level of human experts. One of the biggest difficulties with these explicitly specified rules is that many complex, abstract concepts cannot be implemented in concrete code. For example, the process of human recognition of pictures and understanding of languages cannot be simulated by established rules at all. To solve such problems, a research discipline that allowed machines to automatically learn rules from data, known as machine learning, was born. Machine learning became a popular subject in AI in the 1980s. This is the second stage.

In machine learning, there is a direction to learn complex, abstract logic through neural networks. Research on the direction of neural networks has experienced two ups and downs. Since 2012, the applications of deep neural network technology have made major breakthroughs in fields like computer vision, natural language processing (NLP), and robotics. Some tasks have even surpassed the level of human intelligence. This is the third revival of AI. Deep neural networks eventually have a new name – deep learning. Generally speaking, the essential difference between neural networks and deep learning is not large. Deep learning refers to models or algorithms based on deep neural networks. The relationship between artificial intelligence, machine learning, neural networks, and deep learning is shown in Figure 1-1.

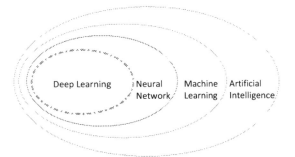

Figure 1-1. *Relationship of artificial intelligence, machine learning, neural networks, and deep learning*

3

1.1.2 Machine Learning

Machine learning can be divided into supervised learning, unsupervised learning, and reinforcement learning, as shown in Figure 1-2.

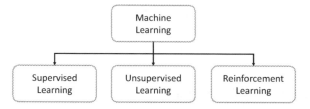

Figure 1-2. *Categories of machine learning*

Supervised Learning. The supervised learning data set contains samples x and sample labels y. The algorithm needs to learn the mapping relationship f_θ: $x \rightarrow y$, where f_θ represents the model function and θ are the parameters of the model. During training, the model parameters θ are optimized by minimizing errors between the model prediction and the real value y, so that the model can have more accurate prediction. Common supervised learning models include linear regression, logistic regression, support vector machines (SVMs), and random forests.

Unsupervised Learning. Collecting labeled data is often more expensive. For a sample-only data set, the algorithm needs to discover the modalities of the data itself. This kind of algorithm is called unsupervised learning. One type of algorithm in unsupervised learning uses itself as a supervised signal, that is, f_θ: $x \rightarrow x$, which is known as self-supervised learning. During training, parameters are optimized by minimizing the error between the model's predicted value $f_\theta(x)$ and itself x. Common unsupervised learning algorithms include self-encoders and generative adversarial networks (GANs).

Reinforcement Learning. This is a type of algorithm that learns strategies for solving problems by interacting with the environment. Unlike supervised and unsupervised learning, reinforcement learning problems do not have a clear "correct" action supervision signal. The algorithm needs to interact with the environment to obtain a lagging reward signal from the environmental feedback. Therefore, it is not possible to calculate the errors between model **Reinforcement Learning** prediction and "correct values" to optimize the network directly. Common reinforcement learning algorithms are Deep Q-Networks (DQNs) and Proximal Policy Optimization (PPO).

1.1.3 Neural Networks and Deep Learning

Neural network algorithms are a class of algorithms that learn from data based on neural networks. They still belong to the category of machine learning. Due to the limitation of computing power and data volume, early neural networks were shallow, usually with around one to four layers. Therefore, the network expression ability was limited. With the improvement of computing power and the arrival of the big data era, highly parallelized graphics processing units (GPUs) and massive data make training of large-scale neural networks possible.

In 2006, Geoffrey Hinton first proposed the concept of deep learning. In 2012, AlexNet, an eight-layer deep neural network, was released and achieved huge performance improvements in the image recognition competition. Since then, neural network models with dozens, hundreds, and even thousands of layers have been developed successively, showing strong learning ability. Algorithms implemented using deep neural networks are generally referred to as deep learning models. In essence, neural networks and deep learning can be considered the same.

Let's simply compare deep learning with other algorithms. As shown in Figure 1-3, rule-based systems usually write explicit logic, which is generally designed for specific tasks and is not suitable for other tasks. Traditional machine learning algorithms artificially design feature detection methods with certain generality, such as SIFT and HOG features. These features are suitable for a certain type of tasks and have certain generality. But the performance highly depends on how to design those features. The emergence of neural networks has made it possible for computers to design those features automatically through neural networks without human intervention. Shallow neural networks typically have limited feature extraction capability, while deep neural networks are capable of extracting high-level, abstract features and have better performance.

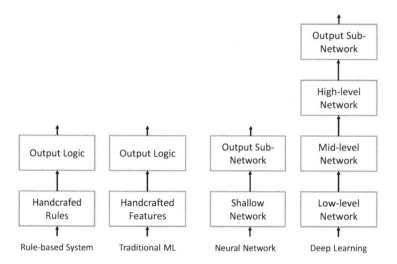

Figure 1-3. *Comparison of deep learning and other algorithms*

1.2 The History of Neural Networks

We divide the development of neural networks into shallow neural network stages and deep learning stages, with 2006 as the dividing point. Before 2006, deep learning developed under the name of neural networks and experienced two ups and two downs. In 2006, Geoffrey Hinton first named deep neural networks as deep learning, which started its third revival.

1.2.1 Shallow Neural Networks

In 1943, psychologist Warren McCulloch and logician Walter Pitts proposed the earliest mathematical model of neurons based on the structure of biological neurons, called MP neuron models after their last name initials. The model $f(x) = h(g(x))$, where $g(x) = \sum_i x_i$, $x_i \in \{0, 1\}$, takes values from $g(x)$ to predict output values as shown in Figure 1-4. If $g(x) \geq 0$, output is 1; if $g(x) < 0$, output is 0. The MP neuron models have no learning ability and can only complete fixed logic judgments.

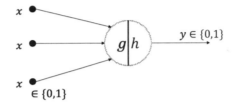

Figure 1-4. *MP neuron model*

In 1958, American psychologist Frank Rosenblatt proposed the first neuron model that can automatically learn weights, called perceptron. As shown in Figure 1-5, the error between the output value o and the true value y is used to adjust the weights of the neurons $\{w_1, w_2, ..., w_n\}$. Frank Rosenblatt then implemented the perceptron model based on the "Mark 1 perceptron" hardware. As shown in Figures 1-6 and 1-7, the input is an image sensor with 400 pixels, and the output has eight nodes. It can successfully identify some English letters. It is generally believed that 1943–1969 is the first prosperous period of artificial intelligence development.

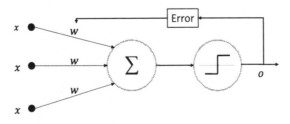

Figure 1-5. *Perceptron model*

7

Figure 1-6. *Frank Rosenblatt and Mark 1 perceptron[1]*

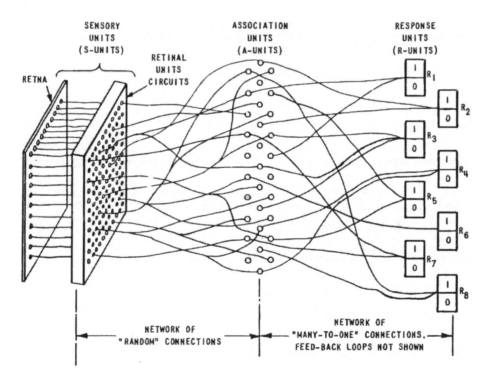

Figure 1-7. *Mark 1 perceptron network architecture[2]*

---

[1] Picture source: https://slideplayer.com/slide/12771753/

[2] Picture source: www.glass-bead.org/article/machines-that-morph-logic/?lang=enview

In 1969, the American scientist Marvin Minsky and others pointed out the main flaw of linear models such as perceptrons in the book *Perceptrons*. They found that perceptrons cannot handle simple linear inseparable problems such as XOR. This directly led to the trough period of perceptron-related research on neural networks. It is generally considered that 1969–1982 was the first winter of artificial intelligence.

Although it was in the trough period of AI, there were still many significant studies published one after another. The most important one is the backpropagation (BP) algorithm, which is still the core foundation of modern deep learning algorithms. In fact, the mathematical idea of the BP algorithm has been derived as early as the 1960s, but it had not been applied to neural networks at that time. In 1974, American scientist Paul Werbos first proposed that the BP algorithm can be applied to neural networks in his doctoral dissertation. Unfortunately, this result has not received enough attention. In 1986, David Rumelhart et al. published a paper using the BP algorithm for feature learning in *Nature*. Since then, the BP algorithm started gaining widespread attention.

In 1982, with the introduction of John Hopfield's cyclically connected Hopfield network, the second wave of artificial intelligence renaissance was started from 1982 to 1995. During this period, convolutional neural networks, recurrent neural networks, and backpropagation algorithms were developed one after another. In 1986, David Rumelhart, Geoffrey Hinton, and others applied the BP algorithm to multilayer perceptrons. In 1989, Yann LeCun and others applied the BP algorithm to handwritten digital image recognition and achieved great success, which is known as LeNet. The LeNet system was successfully commercialized in zip code recognition, bank check recognition, and many other systems. In 1997, one of the most widely used recurrent neural network variants, Long Short-Term Memory (LSTM), was proposed by Jürgen Schmidhuber. In the same year, a bidirectional recurrent neural network was also proposed.

Unfortunately, the study of neural networks has gradually entered a trough with the rise of traditional machine learning algorithms represented by support vector machines (SVMs), which is known as the second winter of artificial intelligence. Support vector machines have a rigorous theoretical foundation, require a small number of training samples, and also have good generalization capabilities. In contrast, neural networks lack theoretical foundation and are hard to interpret. Deep networks are difficult to train, and the performance is normal. Figure 1-8 shows the significant time of AI development between 1943 and 2006.

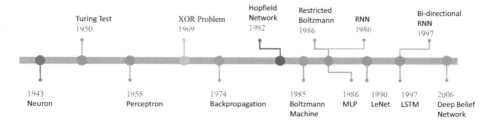

Figure 1-8. *Shallow neural network development timeline*

1.2.2 Deep Learning

In 2006, Geoffrey Hinton et al. found that multilayer neural networks can be better trained through layer-by-layer pre-training and achieved a better error rate than SVM on the MNIST handwritten digital picture data set, turning on the third artificial intelligence revival. In that paper, Geoffrey Hinton first proposed the concept of deep learning. In 2011, Xavier Glorot proposed a Rectified Linear Unit (ReLU) activation function, which is one of the most widely used activation functions now. In 2012, Alex Krizhevsky proposed an eight-layer deep neural network AlexNet, which used the ReLU activation function and Dropout technology to prevent overfitting. At the same time, it abandoned the layer-by-layer pre-training method and directly trained the network on two NVIDIA GTX580 GPUs. AlexNet won the first place in the ILSVRC-2012 picture recognition competition, showing a stunning 10.9% reduction in the top-5 error rate compared with the second place.

Since the AlexNet model was developed, various models have been published successively, including VGG series, GoogleNet series, ResNet series, and DenseNet series. The ResNet series models increase the number of layers in the network to hundreds or even thousands while maintaining the same or even better performance. Its algorithm is simple and universal, and it has significant performance, which is the most representative model of deep learning.

In addition to the amazing results in supervised learning, huge achievements have also been made in unsupervised learning and reinforcement learning. In 2014, Ian Goodfellow proposed generative adversarial networks (GANs), which learned the true distribution of samples through adversarial training to generate samples with higher approximation. Since then, a large number of GAN models have been proposed. The latest image generation models can generate images that reach a degree of fidelity hard to discern from the naked eye. In 2016, DeepMind applied deep neural networks to the field of reinforcement learning and proposed the DQN algorithm, which achieved a level comparable to or even higher than that of humans in 49 games in the Atari game platform. In the field of Go, AlphaGo and AlphaGo Zero intelligent programs from DeepMind have successively defeated human top Go players Li Shishi, Ke Jie, etc. In the multi-agent collaboration Dota 2 game platform, OpenAI Five intelligent programs developed by OpenAI defeated the TI8 champion team OG in a restricted game environment, showing a large number of professional high-level intelligent operations. Figure 1-9 lists the major time points between 2006 and 2019 for AI development.

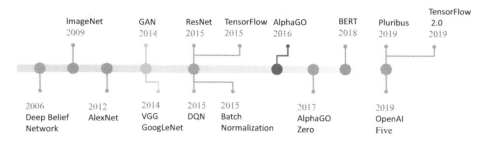

Figure 1-9. *Timeline for deep learning development*

1.3 Deep Learning Characteristics

Compared with traditional machine learning algorithms and shallow neural networks, modern deep learning algorithms usually have the following characteristics.

1.3.1 Data Volume

Early machine learning algorithms are relatively simple and fast to train, and the size of the required dataset is relatively small, such as the Iris flower dataset collected by the British statistician Ronald Fisher in 1936, which contains only three categories of flowers, with each category having 50 samples. With the development of computer technology, the designed algorithms are more and more complex, and the demand for data volume is also increasing. The MNIST handwritten digital picture dataset collected by Yann LeCun in 1998 contains a total of ten categories of numbers from 0 to 9, with up to 7,000 pictures in each category. With the rise of neural networks, especially deep learning networks, the number of network layers is generally large, and the number of model parameters can reach one million, ten million, or even one billion. To prevent overfitting, the size of the training dataset is usually huge. The popularity of modern social media also makes it possible to collect huge amounts of data. For example, the ImageNet dataset released in 2010 included a total of 14,197,122 pictures, and the compressed file size of the entire dataset was 154GB. Figures 1-10 and 1-11 list the number of samples and the size of the data set over time.

Although deep learning has a high demand for large datasets, collecting data, especially collecting labeled data, is often very expensive. The formation of a dataset usually requires manual collection, crawling of raw data and cleaning out invalid samples, and then annotating the data samples with human intelligence, so subjective bias and random errors are inevitably introduced. Therefore, algorithms with small data volume requirement are very hot topics.

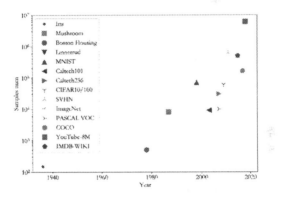

Figure 1-10. *Dataset sample size change over time*

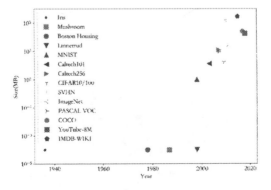

Figure 1-11. *Dataset size change over time*

13

1.3.2 Computing Power

The increase in computing power is an important factor in the third artificial intelligence renaissance. In fact, the basic theory of modern deep learning was proposed in the 1980s, but the real potential of deep learning was not realized until the release of AlexNet based on training on two GTX580 GPUs in 2012. Traditional machine learning algorithms do not have stringent requirements on data volume and computing power like deep learning. Usually, serial training on CPU can get satisfactory results. But deep learning relies heavily on parallel acceleration computing devices. Most of current neural networks use parallel acceleration chips such as NVIDIA GPU and Google TPU to train model parameters. For example, the AlphaGo Zero program needs to be trained on 64 GPUs from scratch for 40 days before surpassing all AlphaGo historical versions. The automatic network structure search algorithm used 800 GPUs to optimize a better network structure.

At present, the deep learning acceleration hardware devices that ordinary consumers can use are mainly from NVIDIA GPU graphics cards. Figure 1-12 illustrates the variation of one billion floating-point operations per second (GFLOPS) of NVIDIA GPU and x86 CPU from 2008 to 2017. It can be seen that the curve of x86 CPU changes relatively slowly, and the floating-point computing capacity of NVIDIA GPU grows exponentially, which is mainly driven by the increasing business of game and deep learning computing.

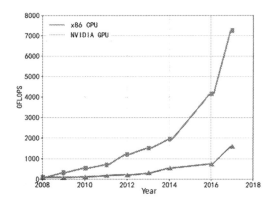

Figure 1-12. *NVIDIA GPU FLOPS change (data source: NVIDIA)*

1.3.3 Network Scale

Early perceptron models and multilayer neural networks only have one or two to four layers, and the network parameters are also around tens of thousands. With the development of deep learning and the improvement of computing capabilities, models such as AlexNet (8 layers), VGG16 (16 layers), GoogleNet (22 layers), ResNet50 (50 layers), and DenseNet121 (121 layers) have been proposed successively, while the size of inputting pictures has also gradually increased from 28×28 to 224×224 to 299×299 and even larger. These changes make the total number of parameters of the network reach ten million levels, as shown in Figure 1-13.

The increase of network scale enhances the capacity of the neural networks correspondingly, so that the networks can learn more complex data modalities and the model performance can be improved accordingly. On the other hand, the increase of the network scale also means that we need more training data and computational power to avoid overfitting.

15

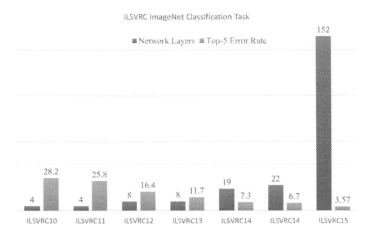

Figure 1-13. *Change of network layers*

1.3.4 General Intelligence

In the past, in order to improve the performance of an algorithm on a certain task, it is often necessary to use prior knowledge to manually design corresponding features to help the algorithm better converge to the optimal solution. This type of feature extraction method is often strongly related to the specific task. Once the scenario changes, these artificially designed features and prior settings cannot adapt to the new scenario, and people often need to redesign the algorithms.

Designing a universal intelligent mechanism that can automatically learn and self-adjust like the human brain has always been the common vision of human beings. Deep learning is one of the algorithms closest to general intelligence. In the computer vision field, previous methods that need to design features for specific tasks and add a priori assumptions have been abandoned by deep learning algorithms. At present, almost all algorithms in image recognition, object detection, and semantic segmentation are based on end-to-end deep learning models, which present good performance and strong adaptability. On the Atari game platform, the DQN algorithm designed by DeepMind can reach human equivalent level in 49 games under the same algorithm, model structure, and hyperparameter settings, showing a certain degree of general intelligence. Figure 1-14 is the network structure of the DQN algorithm. It is not designed for a certain game but can control 49 games on the Atari game platform.

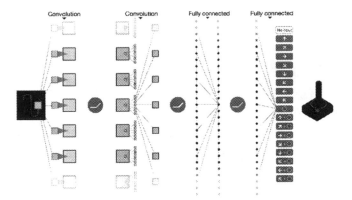

Figure 1-14. *DQN network structure [1]*

1.4 Deep Learning Applications

Deep learning algorithms have been widely used in our daily life, such as voice assistants in mobile phones, intelligent assisted driving in cars, and face payments. We will introduce some mainstream applications of deep learning starting with computer vision, natural language processing, and reinforcement learning.

1.4.1 Computer Vision

Image classification is a common classification problem. The input of the neural network is pictures, and the output value is the probability that the current sample belongs to each category. Generally, the category with the highest probability is selected as the predicted category of the sample.

Image recognition is one of the earliest successful applications of deep learning. Classic neural network models include VGG series, Inception series, and ResNet series.

Object detection refers to the automatic detection of the approximate location of common objects in a picture by an algorithm. It is usually represented by a bounding box and classifies the category information of objects in the bounding box, as shown in Figure 1-15. Common object detection algorithms are RCNN, Fast RCNN, Faster RCNN, Mask RCNN, SSD, and YOLO series.

Semantic segmentation is an algorithm to automatically segment and identify the content in a picture. We can understand semantic segmentation as the classification of each pixel and analyze the category information of each pixel, as shown in Figure 1-16. Common semantic segmentation models include FCN, U-net, SegNet, and DeepLab series.

Figure 1-15. *Object detection example*

Figure 1-16. *Semantic segmentation example*

Video Understanding. As deep learning achieves better results on 2D picture–related tasks, 3D video understanding tasks with temporal dimension information (the third dimension is sequence of frames) are receiving more and more attention. Common video understanding tasks include video classification, behavior detection, and video subject extraction. Common models are C3D, TSN, DOVF, and TS_LSTM.

Image generation learns the distribution of real pictures and samples from the learned distribution to obtain highly realistic generated pictures. At present, common image generation models include VAE series and GAN series. Among them, the GAN series of algorithms have made great progress in recent years. The picture effect produced by the latest GAN model has reached a level where it is difficult to distinguish the authenticity with the naked eye, as shown in Figure 1-17.

In addition to the preceding applications, deep learning has also achieved significant results in other areas, such as artistic style transfer (Figure 1-18), super-resolution, picture de-nosing/hazing, grayscale picture coloring, and many others.

Figure 1-17. *Model-generated image*

Figure 1-18. *Artistic style transfer image*

1.4.2 Natural Language Processing

Machine Translation. In the past, machine translation algorithms were usually based on statistical machine translation models, which were also the technology used by Google's translation system before 2016. In November 2016, Google launched the Google Neural Machine Translation (GNMT) system based on the Seq2Seq model. For the first time, the direct translation technology from source language to target language was realized with 50–90% improvement on multiple tasks. Commonly used machine translation models are Seq2Seq, BERT, GPT, and GPT-2. Among them, the GPT-2 model proposed by OpenAI has about 1.5 billion parameters. At the beginning, OpenAI refused to open-source the GPT-2 model due to technical security reasons.

Chatbot is also a mainstream task of natural language processing. Machines automatically learn to talk to humans, provide satisfactory automatic responses to simple human demands, and improve customer service efficiency and service quality. Chatbot is often used in consulting systems, entertainment systems, and smart homes.

1.4.3 Reinforcement Learning

Virtual Games. Compared with the real environment, virtual game platforms can both train and test reinforcement learning algorithms and can avoid interference from irrelevant factors while also minimizing the cost of experiments. Currently, commonly used virtual game platforms include OpenAI Gym, OpenAI Universe, OpenAI Roboschool, DeepMind OpenSpiel, and MuJoCo, and commonly used reinforcement learning algorithms include DQN, A3C, A2C, and PPO. In the field of Go, the DeepMind AlphaGo program has surpassed human Go experts. In Dota 2 and StarCraft games, the intelligent programs developed by OpenAI and DeepMind have also defeated professional teams under restriction rules.

Robotics. In the real environment, the control of robots has also made some progress. For example, UC Berkeley Lab has made a lot of progress in the areas of imitation learning, meta learning, and few-shot learning in the field of robotics. Boston Dynamics has made gratifying achievements in robot applications. The robots it manufactures perform well on tasks such as complex terrain walking and multi-agent collaboration (Figure 1-19).

Autonomous driving is considered as an application direction of reinforcement learning in the short term. Many companies have invested a lot of resources in autonomous driving, such as Baidu, Uber, and Google. Apollo from Baidu has begun trial operations in Beijing, Xiong'an, Wuhan, and other places. Figure 1-20 shows Baidu's self-driving car Apollo.

Figure 1-19. _Robots from Boston Dynamics[3]_

Figure 1-20. _Baidu's self-driving car Apollo[4]_

1.5 Deep Learning Framework

If a workman wants to be good, he must first sharpen his weapon. After learning about the basic knowledge of deep learning, let's pick the tools used to implement deep learning algorithms.

---

[3] Picture source: www.bostondynamics.com/

[4] Picture source: https://venturebeat.com/2019/01/08/baidu-announces-apollo-3-5-and-apollo-enterprise-says-it-has-over-130-partners/

1.5.1 Major Frameworks

- Theano is one of the earliest deep learning frameworks. It was developed by Yoshua Bengio and Ian Goodfellow. It is a Python-based computing library for positioning low-level operations. Theano supports both GPU and CPU operations. Due to Theano's low development efficiency, long model compilation time, and developers switching to TensorFlow, Theano has now stopped maintenance.

- Scikit-learn is a complete computing library for machine learning algorithms. It has built-in support for common traditional machine learning algorithms, and it has rich documentation and examples. However, scikit-learn is not specifically designed for neural networks. It does not support GPU acceleration, and the implementation of neural network–related layers is also lacking.

- Caffe was developed by Jia Yangqing in 2013. It is mainly used for applications using convolutional neural networks and is not suitable for other types of neural networks. Caffe's main development language is C ++, and it also provides interfaces for other languages such as Python. It also supports GPU and CPU. Due to the earlier development time and higher visibility in the industry, in 2017 Facebook launched an upgraded version of Caffe, Caffe2. Caffe2 has now been integrated into the PyTorch library.

- Torch is a very good scientific computing library, developed based on the less popular programming language Lua. Torch is highly flexible, and it is easy to implement a custom network layer, which is also an excellent gene inherited by PyTorch. However, due to the small number of Lua language users, Torch has been unable to obtain mainstream applications.

- MXNet was developed by Chen Tianqi and Li Mu and is the official deep learning framework of Amazon. It adopts a mixed method of imperative programming and symbolic programming, which has high flexibility, fast running speed, and rich documentation and examples.

- PyTorch is a deep learning framework launched by Facebook based on the original Torch framework using Python as the main development language. PyTorch borrowed the design style of Chainer and adopted imperative programming, which made it very convenient to build and debug the network. Although PyTorch was only released in 2017, due to its sophisticated and compact interface design, PyTorch has received wide acclaim in the academic world. After the 1.0 version, the original PyTorch and Caffe2 were merged to make up for PyTorch's deficiencies in industrial deployment. Overall, PyTorch is an excellent deep learning framework.

- Keras is a high-level framework implemented based on the underlying operations provided by frameworks such as Theano and TensorFlow. It provides a large number of high-level interfaces for rapid training and testing. For common applications, developing with Keras is very efficient. But because there is no low-level implementation, the underlying framework needs to be abstracted, so the operation efficiency is not high, and the flexibility is average.

- TensorFlow is a deep learning framework released by Google in 2015. The initial version only supported symbolic programming. Thanks to its earlier release and Google's influence in the field of deep learning, TensorFlow quickly became the most popular deep learning framework. However, due to frequent changes in the interface design, redundant functional design, and difficulty in symbolic programming development and debugging, TensorFlow 1.x was once criticized by the industry. In 2019, Google launched the official version of TensorFlow 2, which runs in dynamic graph priority mode and can avoid many defects of the TensorFlow 1.x version. TensorFlow 2 has been widely recognized by the industry.

At present, TensorFlow and PyTorch are the two most widely used deep learning frameworks in industry. TensorFlow has a complete solution and user base in the industry. Thanks to its streamlined and flexible interface design, PyTorch can quickly build and debug networks, which has received rave reviews in academia. After TensorFlow 2 was released, it makes it easier for users to learn TensorFlow and seamlessly deploy models to production. This book uses TensorFlow 2 as the main framework to implement deep learning algorithms.

Here are the connections and differences between TensorFlow and Keras. Keras can be understood as a set of high-level API design specifications. Keras itself has an official implementation of the specifications. The same specifications are also implemented in TensorFlow, which is called the tf.keras module, and tf.keras will be used as the unique high-level interface to avoid interface redundancy. Unless otherwise specified, Keras in this book refers to tf.keras.

1.5.2 TensorFlow 2 and 1.x

TensorFlow 2 is a completely different framework from TensorFlow 1.x in terms of user experience. TensorFlow 2 is not compatible with TensorFlow 1.x code. At the same time, it is very different in programming style and functional interface design. TensorFlow 1.x code needs to rely on artificial migration, and automated migration methods are not reliable. Google is about to stop updating TensorFlow 1.x. It is not recommended to learn TensorFlow 1.x now.

TensorFlow 2 supports the dynamic graph priority mode. You can obtain both the computational graph and the numerical results during the calculation. You can debug the code and print the data in real time. The network is built like a building block, stacked layer by layer, which is in line with software development thinking.

Taking simple addition 2.0 + 4.0 as an example, in TensorFlow 1.x, we need to create a calculation graph first as follows:

```
import tensorflow as tf
# 1. Create computation graph with tf 1.x
# Create 2 input variables with fixed name and type
a_ph = tf.placeholder(tf.float32, name='variable_a')
b_ph = tf.placeholder(tf.float32, name='variable_b')
# Create output operation and name
c_op = tf.add(a_ph, b_ph, name='variable_c')
```

The process of creating a computational graph is analogous to the process of establishing a formula $c = a + b$ through symbols. It only records the computational steps of the formula and does not actually calculate the numerical results. The numerical results can only be obtained by running the output c and assigning values $a = 2.0$ and $b = 4.0$ as follows:

```
# 2.Run computational graph with tf 1.x
# Create running environment
sess = tf.InteractiveSession()
# Initialization
init = tf.global_variables_initializer()
sess.run(init) # Run the initialization
# Run the computation graph and return value to c_numpy
c_numpy = sess.run(c_op, feed_dict={a_ph: 2., b_ph: 4.})
# print out the output
print('a+b=',c_numpy)
```

It can be seen that it is so tedious to perform simple addition operations in TensorFlow 1, let alone to create complex neural network algorithms. This programming method of creating a computational graph and then running it later is called symbolic programming.

Next, we use TensorFlow 2 to complete the same operation as follows:

```
import tensorflow as tf
# Use TensorFlow 2 to run
# 1.Create and initialize variable
a = tf.constant(2.)
b = tf.constant(4.)
# 2.Run and get result directly
print('a+b=',a+b)
```

As you can see, the calculation process is very simple, and there are no extra calculation steps.

The method of getting both computation graphs and numerical results at the same time is called imperative programming, also known as dynamic graph mode. TensorFlow 2 and PyTorch are both developed using dynamic graph priority mode, which is easy to debug. In general, the dynamic graph mode is highly efficient for development, but it may not be as efficient as the static graph mode for running. TensorFlow 2 also supports converting the dynamic graph mode to the static graph mode through tf.function, achieving a win-win situation of both development and operation efficiency. In the remaining part of this book, we use TensorFlow to represent TensorFlow 2 in general.

1.5.3 Demo

The core of deep learning is the design idea of algorithms, and deep learning frameworks are just our tools for implementing algorithms. In the following, we will demonstrate the three core functions of the TensorFlow deep learning framework to help us understand the role of frameworks in algorithm design.

a)   **Accelerated Calculation**

The neural network is essentially composed of a large number of basic mathematical operations such as matrix multiplication and addition. One important function of TensorFlow is to use the GPU to conveniently implement parallel computing acceleration functions. In order to demonstrate the acceleration effect of GPU, we can compare mean running time for multiple matrix multiplications on CPU and GPU as follows.

We create two matrices A and B with shape [1, n] and [n, 1], separately. The size of the matrices can be adjusted using parameter n. The code is as follows:

```
# Create two matrices running on CPU
with tf.device('/cpu:0'):
    cpu_a = tf.random.normal([1, n])
    cpu_b = tf.random.normal([n, 1])
    print(cpu_a.device, cpu_b.device)
# Create two matrices running on GPU
with tf.device('/gpu:0'):
    gpu_a = tf.random.normal([1, n])
    gpu_b = tf.random.normal([n, 1])
    print(gpu_a.device, gpu_b.device)
```

Let's implement the functions of the CPU and GPU operations and measure the computation time of the two functions through the timeit. timeit () function. It should be noted that additional environment initialization work is generally required for the first calculation, so this time cannot be counted. We remove this time through the warm-up session and then measure the calculation time as follows:

```
def cpu_run(): # CPU function
    with tf.device('/cpu:0'):
        c = tf.matmul(cpu_a, cpu_b)
    return c

def gpu_run():# GPU function
    with tf.device('/gpu:0'):
        c = tf.matmul(gpu_a, gpu_b)
    return c
# First calculation needs warm-up
cpu_time = timeit.timeit(cpu_run, number=10)
gpu_time = timeit.timeit(gpu_run, number=10)
print('warmup:', cpu_time, gpu_time)
```

```
# Calculate and print mean running time
cpu_time = timeit.timeit(cpu_run, number=10)
gpu_time = timeit.timeit(gpu_run, number=10)
print('run time:', cpu_time, gpu_time)
```

We plot the computation time under CPU and GPU environments at different matrix sizes as shown in Figure 1-21. It can be seen that when the matrix size is small, the CPU and GPU times are almost the same, which does not reflect the advantages of GPU parallel computing. When the matrix size is larger, the CPU computing time significantly increases, and the GPU takes full advantage of parallel computing without almost any change of computation time.

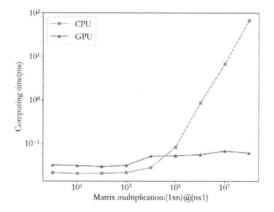

Figure 1-21. *CPU/GPU matrix multiplication time*

b)   **Automatic Gradient Calculation**

When using TensorFlow to construct the forward calculation process, in addition to being able to obtain numerical results, TensorFlow also automatically builds a computational graph. TensorFlow provides automatic differentiation that can calculate the derivative of the output on network parameters without manual derivation. Consider the expression of the following function:

$$y = aw^2 + bw + c$$

The derivative relationship of the output *y* to the variable *w* is

$$\frac{dy}{dw} = 2aw + b$$

Consider the derivative at $(a, b, c, w) = (1, 2, 3, 4)$. We can get $\frac{dy}{dw} = 2 \cdot 1 \cdot 4 + 2 = 10$.

With TensorFlow, we can directly calculate the derivative given the expression of a function without manually deriving the expression of the derivatives. TensorFlow can automatically derive it. The code is implemented as follows:

```
import tensorflow as tf
# Create 4 tensors
a = tf.constant(1.)
b = tf.constant(2.)
c = tf.constant(3.)
w = tf.constant(4.)

with tf.GradientTape() as tape:# Track derivative
    tape.watch([w]) # Add w to derivative watch list
    # Design the function
    y = a * w**2 + b * w + c
# Auto derivative calculation
[dy_dw] = tape.gradient(y, [w])
print(dy_dw) # print the derivative
```

The result of the program is

```
tf.Tensor(10.0, shape=(), dtype=float32)
```

It can be seen that the result of TensorFlow's automatic differentiation is consistent with the result of manual calculation.

c) **Common Neural Network Interface**

In addition to the underlying mathematical functions such as matrix multiplication and addition, TensorFlow also has a series of convenient functions for deep learning systems such as commonly used neural network operation functions, commonly used network layers, network training, model saving, loading, and deployment. Using TensorFlow, you can easily use these functions to complete common production processes, which is efficient and stable.

1.6 Development Environment Installation

After knowing the convenience brought by the deep learning framework, we are now ready to install the latest version of TensorFlow in the local desktop. TensorFlow supports a variety of common operating systems, such as Windows 10, Ubuntu 18.04, and Mac OS. It supports both GPU version running on NVIDIA GPU and CPU version that uses only the CPU to do calculations. We take the most common operating system, Windows 10, NVIDIA GPU, and Python as examples to introduce how to install the TensorFlow framework and other development software.

Generally speaking, the development environment installation is divided into four major steps: the Python interpreter Anaconda, the CUDA acceleration library, the TensorFlow framework, and commonly used editors.

1.6.1 Anaconda Installation

The Python interpreter is the bridge that allows code written in Python to be executed by CPU and is the core software of the Python language. Users can download the appropriate version (Python 3.7 is used here) of the interpreter from `www.python.org/`. After the installation is completed, you can call the python.exe program to execute the source code file written in Python (.py files).

Here we choose to install Anaconda software that integrates a series of auxiliary functions such as the Python interpreter, package management, and virtual environment. We can download Anaconda from `www.anaconda.com/distribution/#download-section` and select the latest version of Python to download and install. As shown in Figure 1-22, check the "Add Anaconda to my PATH environment variable" option, so that you can call the Anaconda program through the command line. As shown in Figure 1-23, the installer asks whether to install the VS Code software together. Select Skip. The entire installation process lasts about 5 minutes, and the specific time depends on the computer performance.

Figure 1-22. *Anaconda installation 1*

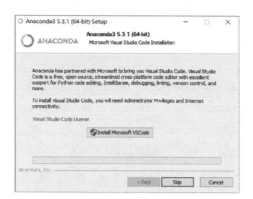

Figure 1-23. *Anaconda installation 2*

After the installation is complete, how can we verify that Anaconda was successfully installed? Pressing the Windows+R key combination on the keyboard, you can bring up the running program dialog box, enter "cmd," and press Enter to open the command-line program "cmd.exe" that comes with Windows. Or click the Start menu and enter "cmd" to find the "cmd.exe" program and open it. Enter the "conda list" command to view the installed libraries in the Python environment. If it is a newly installed Python environment, the listed libraries are all libraries that come with Anaconda, as shown in Figure 1-24. If the "conda list" can pop up a series of library list information normally, the Anaconda software installation is successful. Otherwise, the installation failed, and you need to reinstall.

```
C:\WINDOWS\system32\cmd.exe                                      —   □   ×
Microsoft Windows [Version 10.0.18362.418]
(c) 2019 Microsoft Corporation. All rights reserved.

C:\Users\z390>conda list
# packages in environment at C:\conda:
#
# Name                    Version                   Build  Channel
_ipyw_jlab_nb_ext_conf    0.1.0                     py37_0    defaults
absl-py                   0.8.1                     pypi_0    pypi
adjusttext                0.7.3                     pypi_0    pypi
alabaster                 0.7.12                    py37_0    defaults
anaconda                  2019.07                   py37_0    defaults
anaconda-client           1.7.2                     py37_0    defaults
```

Figure 1-24. *Anaconda installation test*

1.6.2 CUDA Installation

Most of the current deep learning frameworks are based on NVIDIA's GPU graphics card for accelerated calculations, so you need to install the GPU acceleration library CUDA provided by NVIDIA. Before installing CUDA, make sure your computer has an NVIDIA graphics device that supports the CUDA program. If your computer does not have an NVIDIA graphics card – for example, some computer graphics card manufacturers are AMD or Intel – the CUDA program won't work, and you can skip this step and directly install the TensorFlow CPU version.

The installation of CUDA is divided into three steps: CUDA software installation, cuDNN deep neural network acceleration library installation, and environment variable configuration. The installation process is a bit tedious. We will go through them step by step using the Windows 10 system as an example.

CUDA Software Installation Open the official downloading website of the CUDA program: `https://developer.nvidia.com/cuda-10.0-download-archive`. Here we use CUDA 10.0 version: select the Windows platform, x86_64 architecture, 10 system, and exe (local) installation package and then select "Download" to download the CUDA installation software. After the download is complete, open the software. As shown in Figure 1-25, select the "Custom" option and click the "NEXT" button to enter the installation program selection list as shown in Figure 1-26. Here you can select the components that need to be installed and unselect those that do not need to be installed. Under the "CUDA" category, unselect the "Visual Studio Integration" item. Under the "Driver components" category, compare the version number of "Current Version" and "New Version" at the "Display Driver" row. If "Current Version" is greater than "New Version," you need to uncheck the "Display Driver." If "Current Version" is less than or equal to "New Version," leave "Display Driver" checked, as shown in Figure 1-27. After the setup is complete, you can click "NEXT" and follow the instructions to install.

Figure 1-25. *CUDA installation 1*

Figure 1-26. *CUDA installation 2*

After the installation is complete, let's test whether the CUDA software is successfully installed. Open the "cmd" terminal and enter "nvcc -V" to print the current CUDA version information, as shown in Figure 1-28. If the command is not recognized, the installation has failed. We can find the "nvcc.exe" program from the CUDA installation path "C:\Program Files\NVIDIA GPU Computing Toolkit\CUDA\v10.0\bin", as shown in Figure 1-29.

Figure 1-27.   CUDA installation 3

Figure 1-28.   CUDA installation test 1

Figure 1-29.   CUDA installation test 2

cuDNN Neural Network Acceleration Library Installation. CUDA is not a special GPU acceleration library for neural networks; it is designed for a variety of applications that require parallel computing. If you want to accelerate for neural network applications, you need to install an additional cuDNN library. It should be noted that the cuDNN library is not an executable program. You only need to download and decompress the cuDNN file and configure the Path environment variable.

Open the website `https://developer.nvidia.com/cudnn` and select "Download cuDNN." Due to NVIDIA regulations, users need to log in or create a new user to continue downloading. After logging in, enter the cuDNN download interface and check "I Agree To the Terms of the cuDNN Software License Agreement," and the cuDNN version download option will pop up. Select the cuDNN version that matches CUDA 10.0, and click the "cuDNN Library for Windows 10" link to download the cuDNN file, as shown in Figure 1-30. It should be noted that cuDNN itself has a version number, and it also needs to match the CUDA version number.

Figure 1-30. *cuDNN version selection interface*

After downloading the cuDNN file, unzip it and rename the folder "cuda" to "cudnn765". Then copy the "cudnn765" folder to the CUDA installation path "C:\Program Files\NVIDIA GPU Computing Toolkit\ CUDA\v10.0" (Figure 1-31). A dialog box that requires administrator rights may pop up here. Select Continue to paste.

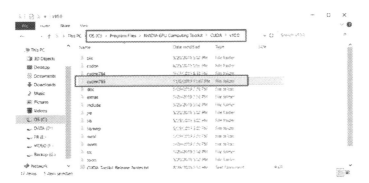

Figure 1-31. *cuDNN installation path*

Environment Variable Configuration. We have completed the installation of cuDNN, but in order for the system to be aware of the location of the cuDNN file, we need to configure the Path environment variable as follows. Open the file browser, right-click "My Computer," select "Properties," select "Advanced system settings," and select "Environment Variables," as shown in Figure 1-32. Select the "Path" environment variable in the "System variables" column and select "Edit," as shown in Figure 1-33. Select "New," enter the cuDNN installation path "C:\Program Files\NVIDIA GPU Computing Toolkit\CUDA\v10.0\cudnn765\bin", and use the "Move Up" button to move this item to the top.

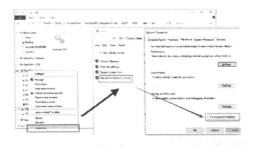

Figure 1-32. *Environment variable configuration 1*

Figure 1-33. *Environment variable configuration 2*

After the CUDA installation is complete, the environment variables should include "C:\Program Files\NVIDIA GPU Computing Toolkit\ CUDA\v10.0\bin", "C:\Program Files\NVIDIA GPU Computing Toolkit\ CUDA\v10.0\libnvvp", and "C:\Program Files\NVIDIA GPU Computing Toolkit\CUDA\v10.0\cudnn765\bin". The preceding path may differ slightly according to the actual path, as shown in Figure 1-34. After confirmation, click "OK" to close all dialog boxes.

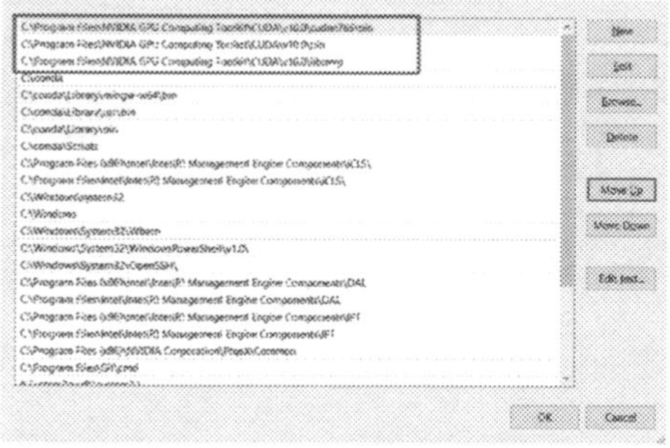

Figure 1-34. *CUDA-related environment variables*

1.6.3 TensorFlow Installation

TensorFlow, like other Python libraries, can be installed using the Python package management tool "pip install" command. When installing TensorFlow, you need to determine whether to install a more powerful

GPU version or a general-performance CPU version based on whether your computer has an NVIDIA GPU graphics card.

```
# Install numpy
pip install numpy
```

With the preceding command, you should be able to automatically download and install the numpy library. Now let's install the latest GPU version of TensorFlow. The command is as follows:

```
# Install TensorFlow GPU version
pip install -U tensorflow
```

The preceding command should automatically download and install the TensorFlow GPU version, which is currently the official version of TensorFlow 2.x. The "-U" parameter specifies that if this package is installed, the upgrade command is executed.

Now let's test whether the GPU version of TensorFlow is successfully installed. Enter "ipython" on the "cmd" command line to enter the ipython interactive terminal, and then enter the "import tensorflow as tf" command. If no errors occur, continue to enter "tf.test.is_gpu_available ()" to test whether the GPU is available. This command will print a series of information. The information beginning with "I" (Information) contains information about the available GPU graphics devices and will return "True" or "False" at the end, indicating whether the GPU device is available, as shown in Figure 1-35. If True, the TensorFlow GPU version is successfully installed; if False, the installation fails. You may need to check the steps of CUDA, cuDNN, and environment variable configuration again or copy the error and seek help from the search engine.

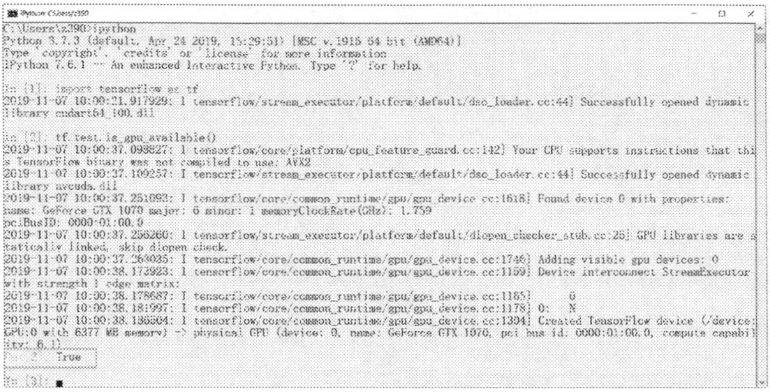

Figure 1-35. *TensorFlow GPU installation test*

If you don't have GPU, you can install the CPU version. The CPU version cannot use the GPU to accelerate calculations, and the computational speed is relatively slow. However, because the models introduced as learning purposes in this book are generally not computationally expensive, the CPU version can also be used. It is also possible to add the NVIDIA GPU device after having a better understanding of deep learning in the future. If the installation of the TensorFlow GPU version fails, we can also use the CPU version directly. The command to install the CPU version is

```
# Install TensorFlow CPU version
pip install -U tensorflow-cpu
```

After installation, enter the "import tensorflow as tf" command in the ipython terminal to verify that the CPU version is successfully installed. After TensorFlow is installed, you can view the version number through "tf.__ version__". Figure 1-36 shows an example. Note that even the code works for all TensorFlow 2.x versions.

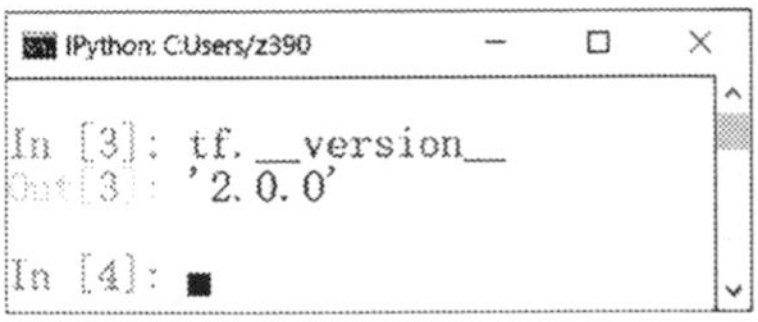

```
IPython: C:Users/z390        —    □    ×

In [3]: tf.__version__
Out[3]: '2.0.0'

In [4]: ▮
```

Figure 1-36. *TensorFlow version test*

The preceding manual process of installing CUDA and cuDNN, configuring the Path environment variable, and installing TensorFlow is the standard installation method. Although the steps are tedious, it is of great help to understand the functional role of each library. In fact, for the novice, you can complete the preceding steps by two commands as follows:

```
# Create virtual environment tf2 with tensorflow-gpu setup
required
# to automatically install CUDA,cuDNN,and TensorFlow GPU
conda create -n tf2 tensorflow-gpu
# Activate tf2 environment
conda activate tf2
```

This quick installation method is called the minimal installation method. This is also the convenience of using the Anaconda distribution. TensorFlow installed through the minimal version requires activation of the corresponding virtual environment before use, which needs to be distinguished from the standard version. The standard version is installed in Anaconda's default environment base and generally does not require manual activation of the base environment.

Common Python libraries can also be installed by default. The command is as follows:

```
# Install common python libraries
pip install -U ipython numpy matplotlib pillow pandas
```

When TensorFlow is running, it will consume all GPU resources by default, which is very computationally unfriendly, especially when the computer has multiple users or programs using GPU resources at the same time. Occupying all GPU resources will make other programs unable to run. Therefore, it is generally recommended to set the GPU memory usage of TensorFlow to the growth mode, that is, to apply for GPU memory resources based on the actual model size. The code implementation is as follows:

```
# Set GPU resource usage method
# Get GPU device list
gpus = tf.config.experimental.list_physical_devices('GPU')
if gpus:
  try:
    # Set GPU usage to growth mode
    for gpu in gpus:
      tf.config.experimental.set_memory_growth(gpu, True)
  except RuntimeError as e:
    # print error
    print(e)
```

1.6.4 Common Editor Installation

There are many ways to write programs in Python. You can use IPython or Jupyter Notebook to write code interactively. You can also use Sublime Text, PyCharm, and VS Code to develop medium and large projects. This book recommends using PyCharm to write and debug code and using VS Code for interactive project development. Both of them are free. Users can download and install them by themselves.

Next, let's start the deep learning journey!

1.7 Summary

1.8 Reference

[1]   V. Mnih, K. Kavukcuoglu, D. Silver, A. A. Rusu,
J. Veness, M. G. Bellemare, A. Graves, M. Riedmiller,
A. K. Fidjeland, G. Ostrovski, S. Petersen, C. Beattie,
A. Sadik, I. Antonoglou, H. King, D. Kumaran,
D. Wierstra, S. Legg, and D. Hassabis, "Human-
level control through deep reinforcement learning,"
Nature, 518, pp. 529–533, 2 2015.

CHAPTER 2

Regression

> Some people worry that artificial intelligence will make us feel inferior, but then, anybody in his right mind should have an inferiority complex every time he looks at a flower.
>
> —Alan Kay

2.1 Neuron Model

An adult brain contains about 100 billion neurons. Each neuron obtains input signals through dendrites and transmits output signals through axons. The neurons are interconnected to form a huge neural network, thus forming the human brain, the basis of perception and consciousness. Figure 2-1 is a typical biological neuron structure. In 1943, the psychologist Warren McCulloch and mathematical logician Walter Pitts proposed a mathematical model of artificial neural networks to simulate the mechanism of biological neurons [1]. This research was further developed by the American neurologist Frank Rosenblatt into the perceptron model [2], which is also the cornerstone of modern deep learning.

© Liangqu Long and Xiangming Zeng 2022
L. Long and X. Zeng, *Beginning Deep Learning with TensorFlow*,
https://doi.org/10.1007/978-1-4842-7915-1_2

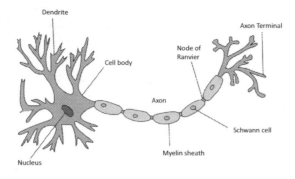

Figure 2-1. *Typical biological neuron structure[1]*

Starting from the structure of biological neurons, we will revisit the exploration of scientific pioneers and gradually unveil the mystery of automatic learning machines.

First, we can abstract the neuron model into the mathematical structure as shown in Figure 2-2 (a). The neuron input vector $x = [x_1, x_2, x_3, ..., x_n]^T$ maps to y through function $f_\theta : x \rightarrow y$, where θ represents the parameters in the function f. Consider a simplified case, such as linear transformation: $f(x) = w^T x + b$. The expanded form is

$$f(x) = w_1 x_1 + w_2 x_2 + w_3 x_3 + ... + w_n x_n + b$$

The preceding calculation logic can be intuitively shown in Figure 2-2 (b).

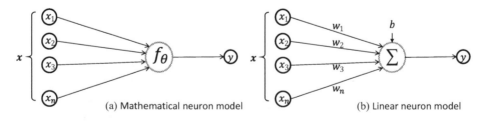

Figure 2-2. *Mathematical neuron model*

---

[1] Source: https://commons.wikimedia.org/wiki/File:Neuron_Hand-tuned.svg

The parameters $\theta = \{w_1, w_2, w_3, ..., w_n, b\}$ determine the state of the neuron, and the processing logic of this neuron can be determined by fixing those parameters. When the number of input nodes $n = 1$ (single input), the neuron model can be further simplified as

$$y = wx + b$$

Then we can plot the change of y as a function of x as shown in Figure 2-3. As the input signal x increases, the output y also increases linearly. Here parameter w can be understood as the slope of the straight line, and b is the bias of the straight line.

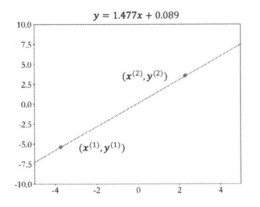

Figure 2-3. *Single-input linear neuron model*

For a certain neuron, the mapping relationship $f_{w, b}$ between x and y is unknown but fixed. Two points can determine a straight line. In order to estimate the value of w and b, we only need to sample any two data points $(x^{(1)}, y^{(1)})$ and $(x^{(2)}, y^{(2)})$ from the straight line in Figure 2-3, where the superscript indicates the data point number:

$$y^{(1)} = wx^{(1)} + b$$

$$y^{(2)} = wx^{(2)} + b$$

If $(x^{(1)}, y^{(1)}) \neq (x^{(2)}, y^{(2)})$, we can solve the preceding equations to get the value of w and b. Let's consider a specific example: $x^{(1)} = 1$, $y^{(1)} = 1.567$, $x^{(2)} = 2$, $y^{(2)} = 3.043$. Substituting the numbers in the preceding formulas gives

$$1.567 = w \cdot 1 + b$$

$$3.043 = w \cdot 2 + b$$

This is the system of binary linear equations that we learned in junior or high school. The analytical solution can be easily calculated using the elimination method, that is, $w = 1.477$, $b = 0.089$.

You can see that we only need two different data points to perfectly solve the parameters of a single-input linear neuron model. For linear neuron models with N input, we only need to sample $N + 1$ different data points. It seems that the linear neuron models can be perfectly resolved. So what's wrong with the preceding method? Considering that there may be observation errors for any sampling point, we assume that the observation error variable ϵ follows a normal distribution $\mathcal{N}(\mu, \sigma^2)$ with μ as mean and σ^2 as variance. Then the samples follow:

$$y = wx + b + \epsilon, \epsilon \sim \mathcal{N}(\mu, \sigma^2)$$

Once the observation error is introduced, even if it is as simple as a linear model, if only two data points are sampled, it may bring a large estimation bias. As shown in Figure 2-4, the data points all have observation errors. If the estimation is based on the two blue rectangular data points, the estimated blue dotted line would have a large deviation from the true orange straight line. In order to reduce the estimation bias introduced by observation errors, we can sample multiple data points $\mathbb{D} = \{(x^{(1)}, y^{(1)}), (x^{(2)}, y^{(2)}), \dots, (x^{(n)}, y^{(n)})\}$ and then find a "best" straight line, so that it minimizes the sum of errors between all sampling points and the straight line.

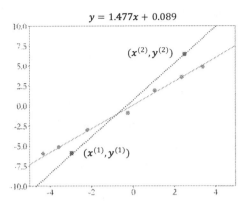

Figure 2-4. *Model with observation errors*

Due to the existence of observation errors, there may not be a straight line that perfectly passes through all the sampling points $\mathbb{D}$. Therefore, we hope to find a "good" straight line close to all sampling points. How to measure "good" and "bad"? A natural idea is to use the mean squared error (MSE) between the predicted value $wx^{(i)} + b$ and the true value $y^{(i)}$ at all sampling points as the total error, that is

$$\mathcal{L} = \frac{1}{n}\sum_{i=1}^{n}\left(wx^{(i)} + b - y^{(i)}\right)^2$$

Then search a set of parameters w^* and b^* to minimize the total error $\mathcal{L}$. The straight line corresponding to the minimal total error is the optimal straight line we are looking for, that is

$$w^*, b^* = \arg\min_{w,b}\frac{1}{n}\sum_{i=1}^{n}\left(wx^{(i)} + b - y^{(i)}\right)^2$$

Here n represents the number of sampling points.

2.2 Optimization Method

Now let's summarize the preceding solution: we need to find the optimal parameters w^* and b^*, so that the input and output meet a linear relationship $y^{(i)} = wx^{(i)} + b$, $i \in [1, n]$. However, due to the existence of observation errors ϵ, it is necessary to sample a data set $\mathbb{D} = \left\{ \left(x^{(1)}, y^{(1)} \right), \left(x^{(2)}, y^{(2)} \right), \ldots, \left(x^{(n)}, y^{(n)} \right) \right\}$, composed of a sufficient number of data samples, to find an optimal set of parameters w^* and b^* to minimize the mean squared error $\mathcal{L} = \frac{1}{n} \sum_{i=1}^{n} \left(wx^{(i)} + b - y^{(i)} \right)^2$.

For a single-input neuron model, only two samples are needed to obtain the exact solution of the equations by the elimination method. This exact solution derived by a strict formula is called an analytical solution. However, in the case of multiple data points ($n \gg 2$), there is probably no analytical solution. We can only use numerical optimization methods to obtain an approximate numerical solution. Why is it called optimization? This is because the computer's calculation speed is very fast. We can use the powerful computing power to "search" and "try" multiple times, thereby reducing the error $\mathcal{L}$ step by step. The simplest optimization method is brute-force search or random experiment. For example, to find the most suitable w^* and b^*, we can randomly sample any w and b from the real number space and calculate the error value $\mathcal{L}$ of the corresponding model. Pick out the smallest error $\mathcal{L}^*$ from all the experiments $\{\mathcal{L}\}$, and its corresponding w^* and b^* are the optimal parameters we are looking for.

This brute-force algorithm is simple and straightforward, but it is extremely inefficient for large-scale, high-dimensional optimization problems. Gradient descent is the most commonly used optimization algorithm in neural network training. With the parallel acceleration capability of powerful graphics processing unit (GPU) chips, it is very suitable for optimizing neural network models with massive data. Naturally it is also suitable for optimizing our simple linear neuron model. Since the gradient descent algorithm is the core algorithm of deep learning, we will first apply the gradient descent algorithm to solve simple neuron models and then detail its application in neural networks in Chapter 7.

With the concept of derivative, if we want to solve the maximum and minimum values of a function, we can simply set the derivative function to be 0 and find the corresponding independent variable values, that is, the stagnation point, and then check the stagnation type. Taking the function $f(x) = x^2 \cdot sin(x)$ as an example, we can plot the function and its derivative in the interval $x \in [-10, 10]$, where the blue solid line is $f(x)$ and the yellow dotted line is $\dfrac{df(x)}{dx}$ as shown in Figure 2-5. It can be seen that the points where the derivative (dashed line) is 0 are the stagnation points, and both the maximum and minimum values of $f(x)$ appear in the stagnation points.

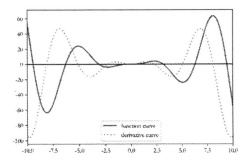

Figure 2-5. *Function f(x) = x² · **sin** (x) and its derivative*

The gradient of a function is defined as a vector of partial derivatives of the function on each independent variable. Considering a three-dimensional function $z = f(x, y)$, the partial derivative of the function with respect to the independent variable x is $\dfrac{\partial z}{\partial x}$, the partial derivative of the function with respect to the independent variable y is recorded as $\dfrac{\partial z}{\partial y}$, and the gradient ∇f is a vector $\left(\dfrac{\partial z}{\partial x}, \dfrac{\partial z}{\partial y} \right)$. Let's look at a specific function $f(x, y) = -(\cos^2 x + \cos^2 y)^2$. As shown in Figure 2-6, the length of the red arrow in the plane represents the modulus of the gradient vector, and the direction of the arrow represents the direction of the gradient vector. It can be seen that the direction of the arrow always points to the function value increasing direction. The steeper the function surface, the longer the length of the arrow, and the larger the modulus of the gradient.

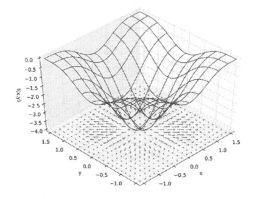

Figure 2-6. *A function and its gradient[2]*

Through the preceding example, we can intuitively feel that the gradient direction of the function always points to the direction in which the function value increases. Then the opposite direction of the gradient should point to the direction in which the function value decreases.

$$x' = x - \eta \cdot \nabla f \qquad (2.1)$$

To take advantage of this property, we just need to follow the preceding equation to iteratively update x'. Then we can get smaller and smaller function values. η is used to scale the gradient vector, which is known as learning rate and generally set to a smaller value, such as 0.01 or 0.001. In particular, for one-dimensional functions, the preceding vector form can be written into a scalar form:

$$x' = x - \eta \cdot \frac{dy}{dx}$$

---

[2] Picture source: https://en.wikipedia.org/wiki/Gradient?oldid=747127712

By iterating and updating x' several times through the preceding formula, the function value y' at x' is always more likely to be smaller than the function value at x.

The method of optimizing parameters by the formula (2.1) is called the gradient descent algorithm. It calculates the gradient ∇f of the function f and iteratively updates the parameters θ to obtain the optimal numerical solution of the parameters θ when the function f reaches its minimum value. It should be noted that model input in deep learning is generally represented as x and the parameters to be optimized are generally represented by θ, w, and b.

Now we will apply the gradient descent algorithm to calculate the optimal parameters w^* and b^* in the beginning of this session. Here the mean squared error function is minimized:

$$\mathcal{L} = \frac{1}{n} \sum_{i=1}^{n} \left(wx^{(i)} + b - y^{(i)} \right)^2$$

The model parameters that need to be optimized are w and b, so we update them iteratively using the following equations:

$$w' = w - \eta \frac{\partial \mathcal{L}}{\partial w}$$

$$b' = b - \eta \frac{\partial \mathcal{L}}{\partial b}$$

2.3 Linear Model in Action

Let's actually train a single-input linear neuron model using the gradient descent algorithm. First, we need to sample multiple data points. For a toy example with a known model, we directly sample from the specified real model:

$$y = 1.477x + 0.089$$

CHAPTER 2   REGRESSION

01. **Sampling data**

In order to simulate the observation errors, we add an independent error variable ϵ to the model, where ϵ follows a Gaussian distribution with a mean value of 0 and a standard deviation of 0.01 (i.e., variance of 0.01^2):

$$y = 1.477x + 0.089 + \epsilon, \epsilon \sim \mathcal{N}\left(0, 0.01^2\right)$$

By randomly sampling $n = 100$ times, we obtain a training data set $\mathbb{D}^{train}$ using the following code:

```
data = [] # A list to save data samples
for i in range(100): # repeat 100 times
    # Randomly sample x from a uniform distribution
    x = np.random.uniform(-10., 10.)
    # Randomly sample from Gaussian distribution
    eps = np.random.normal(0., 0.01)
    # Calculate model output with random errors
    y = 1.477 * x + 0.089 + eps
    data.append([x, y]) # save to data list
data = np.array(data) # convert to 2D Numpy array
```

In the preceding code, we performed 100 samples in a loop, and each time we randomly sampled one data point x from the uniform distribution $U(-10, 10)$ and then randomly sampled noise ϵ from the Gaussian distribution $\mathcal{N}\left(0, 0.1^2\right)$. Finally, we generated the data using the true model and random noise ϵ and save it as a Numpy array.

02. **Calculating the mean squared error**

Now let's calculate the mean squared error on the training set by averaging the squared difference between the predicted value and the true value at each data point. We can achieve this using the following function:

```
def mse(b, w, points):
    # Calculate MSE based on current w and b
    totalError = 0
    # Loop through all points
    for i in range(0, len(points)):
        x = points[i, 0] # Get ith input
        y = points[i, 1] # Get ith output
        # Calculate the total squared error
        totalError += (y - (w * x + b)) ** 2
    # Calculate the mean of the total squared error
    return totalError / float(len(points))
```

03. **Calculating gradient**

According to the gradient descent algorithm, we need to calculate the gradient at each data point $\left(\dfrac{\partial \mathcal{L}}{\partial w}, \dfrac{\partial \mathcal{L}}{\partial b} \right)$. First, consider expanding the mean squared error function $\dfrac{\partial \mathcal{L}}{\partial w}$:

$$\frac{\partial \mathcal{L}}{\partial w} = \frac{\partial \frac{1}{n} \sum_{i=1}^{n} \left(wx^{(i)} + b - y^{(i)} \right)^2}{\partial w} = \frac{1}{n} \sum_{i=1}^{n} \frac{\partial \left(wx^{(i)} + b - y^{(i)} \right)^2}{\partial w}$$

Because

$$\frac{\partial g^2}{\partial w} = 2 \cdot g \cdot \frac{\partial g}{\partial w}$$

we have

$$\frac{\partial \mathcal{L}}{\partial w} = \frac{1}{n} \sum_{i=1}^{n} 2 \left(wx^{(i)} + b - y^{(i)} \right) \cdot \frac{\partial \left(wx^{(i)} + b - y^{(i)} \right)}{\partial w}$$

$$= \frac{1}{n} \sum_{i=1}^{n} 2 \left(wx^{(i)} + b - y^{(i)} \right) \cdot x^{(i)}$$

57

$$= \frac{2}{n} \sum_{i=1}^{n} \left(wx^{(i)} + b - y^{(i)} \right) \cdot x^{(i)} \qquad (2.2)$$

If it is difficult to understand the preceding derivation, you can review the gradient-related courses in mathematics. The details will also be introduced in Chapter 7 of this book. We can remember the final expression of $\frac{\partial \mathcal{L}}{\partial w}$ for now. In the same way, we can derive the expression of the partial derivative $\frac{\partial \mathcal{L}}{\partial b}$:

$$\frac{\partial \mathcal{L}}{\partial b} = \frac{\partial \frac{1}{n} \sum_{i=1}^{n} \left(wx^{(i)} + b - y^{(i)} \right)^2}{\partial b} = \frac{1}{n} \sum_{i=1}^{n} \frac{\partial \left(wx^{(i)} + b - y^{(i)} \right)^2}{\partial b}$$

$$= \frac{1}{n} \sum_{i=1}^{n} 2 \left(wx^{(i)} + b - y^{(i)} \right) \cdot \frac{\partial \left(wx^{(i)} + b - y^{(i)} \right)}{\partial b}$$

$$= \frac{1}{n} \sum_{i=1}^{n} 2 \left(wx^{(i)} + b - y^{(i)} \right) \cdot 1$$

$$= \frac{2}{n} \sum_{i=1}^{n} \left(wx^{(i)} + b - y^{(i)} \right) \qquad (2.3)$$

According to the expressions (2.2) and (2.3), we only need to calculate the mean value of $(wx^{(i)} + b - y^{(i)}) \cdot x^{(i)}$ and $(wx^{(i)} + b - y^{(i)})$ at each data point. The implementation is as follows:

```
def step_gradient(b_current, w_current, points, lr):
    # Calculate gradient and update w and b.
    b_gradient = 0
    w_gradient = 0
    M = float(len(points)) # total number of samples
    for i in range(0, len(points)):
        x = points[i, 0]
        y = points[i, 1]
```

```
    # dL/db:grad_b = 2(wx+b-y) from equation (2.3)
    b_gradient += (2/M) * ((w_current * x + b_current) - y)
    # dL/dw:grad_w = 2(wx+b-y)*x from equation (2.2)
    w_gradient += (2/M) * x * ((w_current * x + b_
    current) - y)
# Update w',b' according to gradient descent algorithm
# lr is learning rate
new_b = b_current - (lr * b_gradient)
new_w = w_current - (lr * w_gradient)
return [new_b, new_w]
```

04. Gradient update

After calculating the gradient of the error function at *w* and *b*, we can update the value of *w* and *b* according to equation (2.1). Training all samples of the data set once is known as one epoch. We can iterate multiple epochs using previous defined functions. The implementation is as follows:

```
def gradient_descent(points, starting_b, starting_w, lr, num_
iterations):
    # Update w, b multiple times
    b = starting_b # initial value for b
    w = starting_w # initial value for w
    # Iterate num_iterations time
for step in range(num_iterations):
        # Update w, b once
        b, w = step_gradient(b, w, np.array(points), lr)
        # Calculate current loss
    loss = mse(b, w, points)
        if step%50 == 0: # print loss and w, b
            print(f"iteration:{step}, loss:{loss},
            w:{w}, b:{b}")
    return [b, w] # return the final value of w and b
```

The main training function is defined as follows:

```python
def main():
    # Load training dataset
    data = []
    for i in range(100):
        x = np.random.uniform(3., 12.)
        # mean=0, std=0.1
        eps = np.random.normal(0., 0.1)
        y = 1.477 * x + 0.089 + eps
        data.append([x, y])
    data = np.array(data)
    lr = 0.01       # learning rate
    initial_b = 0 # initialize b
    initial_w = 0 # initialize w
    num_iterations = 1000
    # Train 1000 times and return optimal w*,b* and
    corresponding loss
    [b, w]= gradient_descent(data, initial_b, initial_w, lr,
    num_iterations)
    loss = mse(b, w, data) # Calculate MSE
    print(f'Final loss:{loss}, w:{w}, b:{b}')
```

After 1000 iterative updates, the final w and b are the "optimal" solution we are looking for. The results are as follows:

```
iteration:0, loss:11.437586448749, w:0.88955725981925,
b:0.02661765516748428
iteration:50, loss:0.111323083882350, w:1.48132089048970,
b:0.58389075913875
iteration:100, loss:0.02436449474995, w:1.479296279074,
b:0.78524532356388
...
```

```
iteration:950, loss:0.01097700897880, w:1.478131231919,
b:0.901113267769968
Final loss:0.010977008978805611, w:1.4781312318924746,
b:0.901113270434582
```

It can be seen that at the 100th iteration, the values of w and b are already close to the real model values. The w and b obtained after 1000 updates are very close to the real model. The mean squared error of the training process is shown in Figure 2-7.

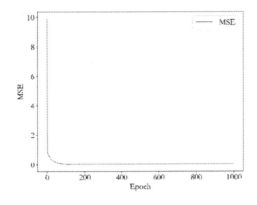

Figure 2-7. *MSE change during the training process*

The preceding example shows the power of the gradient descent algorithm in solving model parameters. It should be noted that for complex nonlinear models, the parameters solved by the gradient descent algorithm may be a local minimum solution instead of a global minimum solution, which is determined by the function non-convexity. However, we found in practice that the performance of the numerical solution obtained by the gradient descent algorithm can often be optimized very well and the corresponding solution can be directly used to approximate the optimal solution.

2.4 Summary

A brief review of our exploration: We first assume that the neuron model with n input is a linear model, and then we can calculate the exact solution of w and b through $n + 1$ samples. After introducing the observation error, we can sample multiple sets of data points and optimize through the gradient descent algorithm to obtain the numerical solution of w and b.

If we look at this problem from another angle, it can actually be understood as a set of continuous value (vector) prediction problems. Given a data set $\mathbb{D}$, we need to learn a model from the data set in order to predict the output value of an unseen sample. After assuming the type of model, the learning process becomes a problem of searching for model parameters. For example, if we assume that the neuron is a linear model, then the training process is the process of searching the linear model parameters w and b. After training, we can use the model output value as an approximation of the real value for any new input. From this perspective, it is a continuous value prediction problem.

In real life, continuous value prediction problems are very common, such as the prediction of stock price trends, the prediction of temperature and humidity in weather forecasts, the prediction of age, the prediction of traffic flow, and so on. We call it a regression problem if its predictions are in a continuous range of real numbers or belong to a certain continuous range of real numbers. In particular, if a linear model is used to approximate the real model, then we call it linear regression, which is a specific implementation of regression problems.

In addition to the continuous value prediction problem, is there a discrete value prediction problem? For example, the prediction of the front and back of a coin can only have two types of prediction: front and back. Given a picture, the type of objects in this picture can only be some discrete categories such as cats or dogs. Problems like those are known as classification problems, which will be introduced in the next chapter.

2.5 References

[1]. W. S. McCulloch and W. Pitts, "A logical calculus of the ideas immanent in nervous activity," *The Bulletin of Mathematical Biophysics*, 5, pp. 115–133, 01 12 1943.

[2]. F. Rosenblatt, *The Perceptron, a Perceiving and Recognizing Automaton Project Para*, Cornell Aeronautical Laboratory, 1957.

CHAPTER 3

Classification

A year spent in artificial intelligence is enough to make one believe in God.

—Alan Perlis

The linear regression model for continuous variable prediction has been introduced previously. Now let's dive into the classification problem. A typical application of the classification problem is to teach computers how to automatically recognize objects in images. Let's consider one of the simplest tasks in image classification: 0–9 digital picture recognition, which is relatively simple and also has a very wide range of applications, such as postal code, courier number, and mobile phone number recognition. We will take 0–9 digital picture recognition as an example to explore how to use machine learning to solve the classification problem.

3.1 Handwritten Digital Picture Dataset

Machine learning needs to learn from the data, so it first needs to collect a large amount of real data. Taking handwritten digital picture recognition as an example, as shown in Figure 3-1, we need to collect a large number of 0–9 digital pictures written by real people. In order to facilitate storage and calculation, the collected pictures are generally scaled to a fixed size, such as 224×224 or 96×96 pixels. These pictures will be used as the input data x.

© Liangqu Long and Xiangming Zeng 2022
L. Long and X. Zeng, *Beginning Deep Learning with TensorFlow*,
https://doi.org/10.1007/978-1-4842-7915-1_3

At the same time, we need to label each image, which will be used as the real value of the image. This label indicates which specific category the image belongs to. For handwritten digital picture recognition, the labels are numbers 0–9 to represent pictures of 0–9.

Figure 3-1. *Handwritten digital pictures*

If we want the model to perform well on new samples, that is, achieve good model generalization ability, then we need to increase the size and diversity of the data set as much as possible, so that the training data set is as close as possible to the real population distribution and the model can also perform well on unseen samples.

In order to facilitate algorithm evaluation, Lecun et al. [1] released a handwritten digital picture data set named MNIST, which contains real handwritten pictures of numbers 0–9. Each number has a total of 7,000 pictures, collected from different writing styles. The total number of pictures is 70,000. Among them, 60,000 pictures are used for training, and the remaining 10,000 pictures are used as a test set.

Because the information in handwritten digital pictures is relatively simple, each picture is scaled to the same size 28 × 28 pixels while retaining only grayscale information, as shown in Figure 3-2. These pictures are written by real people, including rich information such as font size, writing style, and line thickness, to ensure that the distribution of these pictures is as close as possible to the population distribution of real handwritten digital pictures, thereby ensuring model generalization ability.

Figure 3-2. *MNIST dataset examples*

Now let's look at the representation of a picture. A picture contains h rows and w columns with h×w pixel values. Generally, pixel values are integers ranging from 0 to 255 to express color intensity information. For example, 0 represents the lowest intensity, and 255 indicates the highest intensity. If it is a color picture, each pixel contains the intensity information of the three channels R, G, and B, which, respectively, represent the color intensity of colors red, green, and blue. Therefore, unlike a grayscale image, each pixel of a color picture is represented by a one-dimensional vector with three elements, which represent the intensity of R, G, and B colors. As a result, a color image is saved as a tensor with dimension [h, w, 3], while a grayscale picture only needs a two-dimensional matrix with shape [h, w] or a three-dimensional tensor with shape [h, w, 1] to represent its information. Figure 3-3 shows the matrix content of a picture for number 8. It can be seen that the black pixels in the picture are represented by 0 and the grayscale information is represented by 0–255. The whiter pixels in the picture correspond to the larger values in the matrix.

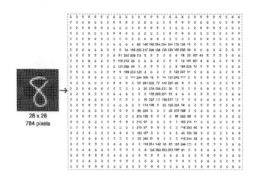

Figure 3-3. *How a picture is represented*[1]

Deep learning frameworks like TensorFlow and PyTorch can easily download, manage, and load the MNIST dataset through a few lines of code. Here we use TensorFlow to automatically download the MNIST dataset and convert it to a Numpy array format:

```
import  os
import  tensorflow as tf
from    tensorflow import keras
from    tensorflow.keras import layers, optimizers, datasets
# load MNIST dataset
(x, y), (x_val, y_val) = datasets.mnist.load_data()
# convert to float type and rescale to [-1, 1]
x = 2*tf.convert_to_tensor(x, dtype=tf.float32)/255.-1
# convert to integer tensor
y = tf.convert_to_tensor(y, dtype=tf.int32)
# one-hot encoding
y = tf.one_hot(y, depth=10)
print(x.shape, y.shape)
```

---

[1] Data source: https://towardsdatascience.com/how-to-teach-a-computer-to-see-with-convolutional-neural-networks-96c120827cd1

```
# create training dataset
train_dataset = tf.data.Dataset.from_tensor_slices((x, y))
# train in batch
train_dataset = train_dataset.batch(512)
```

The load_data () function returns two tuple objects: the first is the training set, and the second is the test set. The first element of the first tuple is the training picture data X, and the second element is the corresponding category number Y. Similar to Figure 3-3, each image in the training set X consists of 28×28 pixels, and there are 60,000 images in the training set X, so the final dimension of X is (60000,28,28). The size of Y is (60,000), representing the 60,000 digital numbers ranging from 0–9. Similarly, the test set contains 10,000 test pictures and corresponding digital numbers with dimensions (10000,28,28) and (10,000) separately.

The MNIST dataset loaded from TensorFlow contains images with values from 0 to 255. In machine learning, it is generally desired that the range of data is distributed in a small range around 0. Therefore, we rescale the pixel range to interval $[-1, 1]$, which will benefit the model optimization process.

The calculation process of each picture is universal. Therefore, we can calculate multiple pictures at once, making full use of the parallel computing power of CPU or GPU. We use a matrix of shape $[h, w]$ to represent a picture. For multiple pictures, we can add one more dimension in front and use a tensor of shape $[b, h, w]$ to represent them. Here b represents the batch size. Color pictures can be represented by a tensor with the shape of $[b, h, w, c]$, where c represents the number of channels, which is 3 for color pictures. TensorFlow's Dataset object can be used to conveniently convert a dataset into batches using the batch() function.

3.2 Build a Model

Recall the biological neuron structure we discussed in the last chapter. We reduce the input vector $x = \left[x_1, x_2, \ldots, x_{d_{in}} \right]^T$ to a single input scalar x, and the model can be expressed as $y = xw + b$. If it is a multi-input, single-output model structure, we need to use the vector form:

$$y = w^T x + b = \left[w_1, w_2, w_3, \ldots, w_{d_{in}} \right] \cdot \left[x_1\ x_2\ x_3 \vdots x_{d_{in}} \right] + b$$

More generally, by combining multiple multi-input, single-output neuron models, we can build a multi-input, multi-output model:

$$y = Wx + b$$

where $x \in R^{d_{in}}$, $b \in R^{d_{out}}$, $y \in R^{d_{out}}$, and $W \in R^{d_{out} \times d_{in}}$.

For multiple-output and batch training, we write the model in batch form:

$$Y = X @ W + b \tag{3.1}$$

where $X \in R^{b \times d_{in}}$, $b \in R^{d_{out}}$, $Y \in R^{b \times d_{out}}$, $W \in R^{d_{in} \times d_{out}}$, d_{in} represents input dimension, and d_{out} indicates output dimension. X has shape $[b, d_{in}]$, b is the number of samples and d_{in} is the length of each sample. W has shape $[d_{in}, d_{out}]$, containing $d_{in} * d_{out}$ parameters. Bias vector b has shape d_{out}. The @ symbol means matrix multiplication. Since the result of the operation $X @ W$ is a matrix of shape $[b, d_{out}]$, it cannot be directly added to the vector b. Therefore, the + sign in batch form needs to support broadcasting, that is, expand the vector b into a matrix of shape $[b, d_{out}]$ by replicating b.

Consider two samples with $d_{in} = 3$ and $d_{out} = 2$. Equation 3.1 is expanded as follows:

$$\left[o_1^{(1)}\ o_2^{(1)}\ o_1^{(2)}\ o_2^{(2)} \right] = \left[x_1^{(1)}\ x_2^{(1)}\ x_3^{(1)}\ x_1^{(2)}\ x_2^{(2)}\ x_3^{(2)} \right] \left[w_{11}\ w_{12}\ w_{21}\ w_{22}\ w_{31}\ w_{32} \right]$$
$$+ \left[b_1\ b_2 \right]$$

where superscripts like (1) and (2) represent the sample index and subscripts such as 1 and 2 indicate the elements of a certain sample vector. The corresponding model structure is shown in Figure 3-4.

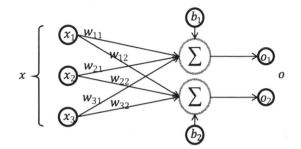

Figure 3-4. *A neural network with three inputs and two outputs*

It can be seen that the matrix form is more concise and clearer, and at the same time, the parallel acceleration capability of matrix calculation can be fully utilized. So how to transform the input and output of the image recognition task into a tensor form?

A grayscale image is stored using a matrix with shape $[h, w]$, and b pictures are stored using a tensor with shape $[b, h, w]$. However, our model can only accept vectors, so we need to flatten the $[h, w]$ matrix into a vector of length $[h \cdot w]$, as shown in Figure 3-5, where the length of the input features $d_{in} = h \cdot w$.

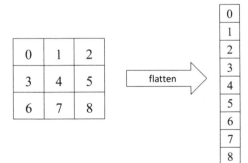

Figure 3-5. *Flatten a matrix*

For the output label y, the digital coding has been introduced previously. It can use a number to represent the label information. The output only needs one number to represent the predicted category value of the network, such as number 1 for cat and number 3 for fish. However, one of the biggest problems with digital coding is that there is a natural order relationship between numbers. For example, if the tags corresponding to 1, 2, and 3 are cat, dog, and fish, there is no order relationship between them, but $1 < 2 < 3$. Therefore, if digital coding is used, it will force the model to learn this unnecessary constraint. In other words, digital coding would change nominal scale (i.e., no specific order) to ordinal scale (i.e., has a specific order), which is not suitable for this case.

So how to solve this problem? The output actually can be set to a set of vectors with length d_{out}, where d_{out} is the same as the number of categories. For example, if the output belongs to the first category, then the corresponding index is set to 1, and the other positions are set to 0. This encoding method is called one-hot encoding. Taking the "cat, dog, fish, and bird" recognition system in Figure 3-6 as an example, all the samples belong to only one of the four categories of "cat, dog, fish, and bird." We use the index positions to indicate the categories of cat, dog, fish, and bird, respectively. For all pictures of cats, their one-hot encoding is $[1, 0, 0, 0]$; for all dog pictures, their one-hot encoding is $[0, 1, 0, 0]$; and so on. One-hot encoding is widely used in classification problems.

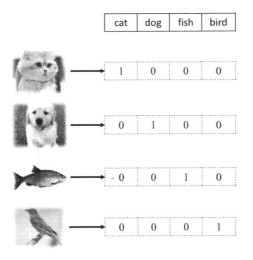

cat	dog	fish	bird

Figure 3-6. *One-hot encoding example*

The total number of categories of handwritten digital pictures is ten, that is, $d_{out} = 10$. For a sample, suppose it belongs to a category *i*, that is, number *i*. Using one-hot encoding, we can represent it using a vector *y* with length 10, where the ith element in this vector is 1 and the rest is 0. For example, the one-hot encoding of picture 0 is $[1, 0, 0, ..., 0]$, and the one-hot encoding of picture 2 is $[0, 0, 1, ..., 0]$, and the one-hot encoding of picture 9 is $[0, 0, 0, ..., 1]$. One-hot encoding is very sparse. Compared with digital encoding, it needs more storage, so digital encoding is generally used for storage. During calculation, digital encoding is converted to one-hot encoding, which can be achieved through the tf.one_hot() function as follows:

```
y = tf.constant([0,1,2,3]) # digits 0-3
y = tf.one_hot(y, depth=10) # one-hot encoding with length 10
print(y)
Out[1]:
tf.Tensor(
```

```
[[1. 0. 0. 0. 0. 0. 0. 0. 0. 0.]  # one-hot encoding of
number 0
 [0. 1. 0. 0. 0. 0. 0. 0. 0. 0.]  # one-hot encoding of
number 1
 [0. 0. 1. 0. 0. 0. 0. 0. 0. 0.]  # one-hot encoding of
number 2
 [0. 0. 0. 1. 0. 0. 0. 0. 0. 0.]], shape=(4, 10),
dtype=float32)
```

Now let's return to the task of handwritten digital picture recognition. The input is a flattened picture vector $x \in R^{784}$, and the output is a vector of length 10 $o \in R^{10}$ corresponding one-hot encoding of a certain number, which forms a multi-input, multi-output linear model $o = W^Tx + b$. We hope that the model output is closer to the real label.

3.3 Error Calculation

For classification problems, our goal is to maximize a certain performance metric, such as accuracy. But when accuracy is used as a loss function, it is in fact indifferentiable. As a result, the gradient descent algorithm cannot be used to optimize the model parameters. The general approach is to establish a smooth and derivable proxy objective function, such as optimizing the distance between the output of the model and the one-hot encoded real label. The model obtained by optimizing the proxy objective function generally also performs well on a testing dataset. Compared with the regression problem, the optimization and evaluation objective functions of the classification problem are inconsistent. The goal of training a model is to find the optimal numerical solution W^* and b^* by optimizing the loss function L:

$$W^*, b^* = arg\ min_{\underset{\smile}{W,b}} L(o, y)$$

For the error calculation of a classification problem, it is more common to use the cross-entropy loss function instead of the mean squared error loss function introduced in the regression problem. We will introduce the cross-entropy loss function in future chapters. Here we still use the mean squared error loss function to solve the handwritten digital picture recognition problem for simplicity. The mean squared error loss function for n samples can be expressed as

$$L(o,y) = \frac{1}{n}\sum_{i=1}^{n}\sum_{j=1}^{10}\left(o_j^{(i)} - y_j^{(i)}\right)^2$$

Now we only need to use the gradient descent algorithm to optimize the loss function to get the optimal solution W and b and then use the obtained model to predict the unknown handwritten digital pictures $x \in D^{test}$.

3.4 Do We Really Solve the Problem?

According to the preceding solution, is the problem of handwritten digital picture recognition really solved perfectly? There are at least two major issues:

- A **linear model** is one of the simplest models in machine learning. It has only a few parameters and can only express linear relationships. The perception and decision-making of complex brains are far more complex than a linear model. Therefore, the linear model is clearly not enough.

- **Complexity** is the model ability to approximate complex distributions. The preceding solution only uses a one-layer neural network model composed of

a small number of neurons. Compared with the 100
billion neuron interconnection structure in the human
brain, its generalization ability is obviously weaker.

Figure 3-7 shows an example of model complexity and data
distribution. The distribution of sampling points with observation errors
is plotted. The actual distribution may be a quadratic parabolic model.
As shown in Figure 3-7 (a), if you use a linear model to fit the data, it is
difficult to learn a good model; if you use a suitable polynomial function
model to learn, such as a quadratic polynomial, you can learn a suitable
model as shown in Figure 3-7 (b). But when the model is too complex,
such as a ten-degree polynomial, it is likely to overfit and hurt the
generalization ability of the model, as shown in Figure 3-7 (c).

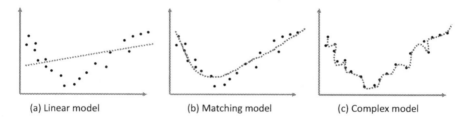

(a) Linear model   (b) Matching model   (c) Complex model

Figure 3-7. *Model complexity*

The multi-neuronal model we currently use is still a linear model with
weak generalization ability. Next, we'll try to solve these two problems.

3.5 Nonlinear Model

Since a linear model is not feasible, we can embed a nonlinear function
in the linear model and convert it to a nonlinear model. We call this
nonlinear function the activation function, which is represented by σ:

$$o = \sigma(Wx + b)$$

Here σ represents a specific nonlinear activation function, such as the Sigmoid function (Figure 3-8 (a)) and the ReLU function (Figure 3-8 (b)).

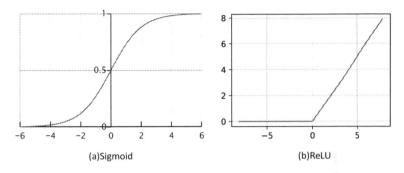

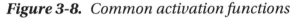

Figure 3-8. *Common activation functions*

The ReLU function only retains the positive part of function $y = x$ and sets the negative part to be zeros. It has a unilateral suppression characteristic. Although simple, the ReLU function has excellent nonlinear characteristics, easy gradient calculation, and stable training process. It is one of the most widely used activation functions for deep learning models. Here we convert the model to a nonlinear model by embedding the ReLU function:

$$o = ReLU(Wx + b)$$

3.6 Model Complexity

To increase the model complexity, we can repeatedly stack multiple transformations such as

$$h_1 = ReLU(W_1 x + b_1)$$

$$h_2 = ReLU(W_2 h_1 + b_2)$$

$$o = W_3 h_2 + b_3$$

In the preceding equations, we take the output value h_1 of the first-layer neuron as the input of the second-layer neuron and then take the output h_2 of the second-layer neuron as the input of the third-layer neuron, and the output of the last-layer neuron is the model output.

As shown in Figure 3-9, the function embedding appears as the connected network one after the other. We call the layer where the input node x is located the input layer. The output of each nonlinear module h_i along with its parameters W_i and b_i is called a network layer. In particular, the layer in the middle of the network is called the hidden layer, and the last layer is called the output layer. This network structure formed by the connection of a large number of neurons is called a neural network. The number of nodes in each layer and the number of layers determine the complexity of the neural network.

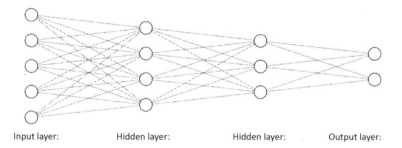

Input layer: Hidden layer: Hidden layer: Output layer:

Figure 3-9. *Three-layer neural network architecture*

Now our network model has been upgraded to a three-layer neural network, which has a descent complexity and good nonlinear generalization ability. Next, let's discuss how to optimize the network parameters.

3.7 Optimization Method

We've introduced the detailed optimization process in Chapter 2 for regression problems. Actually, similar optimization methods can also be used to solve classification problems. For a network model with only one layer, we can directly derive the partial derivative expression of $\dfrac{\partial L}{\partial w}$ and $\dfrac{\partial L}{\partial b}$ and then calculate the gradient for each step and update the parameters w and b using the gradient descent algorithm. However, as complex nonlinear functions are embedded, the number of network layers and the length of data features also increase, the model becomes very complicated, and it is difficult to manually derive the gradient expressions. Besides, once the network structure changes, the model function and corresponding gradient expressions also change. Therefore, it is obviously not feasible to rely on the manual calculation of the gradient.

That is why we have the invention of deep learning frameworks. With the help of autodifferentiation technology, deep learning frameworks can build the neural network's computational graph during the calculation of each layer's output and corresponding loss function and then automatically calculate the gradient $\dfrac{\partial L}{\partial \theta}$ of any parameter θ. Users only need to set up the network structure, and the gradient will automatically be calculated and updated, which is very convenient and efficient to use.

3.8 Hands-On Handwritten Digital Image Recognition

In this section, we will experience the fun of neural networks without introducing too much detail of TensorFlow. The main purpose of this section is not to teach every detail, but to give readers a comprehensive and intuitive experience of neural network algorithms. Let's start experiencing the magical image recognition algorithm!

3.8.1 Build the Network

For the first layer, the input is $x \in R^{784}$, and the output $h_1 \in R^{256}$ is a vector of length 256. We don't need to explicitly write the calculation logic of $h_1 = ReLU(W_1 x + b_1)$. It can be achieved in TensorFlow with a single line of code:

```
# Create one layer with 256 output dimension and ReLU
activation function
layers.Dense(256, activation='relu')
```

Using TensorFlow's Sequential function, we can easily build a multilayer network. For a three-layer network, it can be implemented as follows:

```
# Build a 3-layer network. The output of 1st layer is the input
of 2nd layer.
model = keras.Sequential([
    layers.Dense(256, activation='relu'),
    layers.Dense(128, activation='relu'),
    layers.Dense(10)])
```

The number of output nodes in the three layers is 256, 128, and 10, respectively. Calling model (x) can directly return the output of the last layer.

3.8.2 Model Training

After building the three-layer neural network, given the input x, we can call model(x) to get the model output o and calculate the current loss L:

```
with tf.GradientTape() as tape: # Record the gradient
calculation
    # Flatten x, [b, 28, 28] => [b, 784]
```

```
x = tf.reshape(x, (-1, 28*28))
# Step1. get output [b, 784] => [b, 10]
out = model(x)
# [b] => [b, 10]
y_onehot = tf.one_hot(y, depth=10)
# Calculate squared error, [b, 10]
loss = tf.square(out-y_onehot)
# Calculate the mean squared error, [b]
loss = tf.reduce_sum(loss) / x.shape[0]
```

Then we use the autodifferentiation function from TensorFlow tape.gradient(loss, model.trainable_variables) to calculate all the gradients $\frac{\partial L}{\partial \theta}, \theta \in \{W_1, b_1, W_2, b_2, W_3, b_3\}$。 :

```
# Step3. Calculate gradients w1, w2, w3, b1, b2, b3
grads = tape.gradient(loss, model.trainable_variables)
```

The gradient results are saved using the grads list variable. Then we use the optimizer object to automatically update the model parameters θ according to the gradient update rule.

$$\theta' = \theta - \eta \cdot \frac{\partial L}{\partial \theta}$$

Code is as follows:

```
# Auto gradient calculation
grads = tape.gradient(loss, model.trainable_variables)
# w' = w - lr * grad, update parameters
optimizer.apply_gradients(zip(grads, model.trainable_
variables))
```

After multiple iterations, the learned model f_θ can be used to predict the categorical probability of unknown pictures. The model testing part is not discussed here for now.

The training error curve of the MNIST data set is shown in Figure 3-10. Because the three-layer neural network has relatively strong generalization ability and the task of handwritten digital picture recognition is relatively simple, the training error decreases quickly. In Figure 3-10, the x-axis represents the number of times of iterating over all training samples, which is called epoch. Iterating all training samples once is called one epoch. We can test the model's accuracy and other indicators after several epochs to monitor the model training effect.

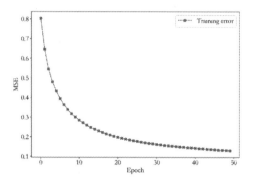

Figure 3-10. *Training error of MNIST dataset*

3.9 Summary

In this chapter, by analogizing a one-layer linear regression model to the classification problem, we proposed a three-layer nonlinear neural network model to solve the problem of handwritten digital picture recognition. After this chapter, everyone should have a good understanding of the (shallow) neural network algorithms. Besides digital picture recognition, classification models also have a variety of applications. For example, classification models are used to separate spam and non-spam emails, conduct sentiment analysis with unstructured text, and process images for segmentation purposes. We will run into more classification problems and applications in future chapters.

Next, we will learn some basic knowledge of TensorFlow and lay a solid foundation for subsequent learning and implementation of deep learning algorithms.

3.10 Reference

[1].   Y. Lecun, L. Bottou, Y. Bengio, and P. Haffner, "Gradient-based learning applied to document recognition," *Proceedings of the IEEE*, 1998.

CHAPTER 4

Basic TensorFlow

I envision that in the future, we may be equivalent to robot
pet dogs, and by then I will also support robots.

—Claude Shannon

TensorFlow is a scientific computing library of deep learning algorithms.
All operations are performed based on tensor objects. Complex neural
network algorithms are essentially a combination of basic operations
such as multiplication and addition of tensors. Therefore, it is important
to get familiar with the basic tensor operation in TensorFlow. Only by
mastering these operations can we realize various complex and novel
network models at will and understand the essence of various models and
algorithms.

4.1 Data Types

The basic data types in TensorFlow include numeric, string, and Boolean.

© Liangqu Long and Xiangming Zeng 2022
L. Long and X. Zeng, *Beginning Deep Learning with TensorFlow*,
https://doi.org/10.1007/978-1-4842-7915-1_4

4.1.1 Numeric

A numeric tensor is the main data format of TensorFlow. According to the dimension, it can be divided into

- Scalar: A single real number, such as 1.2 and 3.4, has a dimension of 0 and a shape of [].

- Vector: An ordered set of real numbers, wrapped by square brackets, such as [1.2] *and* [1.2, 3.4], has a dimension of 1 and a shape of [n] depending on the length.

- Matrix: An ordered set of real numbers in *n* rows and *m* columns, such as [[1, 2], [3, 4]], has a dimension of 2 and a shape of [*n*, *m*].

- Tensor: An array with dimension greater than 2. Each dimension of the tensor is also known as the axis. Generally, each dimension represents specific physical meaning. For example, a tensor with a shape of [2,32,32,3] has four dimensions. If it represents image data, each dimension or axis represents the number of images, image height, image width, and number of color channels, that is, 2 represents two pictures, image height and width are both 32, and 3 represents a total of three color channels, that is, RGB. The number of dimensions of the tensor and the specific physical meaning represented by each dimension need to be defined by users.

In TensorFlow, scalars, vectors, and matrices are also collectively referred to as tensors without distinction. You need to make your own judgment based on the dimension or shape of tensors. The same convention applies in this book.

First, let's create a scalar in TensorFlow. The implementation is as follows:

```
In [1]:
a = 1.2 # Create a scalar in Python
aa = tf.constant(1.2)  # Create a scalar in TensorFlow
type(a), type(aa), tf.is_tensor(aa)
Out[1]:
    (float, tensorflow.python.framework.ops.EagerTensor, True)
```

If we want to use the functions provided by TensorFlow, we must create tensors in the way specified by TensorFlow, not the standard Python language. We can print out the relevant information of tensor x through print (x) or x. The code is as follows:

```
In [2]: x = tf.constant([1,2.,3.3])
x # print out x
Out[2]:
<tf.Tensor: id=165, shape=(3,), dtype=float32,
numpy=array([1. , 2. , 3.3], dtype=float32)>
```

In the output, id is the index of the internal object in TensorFlow, shape represents the shape of the tensor, and dtype represents the numerical precision of the tensor. The numpy() method can return data in the type of Numpy.array, which is convenient for exporting data to other modules in the system.

```
In [3]:  x.numpy()      # Convert TensorFlow (TF) tensor to
numpy array
Out[3]:
array([1. , 2. , 3.3], dtype=float32)
```

Unlike scalars, the definition of a vector must be passed to the tf.constant () function through a list container. For example, here's how to create a vector:

```
In [4]:
a = tf.constant([1.2])  # Create a vector with one element
a, a.shape
Out[4]:
(<tf.Tensor: id=8, shape=(1,), dtype=float32, numpy=array([1.2],
dtype=float32)>,
 TensorShape([1]))
```

Create a vector with three elements:

```
In [5]:
a = tf.constant([1,2, 3.])
a, a.shape
Out[5]:
 (<tf.Tensor: id=11, shape=(3,), dtype=float32,
 numpy=array([1., 2., 3.], dtype=float32)>,
 TensorShape([3]))
```

Similarly, the implementation of a matrix is as follows:

```
In [6]:
a = tf.constant([[1,2],[3,4]])  # Create a 2x2 matrix
a, a.shape
Out[6]:
(<tf.Tensor: id=13, shape=(2, 2), dtype=int32, numpy=
 array([[1, 2],
        [3, 4]])>, TensorShape([2, 2]))
```

A three-dimensional tensor can be defined as

```
In [7]:
a = tf.constant([[[1,2],[3,4]],[[5,6],[7,8]]])
Out[7]:
<tf.Tensor: id=15, shape=(2, 2, 2), dtype=int32, numpy=
array([[[1, 2],
        [3, 4]],
       [[5, 6],
        [7, 8]]])>
```

4.1.2 String

In addition to numeric types, TensorFlow also supports a string type. For example, when processing image data, we can first record the path string of the images and then read the image tensors according to the path through the preprocessing function. A string tensor can be created by passing in a string object, for example:

```
In [8]:
a = tf.constant('Hello, Deep Learning.')
a
Out[8]:
<tf.Tensor: id=17, shape=(), dtype=string, numpy=b'Hello,
Deep Learning.'>
```

The tf.strings module provides common utility functions for strings, such as lower(), join(), length(), and split(). For example, we can convert all strings to lowercase:

```
In [9]:
tf.strings.lower(a)  # Convert string a to lowercase
Out[9]:
<tf.Tensor: id=19, shape=(), dtype=string, numpy=b'hello,
deep learning.'>
```

Deep learning algorithms are mainly based on numerical tensor operations, and string data is used less frequently, so we won't go into too much detail here.

4.1.3 Boolean

In order to facilitate the comparison operation, TensorFlow also supports Boolean tensors. We can easily convert Python standard Boolean data into a TensorFlow internal Boolean as follows:

```
In [10]: a = tf.constant(True)
a
Out[10]:
<tf.Tensor: id=22, shape=(), dtype=bool, numpy=True>
```

Similarly, we can create a Boolean vector as follows:

```
In [1]:
a = tf.constant([True, False])
Out[1]:
<tf.Tensor: id=25, shape=(2,), dtype=bool, numpy=array([ True,
False])>
```

It should be noted that the Tensorflow and standard Python Boolean types are not always equivalent and cannot be used universally, for example:

```
In [1]:
a = tf.constant(True) # Create TF Boolean data
a is True # Whether a is a Python Boolean
Out[1]:
False # TF Boolean is not a Python Boolean
In [2]:
a == True  # Are they numerically the same?
```

```
Out[2]:
<tf.Tensor: id=8, shape=(), dtype=bool, numpy=True> # Yes,
numerically, they are equal.
```

4.2 Numerical Precision

For a numeric tensor, it can be saved with a different byte length corresponding to a different precision. For example, a floating-point number 3.14 can be saved with 16-bit, 32-bit, or 64-bit precision. The longer the bit, the higher the accuracy and, of course, the larger memory space the number occupies. Commonly used precision types in TensorFlow are tf.int16, tf.int32, tf.int64, tf.float16, tf.float32, and tf.float64 where tf.float64 is known as tf.double.

When creating a tensor, we can specify its precision, for example:

```
In [12]:
tf.constant(123456789, dtype=tf.int16)
tf.constant(123456789, dtype=tf.int32)
Out[12]:
<tf.Tensor: id=33, shape=(), dtype=int16, numpy=-13035>
<tf.Tensor: id=35, shape=(), dtype=int32, numpy=123456789>
```

Note that when precision is too low, the data 123456789 overflows, and the wrong result is returned. Generally, tf.int32 and tf.int64 precisions are used more often for integers. For floating-point numbers, high-precision tensors can represent data more accurately. For example, when tf.float32 is used for π, the actual data saved is 3.1415927:

```
In [1]:
import numpy as np
tf.constant(np.pi, dtype=tf.float32)  # Save pi with 32 byte
Out[1]:
<tf.Tensor: id=29, shape=(), dtype=float32, numpy=3.1415927>
```

If we use tf.float64, we can get higher precision:

```
In [2]:
tf.constant(np.pi, dtype=tf.float64)  # Save pi with 64 byte
Out[2]:
<tf.Tensor: id=31, shape=(), dtype=float64,
numpy=3.141592653589793>
```

For most deep learning algorithms, tf.int32 and tf.float32 are able to generally meet the accuracy requirements. Some algorithms that require higher accuracy, such as reinforcement learning, can use tf.int64 and tf.float64.

The tensor precision can be accessed through the dtype property. For some operations that can only handle a specified precision type, the precision type of the input tensor needs to be checked in advance, and the tensor that does not meet the requirements should be converted to the appropriate type using the tf.cast function, for example:

```
In [3]:
a = tf.constant(3.14, dtype=tf.float16)
print('before:',a.dtype)  # Get a's precision
if a.dtype != tf.float32:  # If a is not tf.float32, convert it
to tf.float32.
    a = tf.cast(a,tf.float32)  # Convert a to tf.float32
print('after :',a.dtype)  # Get a's current precision
Out[3]:
before: <dtype: 'float16'>
after : <dtype: 'float32'>
```

When performing type conversion, you need to ensure the legality of the conversion operation. For example, when converting a high-precision tensor into a low-precision tensor, hidden data overflow risks may occur:

```
In [4]:
a = tf.constant(123456789, dtype=tf.int32)
tf.cast(a, tf.int16)  # Convert a to lower precision and we
have overflow
Out[4]:
<tf.Tensor: id=38, shape=(), dtype=int16, numpy=-13035>
```

Conversions between Boolean and integer types are also legal and are common:

```
In [5]:
a = tf.constant([True, False])
tf.cast(a, tf.int32)  # Convert boolean to integers
Out[5]:
<tf.Tensor: id=48, shape=(2,), dtype=int32,
numpy=array([1, 0])>
```

In general, 0 means False and 1 means True during type conversion. In TensorFlow, non-zero numbers are treated as True, for example:

```
In [6]:
a = tf.constant([-1, 0, 1, 2])
tf.cast(a, tf.bool)  # Convert integers to booleans
Out[6]:
<tf.Tensor: id=51, shape=(4,), dtype=bool, numpy=array([ True,
False,  True,  True])>
```

4.3 Tensors to Be Optimized

In order to distinguish tensors that need to calculate gradient information from tensors that do not need to calculate gradient information, TensorFlow adds a special data type to support the recording of gradient information: tf.Variable. tf.Variable adds attributes such as name and trainable on the basis of ordinary tensors to support the construction of computational graphs. Since the gradient operation consumes a large amount of computing resources and automatically updates related parameters, tf.Variable does not need to be encapsulated for tensors that don't need gradient information, such as the input X of a neural network. Instead, tensors that need to calculate the gradient, such as the W and b of neural network layers, need to be wrapped by tf.Variable in order for TensorFlow to track relevant gradient information.

The tf.Variable() function can be used to convert an ordinary tensor into a tensor with gradient information, for example:

```
In [20]:
a = tf.constant([-1, 0, 1, 2])  # Create TF tensor
aa = tf.Variable(a)  # Convert to tf.Variable type
aa.name, aa.trainable # Get tf.Variable properties
Out[20]:
 ('Variable:0', True)
```

The name and trainable attributes are specific for the tf.Variable type. The name attribute is used to name the variables in the computational graph. This naming system is maintained internally by TensorFlow and generally does not require users to do anything about it. The trainable attribute indicates whether the gradient information needs to be recorded for the tensor. When the Variable object is created, the trainable flag is enabled by default. You can set the trainable attribute to be False to avoid recording the gradient information.

In addition to creating tf.Variable tensors through ordinary tensors, you can also create them directly, for example:

```
In [21]:
a = tf.Variable([[1,2],[3,4]])  # Directly create Variable
type tensor
a
```

```
Out[21]:
<tf.Variable 'Variable:0' shape=(2, 2) dtype=int32, numpy=
array([[1, 2],
       [3, 4]])>
```

The tf.Variable tensors can be considered as a special type of ordinary tensors. In fact, ordinary tensors can also be temporarily added to a list of tracking gradient information through the GradientTape.watch() method in order to support the automatic differentiation function.

4.4 Create Tensors

In TensorFlow, you can create tensors in a variety of ways, such as from a Python list, from a Numpy array, or from a known distribution.

4.4.1 Create Tensors from Arrays and Lists

Numpy array and Python list are very important data containers in Python. Many data are loaded into arrays or lists before being converted to tensors. The output data of TensorFlow are also usually exported to arrays or lists, which makes them easy to use for other modules.

The tf.convert_to_tensor function can be used to create a new tensor from a Python list or Numpy array, for example:

```
In [22]:
# Create a tensor from a Python list
tf.convert_to_tensor([1,2.])
Out[22]:
<tf.Tensor: id=86, shape=(2,), dtype=float32, numpy=array([1.,
2.], dtype=float32)>
In [23]:
# Create a tensor from a Numpy array
tf.convert_to_tensor(np.array([[1,2.],[3,4]]))
```

```
Out[23]:
<tf.Tensor: id=88, shape=(2, 2), dtype=float64, numpy=
array([[1., 2.],
       [3., 4.]])>
```

Note that Numpy floating-point arrays store data with 64-bit precision by default. When converting to a tensor type, the precision is tf.float64. You can convert it to tf.float32 when needed. In fact, both tf.constant() and tf.convert_to_tensor() can automatically convert Numpy arrays or Python lists to tensor types.

4.4.2 Create All-0 or All-1 Tensors

Creating tensors with all 0s or 1s is a very common tensor initialization method. Consider linear transformation $y = Wx + b$. The weight matrix W can be initialized with a matrix of all 1s, and b can be initialized with a vector of all 0s. So the linear transformation changes to $y = x$. We can use tf.zeros() or tf.ones() to create all-zero or all-one tensors with arbitrary shapes:

```
In [24]: tf.zeros([]),tf.ones([])
Out[24]:
 (<tf.Tensor: id=90, shape=(), dtype=float32, numpy=0.0>,
 <tf.Tensor: id=91, shape=(), dtype=float32, numpy=1.0>)
```

Create a vector of all 0s and all 1s:

```
In [25]: tf.zeros([1]),tf.ones([1])
Out[25]:
(<tf.Tensor: id=96, shape=(1,), dtype=float32,
numpy=array([0.], dtype=float32)>,
 <tf.Tensor: id=99, shape=(1,), dtype=float32,
 numpy=array([1.], dtype=float32)>)
```

Create a matrix of all zeros:

```
In [26]: tf.zeros([2,2])
Out[26]:
<tf.Tensor: id=104, shape=(2, 2), dtype=float32, numpy=
array([[0., 0.],
       [0., 0.]], dtype=float32)>
```

Create a matrix of all 1s:

```
In [27]: tf.ones([3,2])
Out[27]:
<tf.Tensor: id=108, shape=(3, 2), dtype=float32, numpy=
array([[1., 1.],
       [1., 1.],
       [1., 1.]], dtype=float32)>
```

With tf.zeros_like and tf.ones_like, you can easily create a tensor with all 0s or 1s that is consistent with the shape of another tensor. For example, here's how to create an all-zero tensor with the same shape as the tensor a:

```
In [28]: a = tf.ones([2,3])  # Create a 2x3 tensor with all 1s
tf.zeros_like(a)  # Create a all zero tensor with the same
shape of a
Out[28]:
<tf.Tensor: id=113, shape=(2, 3), dtype=float32, numpy=
array([[0., 0., 0.],
       [0., 0., 0.]], dtype=float32)>
```

Create an all-one tensor with the same shape as the tensor a:

```
In [29]: a = tf.zeros([3,2])  # Create a 3x2 tensor with all
0s tf.ones_like(a)  # Create a all 1 tensor with the same
shape of a
```

```
Out[29]:
<tf.Tensor: id=120, shape=(3, 2), dtype=float32, numpy=
array([[1., 1.],
       [1., 1.],
       [1., 1.]], dtype=float32)>
```

4.4.3 Create a Customized Numeric Tensor

In addition to initializing a tensor with all 0s or 1s, sometimes it is also necessary to initialize the tensor with a specific value, such as –1. With tf.fill(shape, value), we can create a tensor with a specific numeric value, where the dimension is specified by the shape parameter. For example, here's how to create a scalar with element –1:

```
In [30]:tf.fill([], -1)  #
Out[30]:
<tf.Tensor: id=124, shape=(), dtype=int32, numpy=-1>
```

Create a vector with all elements –1:

```
In [31]:tf.fill([1], -1)
Out[31]:
<tf.Tensor: id=128, shape=(1,), dtype=int32, numpy=array([-1])>
```

Create a matrix with all elements 99:

```
In [32]:tf.fill([2,2], 99)  # Create a 2x2 matrix with all 99s
Out[32]:
<tf.Tensor: id=136, shape=(2, 2), dtype=int32, numpy=
array([[99, 99],
       [99, 99]])>
```

4.4.4 Create a Tensor from a Known Distribution

Sometimes, it is very useful to create tensors sampled from common distributions such as normal (or Gaussian) and uniform distributions. For example, in convolutional neural networks, the convolution kernel *W* is usually initialized from a normal distribution to facilitate the training process. In adversarial networks, hidden variables *z* are generally sampled from a uniform distribution.

With tf.random.normal(shape, mean=0.0, stddev=1.0), we can create a tensor with dimension defined by the shape parameter and values sampled from a normal distribution $N(mean, stddev^2)$. For example, here's how to create a tensor from a normal distribution with mean 0 and standard deviation of 1:

```
In [33]: tf.random.normal([2,2])  # Create a 2x2 tensor from a
normal distribution
Out[33]:
<tf.Tensor: id=143, shape=(2, 2), dtype=float32, numpy=
array([[-0.4307344 ,  0.44147003],
       [-0.6563149 , -0.30100572]], dtype=float32)>
```

Create a tensor from a normal distribution with mean of 1 and standard deviation of 2:

```
In [34]: tf.random.normal([2,2], mean=1,stddev=2)
Out[34]:
<tf.Tensor: id=150, shape=(2, 2), dtype=float32, numpy=
array([[-2.2687864, -0.7248812],
       [ 1.2752185,  2.8625617]], dtype=float32)>
```

With tf.random.uniform(shape, minval=0, maxval=None, dtype=tf. float32), we can create a uniformly distributed tensor sampled from the interval [*minval, maxval*). For example, here's how to create a matrix uniformly sampled from the interval [0, 1) with shape of [2, 2]:

```
In [35]: tf.random.uniform([2,2])
Out[35]:
<tf.Tensor: id=158, shape=(2, 2), dtype=float32, numpy=
array([[0.65483284, 0.63064325],
       [0.008816  , 0.81437767]], dtype=float32)>
```

Create a matrix uniformly sampled from an interval [0, 10) with shape of [2, 2]:

```
In [36]: tf.random.uniform([2,2],maxval=10)
Out[36]:
<tf.Tensor: id=166, shape=(2, 2), dtype=float32, numpy=
array([[4.541913  , 0.26521802],
       [2.578913  , 5.126876  ]], dtype=float32)>
```

If we need to uniformly sample integers, we must specify the maxval parameter and set the data type as tf.int*:

```
In [37]:
# Create a integer tensor from a uniform distribution with
interval [0,100)
tf.random.uniform([2,2],maxval=100,dtype=tf.int32)
Out[37]:
<tf.Tensor: id=171, shape=(2, 2), dtype=int32, numpy=
array([[61, 21],
       [95, 75]])>
```

Please notice that these outputs from all random functions may be distinct. However, it does not affect the usage of these functions.

4.4.5 Create a Sequence

When looping or indexing a tensor, it is often necessary to create a
continuous sequence of integers, which can be implemented by the
tf.range() function. The function tf.range(limit, delta=1) can create integer
sequences with delta steps and within interval [0, *limit*). For example,
here's how to create an integer sequence of 0–10 with step of 1:

```
In [38]: tf.range(10)  # 0~10, 10 is not included
Out[38]:
<tf.Tensor: id=180, shape=(10,), dtype=int32, numpy=array([0,
1, 2, 3, 4, 5, 6, 7, 8, 9])>
```

Create an integer sequence between 0 and 10 with step of 2:

```
In [39]: tf.range(10,delta=2) # 10 is not included
Out[39]:
<tf.Tensor: id=185, shape=(5,), dtype=int32, numpy=array([0, 2,
4, 6, 8])>
```

With tf.range(start, limit, delta=1), we can create an integer sequence
within interval [*start, limit*) and step of delta:

```
In [40]: tf.range(1,10,delta=2)  # 1~10, 10 is not included
Out[40]:
<tf.Tensor: id=190, shape=(5,), dtype=int32, numpy=array([1, 3,
5, 7, 9])>
```

4.5 Typical Applications of Tensors

After introducing the properties and creation methods of tensors,
the following will introduce the typical application of tensors in each
dimension, so that readers can intuitively think of their main physical

meaning and purpose and lay the foundation for the study of a series of abstract operations such as the dimensional transformation of subsequent tensors.

This section will inevitably mention the network models or algorithms that will be learned in future chapters. You don't need to fully understand them now, but can have a preliminary impression.

4.5.1 Scalar

In TensorFlow, a scalar is the easiest to understand. It is a simple number with 0 dimension and a shape of []. Typical uses of scalars are the representation of error values and various metrics, such as accuracy, precision, and recall.

Consider the training curve of a model. As shown in Figure 4-1, the x-axis is the number of training steps, and the y-axis is Loss per Query Image error change (Figure 4-1 (a)) and accuracy change (Figure 4-1 (b)), where the loss value and accuracy are scalars generated by tensor calculation.

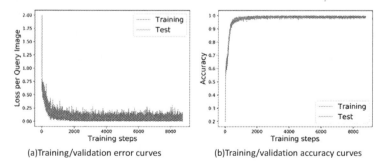

(a)Training/validation error curves   (b)Training/validation accuracy curves

Figure 4-1. *Loss and accuracy curves*

Take the mean squared error function as an example. After tf.keras. losses.mse (or tf.keras.losses.MSE, the same function) returns the error value on each sample and finally takes the average value of the error as the error of the current batch, it automatically becomes a scalar:

```
In [41]:
out = tf.random.uniform([4,10]) # Create a model output example
y = tf.constant([2,3,2,0]) # Create a real observation
y = tf.one_hot(y, depth=10) # one-hot encoding
loss = tf.keras.losses.mse(y, out) # Calculate MSE for
each sample
loss = tf.reduce_mean(loss) # Calculate the mean of MSE
print(loss)
Out[41]:
tf.Tensor(0.19950335, shape=(), dtype=float32)
```

4.5.2 Vector

Vectors are very common in neural networks. For example, in fully connected networks and convolutional neural networks, bias tensors b are represented by vectors. As shown in Figure 4-2, a bias value is added to the output nodes of each fully connected layer, and the bias of all output nodes is represented as a vector form $b = [b_1, b_2]^T$:

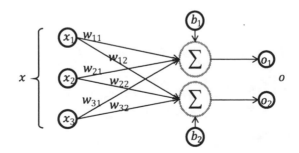

Figure 4-2. *Application of bias vectors*

Considering a network layer of two output nodes, we create a bias vector of length 2 and add back on each output node:

```
In [42]:
# Suppose z is the output of an activation function
z = tf.random.normal([4,2])
b = tf.zeros([2]) # Create a bias vector
z = z + b
Out[42]:
<tf.Tensor: id=245, shape=(4, 2), dtype=float32, numpy=
array([[ 0.6941646 ,  0.4764454 ],
       [-0.34862405, -0.26460952],
       [ 1.5081744 , -0.6493869 ],
       [-0.26224667, -0.78742725]], dtype=float32)>
```

Note that the tensor z with shape [4, 2] and the vector b with shape [2] can be added directly. Why is this? We will reveal it in the "Broadcasting" section later.

For a network layer created through the high-level interface class Dense(), the tensors W and b are automatically created and managed by the class internally. The bias variable b can be accessed through the bias member of the fully connected layer. For example, if a linear network layer with four input nodes and three output nodes is created, then its bias vector b should have length of 3 as follows:

```
In [43]:
fc = layers.Dense(3) # Create a dense layer with output
length of 3
# Create W and b through build function with input nodes of 4
fc.build(input_shape=(2,4))
fc.bias # Print bias vector
```

```
Out[43]:
<tf.Variable 'bias:0' shape=(3,) dtype=float32,
numpy=array([0., 0., 0.], dtype=float32)>
```

It can be seen that the bias member of the class is a vector of length 3 and is initialized to all 0s. This is also the default initialization scheme of the bias b. Besides, the type of the bias vector is Variable, because gradient information is needed for both W and b.

4.5.3 Matrix

A matrix is also a very common type of tensor. For example, the shape of a batch input tensor X of a fully connected layer is $[b, d_{in}]$, where b represents the number of input samples, that is, batch size, and d_{in} represents the length of the input feature. For example, the feature length 4 and the input containing a total of two samples can be expressed as a matrix:

```
x = tf.random.normal([2,4])  # A tensor with 2 samples and 4
features
```

Let the number of output nodes of the fully connected layer be three and then the shape of its weight tensor W [4,3]. We can directly implement a network layer using the tensors X, W and vector b. The code is as follows:

```
In [44]:
w = tf.ones([4,3])
b = tf.zeros([3])
o = x@w+b # @ means matrix multiplication
Out[44]:
<tf.Tensor: id=291, shape=(2, 3), dtype=float32, numpy=
array([[ 2.3506963,  2.3506963,  2.3506963],
       [-1.1724043, -1.1724043, -1.1724043]], dtype=float32)>
```

In the preceding code, both X and W are matrices. The preceding code implements a linear transformation network layer, and the activation function is empty. In general, the network layer $\sigma(X @ W + b)$ is called a fully connected layer, which can be directly implemented by the Dense() class in TensorFlow. In particular, when the activation function σ is empty, the fully connected layer is also called a linear layer. We can create a network layer with four input nodes and three output nodes through the Dense() class and view its weight matrix W through the kernel member of the fully connected layer:

```
In [45]:
fc = layers.Dense(3) # Create fully-connected layer with 3
output nodes
fc.build(input_shape=(2,4)) # Define the input nodes to be 4
fc.kernel # Check kernel matrix W
Out[45]:
<tf.Variable 'kernel:0' shape=(4, 3) dtype=float32, numpy=
array([[ 0.06468129, -0.5146048 , -0.12036425],
       [ 0.71618867, -0.01442951, -0.5891943 ],
       [-0.03011459,  0.578704  ,  0.7245046 ],
       [ 0.73894167, -0.21171576,  0.4820758 ]],
      dtype=float32)>
```

4.5.4 Three-Dimensional Tensor

A typical application of a three-dimensional tensor is to represent a sequence signal. Its format is

$$X = [b, sequence\ length, feature\ length]$$

where the number of sequence signals is b, sequence length represents the number of sampling points or steps in the time dimension, and feature length represents the feature length of each point.

Consider the representation of sentences in natural language processing (NLP), such as the sentiment classification network that evaluates whether a sentence is a positive sentiment or not, as shown in Figure 4-3. In order to facilitate the processing of strings by neural networks, words are generally encoded into vectors of fixed length through the embedding layer. For example, "a" is encoded as a vector of length 3. Then two sentences with equal length (each sentence has five words) can be expressed as a three-dimensional tensor with shape of [2,5,3], where 2 represents the number of sentences, 5 represents the number of words, and 3 represents the length of the encoded word vector. We demonstrate how to represent sentences through the IMDB dataset as follows:

```
In [46]:  # Load IMDB dataset
from tensorflow import keras
(x_train,y_train),(x_test,y_test)=keras.datasets.imdb.load_
data(num_words=10000)
# Convert each sentence to length of 80 words
x_train = keras.preprocessing.sequence.pad_sequences(x_
train,maxlen=80)
x_train.shape
Out [46]: (25000, 80)
```

We can see that the shape of the x_train is [25000, 80], where 25000 represents the number of sentences, 80 represents a total of 80 words in each sentence, and each word is represented by a numeric encoding method. Next, we use the layers.Embedding function to convert each numeric encoded word into a vector of length 100:

```
In [47]: # Create Embedding layer with 100 output length
embedding=layers.Embedding(10000, 100)
# Convert numeric encoded words to word vectors
out = embedding(x_train)
out.shape
Out[47]: TensorShape([25000, 80, 100])
```

Through the embedding layer, the shape of the sentence tensor becomes [25000,80,100], where 100 represents that each word is encoded as a vector of length 100.

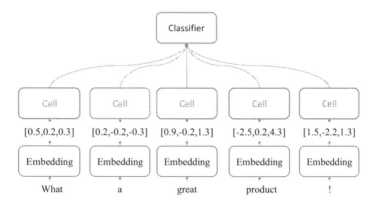

Figure 4-3. *Sentiment classification network*

For a sequence signal with one feature, such as the price of a product within 60 days, only one scalar is required to represent the product price, so the price change of two products can be expressed using a tensor of shape [2, 60]. In order to facilitate the uniform format, the price change can also be expressed as a tensor of shape [2,60,1], where 1 represents the feature length of 1.

4.5.5 Four-Dimensional Tensor

Most times we only use tensors with dimension less than five. For larger-dimension tensors, such as five-dimensional tensor representation in meta learning, a similar principle can be applied. Four-dimensional tensors are widely used in convolutional neural networks. They are used to save feature maps. The format is generally defined as

$$[b,h,w,c]$$

where b indicates the number of input samples; h and w represent the height and width of the feature map, respectively; and c is the number of channels. Some deep learning frameworks also use the format of $[b, c, h, w]$, such as PyTorch. Image data is a type of feature map. A color image with three channels of RGB contains h rows and w columns of pixels. Each point requires three values to represent the color intensity of the RGB channel, so a picture can be expressed using a tensor of shape $[h, w, 3]$. As shown in Figure 4-4, the top picture represents the original image, which contains the intensity information of the three lower channels.

Figure 4-4. *Feature maps of RGB images*

In neural networks, multiple inputs are generally calculated in parallel to improve the computation efficiency, so the tensor of b pictures can be expressed as $[b, h, w, 3]$:

```
In [48]:
# Create 4 32x32 color images
x = tf.random.normal([4,32,32,3])
# Create convolutional layer
layer = layers.Conv2D(16,kernel_size=3)
out = layer(x)
out.shape
Out[48]: TensorShape([4, 30, 30, 16])
```

The convolution kernel tensor is also a four-dimensional tensor, which can be accessed through the kernel member variable:

```
In [49]: layer.kernel.shape
Out[49]: TensorShape([3, 3, 3, 16])
```

4.6 Indexing and Slicing

Part of the tensor data can be extracted through indexing and slicing operations, which are used very frequently.

4.6.1 Indexing

In TensorFlow, the standard Python indexing method is supported, such as $[i][j]$ and comma and ":". Consider four color pictures with 32×32 size (for convenience, most of the tensors are generated by random normal distribution, the same hereinafter). The corresponding tensor has shape [4,32,32,3] as follows:

```
x = tf.random.normal([4,32,32,3])
```

Next, we use the indexing method to read part of the data from the tensor.

- Read the first image data:

```
x = tf.random.normal ([4,32,32,3]) # Create a 4D tensor
In [51]: x[0]   # Index 0 indicates the 1st element in Python
Out[51]:<tf.Tensor: id=379, shape=(32, 32, 3),
dtype=float32, numpy=
array([[[ 1.3005302 ,   1.5301839 ,  -0.32005513],
       [-1.3020388 ,   1.7837263 ,  -1.0747638 ], ...
       [-1.1092019 ,  -1.045254   ,  -0.4980363 ],
       [-0.9099222 ,   0.3947732 ,  -0.10433522]]], dtype=float32)>
```

- Read the second row of the first picture:

```
In [52]: x[0][1]
Out[52]:
<tf.Tensor: id=388, shape=(32, 3), dtype=float32, numpy=
array([[ 4.2904025e-01,  1.0574218e+00,  3.1540772e-01],
       [ 1.5800388e+00, -8.1637271e-02,  6.3147342e-01], ...,
       [ 2.8893018e-01,  5.8003378e-01, -1.1444757e+00],
       [ 9.6100050e-01, -1.0985689e+00,  1.0827581e+00]],
      dtype=float32)>
```

- Read the second row and third column of the first
  picture:

```
In [53]: x[0][1][2]
Out[53]:
<tf.Tensor: id=401, shape=(3,), dtype=float32,
numpy=array([-0.55954427,  0.14497331,  0.46424514],
dtype=float32)>
```

- Select the second row, first column, and second (B)
  channel of the third picture:

```
In [54]: x[2][1][0][1]
Out[54]:
<tf.Tensor: id=418, shape=(), dtype=float32, numpy=-0.84922135>
```

When the number of dimensions is large, the way of using $[i][j]...[k]$ is inconvenient. Instead, we can use the $[i, j, ..., k]$ for indexing. They are equivalent.

- Read the tenth row and third column of the second picture:

```
In [55]: x[1,9,2]
Out[55]:
<tf.Tensor: id=436, shape=(3,), dtype=float32, numpy=array([
1.7487534 , -0.41491988, -0.2944692 ], dtype=float32)>
```

4.6.2 Slicing

A slice of data can be easily extracted using the format *start* : *end* : *step*, where start is the index of the starting position, end is the index of the ending position (excluding), and step is the sampling step size.

Taking the image tensor with shape [4,32,32,3] as an example, we'll explain how to use slicing to obtain data at different positions. For example, read the second and third pictures as follows:

```
In [56]: x[1:3]
Out[56]:
<tf.Tensor: id=441, shape=(2, 32, 32, 3), dtype=float32, numpy=
array([[[[ 0.6920027 ,  0.18658352,  0.0568333 ],
         [ 0.31422952,  0.75933754,  0.26853144],
         [ 2.7898    , -0.4284912 , -0.26247284],...
```

There are many abbreviations for the *start* : *end* : *step* slicing method. The start, end, and step parameters can be selectively omitted as needed. When all of them are omitted like ::, it indicates that the reading is from the beginning to the end and the step size is 1. For example, x [0, ::] means read all the rows of the first picture, where :: means all the rows in the row dimension, which is equivalent to x [0]:

```
In [57]: x[0,::]        # Read 1st picture
Out[57]:
<tf.Tensor: id=446, shape=(32, 32, 3), dtype=float32, numpy=
array([[[ 1.3005302 ,   1.5301839 , -0.32005513],
        [-1.3020388 ,   1.7837263 , -1.0747638 ],
        [-1.1230233 , -0.35004002,  0.01514002],
        ...
```

For brevity, :: can be shortened to a single colon :, for example:

```
In [58]: x[:,0:28:2,0:28:2,:]
Out[58]:
<tf.Tensor: id=451, shape=(4, 14, 14, 3), dtype=float32, numpy=
array([[[[ 1.3005302 ,   1.5301839 , -0.32005513],
         [-1.1230233 , -0.35004002,  0.01514002],
         [ 1.3474811 ,   0.639334  , -1.0826371 ],
         ...
```

The preceding code represents reading all pictures, interlaced sampling, and reading all channel data, which is equivalent to scaling 50% of the original height and width of the picture.

Let's summarize different ways of slicing, where "start" can be omitted when reading from the first element, that is, "start = 0" can be omitted, "end" can be omitted when the last element is taken, and "step" can be omitted when the step length is 1. The details are summarized in Table 4-1.

Table 4-1. *Summary of slicing methods*

Method	Meaning
start:end:step	Read from "start" to "end" (excluding) with step length of "step."
start:end	Read from "start" to "end" (excluding) with step length of 1.
start:	Read from "start" to the end of object with step length of 1.
start::step	Read from "start" to the end of object with step length of "step."
:end:step	Read from the 0th item to "end" (excluding) with step length of "step."
:end	Read from 0th item to "end" (excluding) with step length of 1.
::step	Read from 0th item to the last item with step length of "step."
::	Read all items.
:	Read all items.

In particular, step can be negative. For example, $start : end : -1$ means starting from "start," reading in reverse order, and ending with "end" (excluding), and the index "end" is smaller than "start." Consider a simple sequence vector from 0 to 9, and take the first element in reverse order, excluding the first element:

```
In [59]: x = tf.range(9)  # Create the vector
x[8:0:-1]  # Reverse slicing
Out[59]:
<tf.Tensor: id=466, shape=(8,), dtype=int32, numpy=array([8, 7,
6, 5, 4, 3, 2, 1])>
```

Fetch all elements in reverse order as follows:

```
In [60]: x[::-1]
Out[60]:
<tf.Tensor: id=471, shape=(9,), dtype=int32, numpy=array([8, 7,
6, 5, 4, 3, 2, 1, 0])>
```

Reverse sampling every two items is implemented as follows:

```
In [61]: x[::-2]
Out[61]:
<tf.Tensor: id=476, shape=(5,), dtype=int32, numpy=array([8, 6,
4, 2, 0])>
```

Read all the channels of each picture, where both rows and columns are sampled every two elements in reverse order. The implementation is as follows:

```
In [62]: x = tf.random.normal([4,32,32,3])
x[0,::-2,::-2]
Out[62]:
<tf.Tensor: id=487, shape=(16, 16, 3), dtype=float32, numpy=
array([[[ 0.63320625,  0.0655185 ,   0.19056146],
        [-1.0078577 , -0.61400175,  0.61183935],
        [ 0.9230892 , -0.6860094 , -0.01580668],
        ...
```

When the tensor has large dimensions, the dimensions that do not need to be sampled generally use a single colon ":" to indicate that all elements are selected. As a result, a lot of ":" may appear. Consider the image tensor with shape [4,32,32,3]. When the data on the green channel needs to be read, all the previous dimensions are extracted as

```
In [63]: x[:,:,:,1]  # Read data on Green channel
Out[63]:
<tf.Tensor: id=492, shape=(4, 32, 32), dtype=float32, numpy=
array([[[ 0.575703  ,  0.11028383, -0.9950867 ,
..., 0.38083118, -0.11705163, -0.13746642],
        ...
```

In order to avoid the situation of too many colons like $x[:, :, :, 1]$, we can use the symbol "⋯" to take all the data in multiple dimensions, where the number of dimensions needs to be automatically inferred according to the rules: When the symbol ⋯ appears in slice mode, the dimension to the left of "⋯" will be automatically aligned to the left maximum. The dimension to the right of the symbol "⋯" will be automatically aligned to the far right. The system will automatically infer the number of dimensions represented by the symbol "⋯". The details are summarized in Table 4-2.

Table 4-2. *"..." slicing method summary*

Method	Meaning
a,⋯,b	Select 0 to a for dimension a, b to end for dimension b, and all elements for other dimensions.
a,⋯	Select 0 to a for dimension a and all elements for other dimensions.
⋯,b	Select b to end for dimension b and all elements for other dimensions.
⋯	Read all elements.

We list more examples as follows:

- Read the green and blue channel data of the first and second pictures:

```
In [64]: x[0:2,...,1:]
Out[64]:
<tf.Tensor: id=497, shape=(2, 32, 32, 2), dtype=float32, numpy=
array([[[[ 0.575703  ,  0.8872789 ],
        [ 0.11028383, -0.27128693],
        [-0.9950867 , -1.7737272 ],
        ...
```

- Read the last two pictures:

```
In [65]: x[2:,...]  # equivalent to x[2:]
Out[65]:
<tf.Tensor: id=502, shape=(2, 32, 32, 3), dtype=float32, numpy=
array([[[[-8.10753584e-01,  1.10984087e+00,  2.71821529e-01],
        [-6.10031188e-01, -6.47952318e-01, -4.07003373e-01],
        [ 4.62206364e-01, -1.03655539e-01, -1.18086267e+00],
        ...
```

- Read red and green channel data:

```
In [66]: x[...,:2]
Out[66]:
<tf.Tensor: id=507, shape=(4, 32, 32, 2), dtype=float32, numpy=
array([[[[-1.26881   ,  0.575703  ],
        [ 0.98697686,  0.11028383],
        [-0.66420585, -0.9950867 ],
        ...
```

4.6.3 Slicing Summary

Tensor indexing and slicing methods are various, especially the slicing operation, which is easy for beginners to get confused. In essence, the slicing operation has only this basic form of *start* : *end* : *step*. Through this basic form, some default parameters are purposefully omitted, and multiple abbreviated methods are derived. So it is easier and faster to write. Since the number of dimensions that deep learning generally deals with is within four dimensions, you will find that the tensor slice operation is not that complicated in deep learning.

4.7 Dimensional Transformation

In neural networks, dimensional transformation is the core tensor operation. Through dimensional transformation, the data can be arbitrarily switched to meet the computing needs of different situations. Consider the batch form of the linear layer:

$$Y = X @ W + b$$

Assume that two samples, each of which has a feature length of 4, are included in X, with a shape of $[2, 4]$. The number of output nodes of the linear layer is three, that is, the shape of W is $[4, 3]$ and the shape of b is defined $[3]$. Then the result of $X @ W$ has shape of $[2, 3]$. Note that we also need to add b with shape $[3]$. How to add two tensors of different shapes directly?

Recall that what we want to do is adding a bias to each output node of each layer. This bias is shared by all samples at each node. In other words, each sample should add the same bias at each node as shown in Figure 4-5.

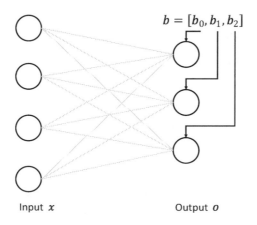

Figure 4-5. *Bias of a linear layer*

Therefore, for the input X of two samples, we need to copy the bias

$$b = \begin{bmatrix} b_1 & b_2 & b_3 \end{bmatrix}$$

to the number of samples into the following matrix form

$$B' = \begin{bmatrix} b_1 & b_2 & b_3 & b_1 & b_2 & b_3 \end{bmatrix}$$

and then add $X' = X \ @\ W$

$$X' = \begin{bmatrix} x'_{11} & x'_{12} & x'_{13} & x'_{21} & x'_{22} & x'_{23} \end{bmatrix}$$

Because they have the same shape at this time, this satisfies the requirement of matrix addition:

$$Y = X' + B' = \begin{bmatrix} x'_{11} & x'_{12} & x'_{13} & x'_{21} & x'_{22} & x'_{23} \end{bmatrix} + \begin{bmatrix} b_1 & b_2 & b_3 & b_1 & b_2 & b_3 \end{bmatrix}$$

In this way, it not only satisfies the requirement that the matrix addition needs to be consistent in shape but also achieves the logic of sharing the bias vector to the output nodes of each input sample. In order to achieve this, we insert a new dimension, batch, to the bias vector b and then copy the data in the batch dimension to get a transformed version B' with shape of [2, 3]. This series of operations is called dimensional transformation.

Each algorithm has different logical requirements for tensor format. When the existing tensor format does not meet the algorithm requirements, the tensor needs to be adjusted to the correct format through dimensional transformation. Basic dimensional transformation includes functions such as changing the view (reshape()), inserting new dimensions (expand_dims()), deleting dimensions (squeeze()), and exchanging dimensions (transpose()).

4.7.1 Reshape

Before introducing the reshape operation, let's first understand the concept of tensor storage and view. The view of the tensor is the way we understand the tensor. For example, the tensor of shape $[2, 4, 4, 3]$ is logically understood as two pictures, each picture having four rows and four columns and each pixel having three channels of RGB data. The storage of a tensor is reflected in that the tensor is stored in the memory as a continuous area. For the same storage, we can have different ways of view. For the $[2, 4, 4, 3]$ tensor, we can consider it as two samples, each of which is characterized by a vector of length 48. The same tensor can produce different views. This is the relationship between storage and view. View generation is very flexible, but needs to be reasonable.

We can generate a vector through tf.range() and generate different views through the tf.reshape() function, for example:

```
In [67]: x=tf.range(96)
x=tf.reshape(x,[2,4,4,3])  # Change view to [2,4,4,3] without
change storage
Out[67]:  # Data is not changed, only view is changed.
<tf.Tensor: id=11, shape=(2, 4, 4, 3), dtype=int32, numpy=
array([[[[ 0,  1,  2],
         [ 3,  4,  5],
         [ 6,  7,  8],
         [ 9, 10, 11]],...
```

When storing data, memory does not support this dimensional hierarchy concept, and data can only be written to memory in a tiled and sequential manner. Therefore, this hierarchical relationship needs to be managed manually, that is, the storage order of each tensor needs to be manually tracked. For ease of expression, we refer to the dimension on the left side of the tensor shape list as the large dimension and the dimension on the right side of the shape list as the small dimension. For example, in a tensor of shape $[2, 4, 4, 3]$, the number of images 2 is called the large dimension, and the number of channels 3 is called the small dimension. Under the setting of priority to write in small dimension first, the memory layout of the preceding tensor **x** is

1	2	3	4	5	6	7	8	9	...	...	...	93	94	95

Changing the view of the tensor only changes the way the tensor is understood. It does not change the storage order. Because the writing of a large amount of data consumes more computing resources, this is done to increase the computation efficiency. Because the data has only a flat structure when stored and it is separate from the logical structure, the new logical structure (view) does not need to change the data storage mode, which can save a lot of computing resources. While changing the view operation provides convenience, it also brings a lot of logical dangers. The default premise of changing the view operation is that the storage does not change; otherwise, changing the view operation is illegal. We first introduce legal view transformation operations and then introduce some illegal view transformations.

For example, tensor A is written into the memory according to the initial view of $[b, h, w, c]$. If we change the way of understanding, it can have the following format:

- Tensor $[b, h \cdot w, c]$ represents b pictures with $h \cdot w$ pixels and c channels.

- Tensor $[b, h, w \cdot c]$ represents b pictures with h lines, and the feature length of each line is $w \cdot c$.

- Tensor $[b, h \cdot w \cdot c]$ represents b pictures, and the feature length of each picture is $h \cdot w \cdot c$.

The storage of the preceding views does not need to be changed, so it is all correct.

Syntactically, the view transformation only needs to make sure the total number of elements of the new view and the size of the storage area are equal, that is, the element number of the new view is equal to

$$b \cdot h \cdot w \cdot c$$

It is precisely because the view design has very few grammatical constraints and is completely defined by the user, which makes it prone to logical risks when changing the view.

Now let's consider illegal view transformations. For example, if the new view is defined as $[b, w, h, c]$, $[b, c, h * w]$, or $[b, c, h, w]$, the storage order of the tensor needs to be changed. If the storage order is not updated synchronously, the recovered data will be inconsistent with the new view, resulting in data disorder. This requires the user to understand the data in order to determine whether the operation is legal. We will show how to change the storage of tensors in the "Swap Dimensions" section.

One technique for using view transformation operations correctly is to track the order of the stored dimensions. For example, for tensors saved in the initial view of "number of pictures-rows-columns-channels," the storage is also written in the order of "number of pictures-rows-columns-channels." If the view is restored in the "number of pictures-pixels-channels" method, it does not conflict with the "number of pictures-rows-columns-channels," so correct data can be obtained. However, if the data is restored in the "number of pictures-channels-pixels" method, because the memory layout is in the order of "number of pictures-rows-columns-channels," the order of the view dimensions is inconsistent with the order of the storage dimensions, which leads to disordered data.

Changing views is a very common operation in neural networks. You can implement complex logic by concatenating multiple reshape operations. However, when changing views through reshape, you must always remember the storage order of the tensor. The dimensional order of the new view must be the same as the storage order. Otherwise, you need to synchronize the storage order through the swap dimension operation. For example, for image data with shape [4,32,32,3], shape can be adjusted to [4,1024,3] by reshape operations. The view's dimensional order is $b - pixel - c$, and the tensor's storage order is $[b, h, w, c]$. The tensor with shape [4,1024,3] can be restored to the following:

- When $[b, h, w, c] = [4,32,32,3]$, the dimensional order of the new view and the storage order are consistent, and data can be recovered without disorders.

- When $[b, w, h, c] = [4,32,32,3]$, the dimensional order of the new view conflicts with the storage order.

- When $[h \cdot w \cdot c, b] = [3072, 4]$, the dimensional order of the new view conflicts with the storage order.

In TensorFlow, we can obtain the number of dimensions and shape of a tensor through the tensor's ndim and shape attributes:

```
In [68]: x.ndim,x.shape # Get the tensor's dimension and shape
Out[68]:(4, TensorShape([2, 4, 4, 3]))
```

With tf.reshape (x, new_shape), we can legally change the view of the tensor arbitrarily, for example:

```
In [69]: tf.reshape(x,[2,-1])
Out[69]:<tf.Tensor: id=520, shape=(2, 48), dtype=int32, numpy=
array([[ 0,  1,  2,  3,  4,  5,  6,  7,  8,  9, 10, 11, 12,
        13, 14, 15,
        16, 17, 18, 19, 20, 21, 22, 23, 24, 25, 26, 27, 28,
        29, 30, 31,...
        80, 81, 82, 83, 84, 85, 86, 87, 88, 89, 90, 91, 92,
        93, 94, 95]])>
```

The parameter –1 indicates that the length on the current axis needs to be automatically derived according to the rule that the total elements of the tensor are not changed. For example, the preceding –1 can be derived as

$$\frac{2 \cdot 4 \cdot 4 \cdot 3}{2} = 48$$

Change the view of the data again to $[2, 4, 12]$ as follows:

```
In [70]: tf.reshape(x,[2,4,12])
Out[70]:<tf.Tensor: id=523, shape=(2, 4, 12),
dtype=int32, numpy=
array([[[ 0,  1,  2,  3,  4,  5,  6,  7,  8,  9, 10, 11],...
        [36, 37, 38, 39, 40, 41, 42, 43, 44, 45, 46, 47]],
       [[48, 49, 50, 51, 52, 53, 54, 55, 56, 57, 58, 59], ...
        [84, 85, 86, 87, 88, 89, 90, 91, 92, 93, 94, 95]]])>
```

Change the view of the data to $[2,16,3]$ again as follows:

```
In [71]: tf.reshape(x,[2,-1,3])
Out[71]:<tf.Tensor: id=526, shape=(2, 16, 3),
dtype=int32, numpy=
array([[[ 0,  1,  2], ...
        [45, 46, 47]],
       [[48, 49, 50],...
        [93, 94, 95]]])>
```

Through the preceding series of continuous view transformation operations, we need to be aware that the storage order of the tensor has not changed and the data is still stored in the order of the initial order of $0, 1, 2, \cdots, 95$ in memory.

4.7.2 Add and Delete Dimensions

Add a Dimension. Adding a dimension with a length of 1 is equivalent to adding the concept of a new dimension to the original data. The dimension length is 1, so the data does not need to be changed; it is only a change of view.

Consider a specific example. The data of a large grayscale image is saved as a tensor of shape 28 × 28. At the end, a new dimension is added to the tensor, which is defined as the number of channels. Then the shape of the tensor becomes [28,28,1] as follows:

```
In [72]:  # Generate a 28x28 matrix
x = tf.random.uniform([28,28],maxval=10,dtype=tf.int32)
Out[72]:
<tf.Tensor: id=11, shape=(28, 28), dtype=int32, numpy=
array([[6, 2, 0, 0, 6, 7, 3, 3, 6, 2, 6, 2, 9, 3, 0, 3, 2, 8,
1, 3, 6, 2, 3, 9, 3, 6, 1, 7],...
```

With tf.expand_dims (x, axis), we can insert a new dimension before the specified axis:

```
In [73]:  x = tf.expand_dims(x,axis=2)
Out[73]:
<tf.Tensor: id=13, shape=(28, 28, 1), dtype=int32, numpy=
array([[[6],
        [2],
        [0],
        [0],
        [6],
        [7],
        [3],...
```

It can be seen that after inserting a new dimension, the storage order of the data has not changed. Only the view of the data is changed after inserting a new dimension.

In the same way, we can insert a new dimension at the front indicating the number of images dimension with a length of 1. At this time, the shape of the tensor becomes [1,28,28,1]:

```
In [74]: x = tf.expand_dims(x,axis=0)  # Insert a dimension at
the beginning
Out[74]:
<tf.Tensor: id=15, shape=(1, 28, 28, 1), dtype=int32, numpy=
array([[[[6],
        [2],
        [0],
        [0],
        [6],
        [7],
        [3],...
```

Note that when the axis of tf.expand_dims is positive, it means that a new dimension is inserted before the current dimension; when it is negative, it means that a new dimension is inserted after the current dimension. Taking tensor $[b, h, w, c]$ as an example, the actual insertion position of different axis parameters is shown in Figure 4-6.

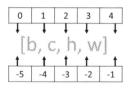
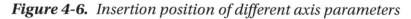

Figure 4-6. *Insertion position of different axis parameters*

Delete a Dimension. Deleting a dimension is the inverse operation of adding a dimension. As with adding a dimension, deleting a dimension can only delete a dimension of length 1, and it does not change the storage order of the tensor. Continue to consider the example of the shape [1,28,28,1]. If we want to delete the number of pictures dimension, we can use the tf.squeeze (x, axis) function. The axis parameter is the index number of the dimension to be deleted:

```
In [75]: x = tf.squeeze(x, axis=0)  # Delete the image number
dimension
Out[75]:
<tf.Tensor: id=586, shape=(28, 28, 1), dtype=int32, numpy=
array([[[8],
        [2],
        [2],
        [0],...
```

Continue to delete the channel number dimension. Since the image number dimension has been deleted, the shape of x at this time is [28,28,1]. When deleting the channel number dimension, we should specify axis = 2 as follows:

```
In [76]: x = tf.squeeze(x, axis=2)  # Delete channel dimension
Out[76]:
<tf.Tensor: id=588, shape=(28, 28), dtype=int32, numpy=
array([[8, 2, 2, 0, 7, 0, 1, 4, 9, 1, 7, 4, 8, 2, 7, 4, 8, 2,
        9, 8, 8, 0, 9, 9, 7, 5, 9, 7],
       [3, 4, 9, 9, 0, 6, 5, 7, 1, 9, 9, 1, 2, 7, 2, 7, 5, 3,
        3, 7, 2, 4, 5, 2, 7, 3, 8, 0],...
```

If we do not specify the dimension parameter axis, that is, tf.squeeze(x), it will delete all dimensions with a length of 1 by default, for example:

```
In [77]:
x = tf.random.uniform([1,28,28,1],maxval=10,dtype=tf.int32)
tf.squeeze(x)     # Delete all dimensions with length 1
Out[77]:
<tf.Tensor: id=594, shape=(28, 28), dtype=int32, numpy=
array([[9, 1, 4, 6, 4, 9, 0, 0, 1, 4, 0, 8, 5, 2, 5, 0, 0, 8,
9, 4, 5, 0, 1, 1, 4, 3, 9, 9],...
```

It is recommended to specify the dimension parameters to be deleted one by one, in order to prevent TensorFlow from accidentally deleting certain dimensions with length of 1, resulting in invalid calculation results.

4.7.3 Swap Dimensions

Changing the view or adding or deleting dimensions will not affect the storage of the tensor. Sometimes it is not enough to change the understanding of the tensor without changing the order of the dimensions. That is, the storage order needs to be adjusted directly. By swapping the dimensions, both the storage order and the view of the tensor are changed.

Swapping dimension operations are very common. For example, the default storage format of an image tensor is the $[b, h, w, c]$ format in TensorFlow, but the image format of some libraries is the $[b, c, h, w]$ format. We take transformation from $[b, h, w, c]$ to $[b, c, h, w]$ as an example to introduce how to use the tf.transpose(x, perm) function to complete the dimension swap operation, where the parameter perm represents the order of the new dimensions. Considering the image tensor with shape [2,32,32,3], the dimensional indexes of "number of pictures, rows, columns, and channels" are 0, 1, 2, and 3, respectively. If the order of the new dimensions is "number of pictures, number of channels, rows, and columns," the corresponding index number becomes $[0, 3, 1, 2]$, so the parameter perm needs to be set to $[0, 3, 1, 2]$. The implementation is as follows:

```
In [78]: x = tf.random.normal([2,32,32,3])
tf.transpose(x,perm=[0,3,1,2])  # Swap dimension
Out[78]:
<tf.Tensor: id=603, shape=(2, 3, 32, 32), dtype=float32, numpy=
array([[[[-1.93072677e+00, -4.80163872e-01, -8.85614634e-01, ...,
              1.49124235e-01,  1.16427064e+00, -1.47740364e+00],
          [-1.94761145e+00,  7.26879001e-01, -4.41877693e-01, ...
```

If we want to change $[b, h, w, c]$ to $[b, w, h, c]$, that is, exchange the height and width dimensions, the new dimension index becomes $[0, 2, 1, 3]$ as follows:

```
In [79]:
x = tf.random.normal([2,32,32,3])
tf.transpose(x,perm=[0,2,1,3]) # Swap dimension
Out[79]:
<tf.Tensor: id=612, shape=(2, 32, 32, 3), dtype=float32, numpy=
array([[[[ 2.1266546 , -0.64206547,  0.01311932],
          [ 0.918484  ,  0.9528751 ,  1.1346699 ],
          ...,
```

It should be noted that after the dimension swap is completed through tf.transpose, the storage order of the tensor has changed, and the view has changed accordingly. All subsequent operations must be based on the new order and view. Compared with the changing view operation, the dimension swap operation is more computationally expensive.

4.7.4 Copy Data

After inserting a new dimension, we may want to copy data on the new dimension to meet the requirements of subsequent calculations. Consider the example $Y = X @ W + b$. After inserting a new dimension with the number of samples for b, we need to copy the batch size data in the new dimension and change the shape of b to be consistent with $X @ W$ to complete the tensor addition operation.

We can use the tf.tile(x, multiples) function to complete the data replication operation in the specified dimensions. The parameter multiples specifies the replication number for each dimension, respectively. For example, 1 indicates that the data will not be copied, and 2 indicates that the new length is twice of the original length.

Taking the input $[2, 4]$ and a three–output node linear transformation layer as an example, the bias b is defined as

$$b = \begin{bmatrix} b_1 & b_2 & b_3 \end{bmatrix}$$

Insert a new dimension through tf.expand_dims(b, axis = 0) and turn it into a matrix:

$$B = \begin{bmatrix} b_1 & b_2 & b_3 \end{bmatrix}$$

Now the shape of B becomes $[1, 3]$. We need to copy data in the dimension of axis = 0 according to the number of input samples. The batch size here is 2, that is, a copy is made and it becomes

$$B = \begin{bmatrix} b_1 & b_2 & b_3 & b_1 & b_2 & b_3 \end{bmatrix}$$

Through tf.tile(b, multiples = [2,1]), it can be copied once in the axis = 0 dimension and not copied in the axis = 1 dimension. First, insert a new dimension as follows:

```
In [80]:
b = tf.constant([1,2])  # Create tensor b
b = tf.expand_dims(b, axis=0)  # Insert new dimension
b
Out[80]:
<tf.Tensor: id=645, shape=(1, 2), dtype=int32,
numpy=array([[1, 2]])>
```

Copy one replicate of the data in the batch dimension to achieve the following:

```
In [81]: b = tf.tile(b, multiples=[2,1])
Out[81]:
<tf.Tensor: id=648, shape=(2, 2), dtype=int32, numpy=
array([[1, 2],
       [1, 2]])>
```

Now the shape of B becomes [2, 3], and B can be directly added to $X @ W$. Consider another example with a 2×2 matrix. The implementation is as follows:

```
In [82]: x = tf.range(4)
x=tf.reshape(x,[2,2])  # Create 2x2 matrix
Out[82]:
<tf.Tensor: id=655, shape=(2, 2), dtype=int32, numpy=
array([[0, 1],
       [2, 3]])>
```

First, copy one replicate of the data in the column dimension as follows:

```
In [83]: x = tf.tile(x,multiples=[1,2])
Out[83]:
<tf.Tensor: id=658, shape=(2, 4), dtype=int32, numpy=
array([[0, 1, 0, 1],
       [2, 3, 2, 3]])>
```

Then copy one replicate of the data in the row dimension:

```
In [84]: x = tf.tile(x,multiples=[2,1])
Out[84]:
<tf.Tensor: id=672, shape=(4, 4), dtype=int32, numpy=
array([[0, 1, 0, 1],
       [2, 3, 2, 3],
       [0, 1, 0, 1],
       [2, 3, 2, 3]])>
```

After the replication operation in two dimensions, we can see the shape of the data has doubled. This example helps us understand the process of data replication more intuitively.

It should be noted that tf.tile will create a new tensor to save the copied tensor. Since the copy operation involves a large amount of data reading and writing operations, the computational cost is relatively high. The tensor operations between different shapes in the neural network are very common, so is there a lightweight copy operation? This is the broadcasting operation to be introduced next.

4.8 Broadcasting

Broadcasting is a lightweight tensor copying method, which logically expands the shape of the tensor data, but only performs the actual storage copy operation when needed. For most scenarios, the broadcasting mechanism can complete logical operations by avoiding the actual data copying, thereby reducing a large amount of computational cost compared with the tf.tile function.

For all dimensions of length 1, broadcasting has the same effect as tf.tile. The difference is that tf.tile creates a new tensor by performing the copy IO operation. Broadcasting does not immediately copy the data; instead, it will logically change the shape of the tensor, so that the view becomes the copied shape. Broadcasting will use the optimization methods of the deep learning framework to avoid the actual copying of data and complete the logical operations. For the user, the final effect of broadcasting and tf.tile copy is the same, but the broadcasting mechanism saves a lot of computational resources. It is recommended to use broadcasting as much as possible in the calculation process to improve efficiency.

Continuing to consider the preceding example $Y = X @ W + b$, the shape of $X @ W$ is $[2, 3]$, and the shape of b is $[3]$. We can manually complete the copy data operation by combining tf.expand_dims and tf.tile, that is, transform b to shape $[2, 3]$ and then add it to $X @ W$. But in fact, it is also correct to add $X @ W$ directly to b with shape $[3]$, for example:

```
x = tf.random.normal([2,4])
w = tf.random.normal([4,3])
b = tf.random.normal([3])
y = x@w+b # Add tensors with different shapes directly
```

The preceding addition does not throw a logical error. This is because it automatically calls the broadcasting function tf.broadcast_to(x, new_shape), expanding the shape of b to $[2,3]$. The preceding operation is equivalent to

```
y = x@w + tf.broadcast_to(b,[2,3])
```

In other words, when the operator + encounters two tensors with inconsistent shapes, it will automatically consider expanding the two tensors to a consistent shape and then call tf.add to complete the tensor addition operation. By automatically calling tf.broadcast_to(b, [2,3]), it not only achieves the purpose of increasing dimension but also avoids the expensive computational cost of actually copying the data.

The core idea of the broadcasting mechanism is universality. That is, the same data can be generally suitable for other locations. Before verifying universality, we need to align the tensor shape to the right first and then perform universality check: for a dimension of length 1, by default this data is generally suitable for other positions in the current dimension; for dimensions that do not exist, after adding a new dimension, the default current data is also universally applicable to the new dimension, so that it can be expanded into a tensor shape of any number of dimensions.

Considering the tensor A with shape $[w, 1]$, it needs to be extended to shape $[b, h, w, c]$. As shown in Figure 4-7, the first line is the expanded shape, and the second line is the existing shape.

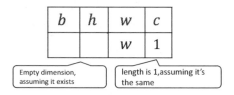

Figure 4-7. *Broadcasting example 1*

First, align the two shapes to the right. For the channel dimension c, the current length of the tensor is 1. By default, this data is also suitable for other positions in the current dimension. The data is logically copied, and the length becomes c; for the nonexisting dimensions b and h, a new dimension is automatically inserted, the length of the new dimension is 1, and at the same time, the current data is generally suitable for other positions in the new dimension, that is, for other pictures and other rows, it is completely consistent with the data of the current row. This automatically expands the corresponding dimensions to b and h, as shown in Figure 4-8.

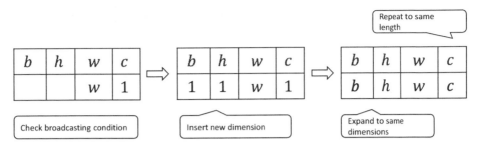

Figure 4-8. *Broadcasting example 2*

The tf.broadcast_to(x, new_shape) function can be used to explicitly perform the automatic expansion function to expand the existing shape to new_shape. The implementation is as follows:

```
In [87]:
A = tf.random.normal([32,1])  # Create a matrix
tf.broadcast_to(A, [2,32,32,3])  # Expand to 4 dimensions
Out[87]:
<tf.Tensor: id=13, shape=(2, 32, 32, 3), dtype=float32, numpy=
array([[[[-1.7571245 , -1.7571245 , -1.7571245 ],
         [ 1.580159  ,  1.580159  ,  1.580159  ],
         [-1.5324328 , -1.5324328 , -1.5324328 ],...
```

It can be seen that, under the guidance of the universality principle, the broadcasting mechanism has become intuitive and easy to understand.

Let us consider an example that does not satisfy the principle of universality, as shown in Figure 4-9.

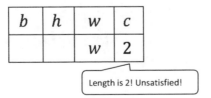

Figure 4-9. *Broadcasting bad example*

In the c dimension, the tensor already has two features, and the length of the corresponding dimension of the new shape is $c(c \neq 2$, such as $c = 3)$. Then these two features in the current dimension cannot be universally applied to other positions, so it does not meet the universality principle. If we apply broadcasting, it will trigger errors, such as

```
In [88]:
A = tf.random.normal([32,2])
tf.broadcast_to(A, [2,32,32,4])
Out[88]:
InvalidArgumentError: Incompatible shapes: [32,2] vs.
[2,32,32,4] [Op:BroadcastTo]
```

When performing tensor operations, some operations will automatically call the broadcasting mechanism when processing tensors of different shapes, such as +,-, *, and /, to broadcast the corresponding tensors into a common shape and then do the calculation accordingly. Figure 4-10 demonstrates some examples of tensor addition in three different shapes.

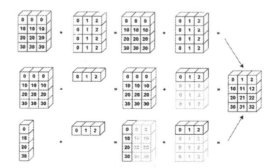

Figure 4-10. *Automatic broadcasting example*

Let's test the automatic broadcasting mechanism of basic operators, for example:

```
a = tf.random.normal([2,32,32,1])
b = tf.random.normal([32,32])
a+b,a-b,a*b,a/b # Test automatic broadcasting for operations +,
-, *, and /
```

These operations can be broadcasted into a common shape before the actual calculation. Using the broadcasting mechanism can make code more concise and efficient.

4.9 Mathematical Operations

We've used some basic mathematical operations such as addition, subtraction, multiplication, and division in previous chapters. In this section, we will systematically introduce the common mathematical operations in TensorFlow.

4.9.1 Addition, Subtraction, Multiplication and Division

Addition, subtraction, multiplication, and division are the most basic mathematical operations. They are implemented by the tf.add, tf.subtract, tf.multiply, and tf.divide functions, respectively, in TensorFlow. TensorFlow has overloaded operators +, − , ∗ , and/. It is generally recommended to use those operators directly. Floor dividing and remainder dividing are two other common operations, implemented by the //and % operators, respectively. Let's demonstrate the division operations, for example:

```
In [89]:
a = tf.range(5)
b = tf.constant(2)
a//b # Floor dividing
Out[89]:
<tf.Tensor: id=115, shape=(5,), dtype=int32, numpy=array([0, 0,
1, 1, 2])>
In [90]: a%b # Remainder dividing
Out[90]:
<tf.Tensor: id=117, shape=(5,), dtype=int32, numpy=array([0, 1,
0, 1, 0])>
```

4.9.2 Power Operations

The power operation can be conveniently completed through the tf.pow(x, a) function, or the operator ** as x**a:

```
In [91]:
x = tf.range(4)
tf.pow(x,3)
Out[91]:
<tf.Tensor: id=124, shape=(4,), dtype=int32, numpy=array([
0,   1,   8, 27])>
In [92]: x**2
Out[92]:
<tf.Tensor: id=127, shape=(4,), dtype=int32, numpy=array([0,
1, 4, 9])>
```

Set the exponent to the form of $\dfrac{1}{a}$ to implement the root operation $\sqrt[a]{x}$, for example:

```
In [93]: x=tf.constant([1.,4.,9.])
x**(0.5)   # square root
```

```
Out[93]:
<tf.Tensor: id=139, shape=(3,), dtype=float32, numpy=array([1.,
2., 3.], dtype=float32)>
```

In particular, for common square and square root operations, tf.square(x) and tf.sqrt(x) can be used. The square operation is implemented as follows:

```
In [94]:x = tf.range(5)
x = tf.cast(x, dtype=tf.float32)  # convert to float type
x = tf.square(x)
Out[94]:
<tf.Tensor: id=159, shape=(5,), dtype=float32, numpy=array([
0.,   1.,   4.,   9.,  16.], dtype=float32)>
```

The square root operation is implemented as follows:

```
In [95]:tf.sqrt(x)
Out[95]:
<tf.Tensor: id=161, shape=(5,), dtype=float32, numpy=array([0.,
1., 2., 3., 4.], dtype=float32)>
```

4.9.3 Exponential and Logarithmic Operations

Exponential operations can also be easily implemented using tf.pow(a, x) or the ** operator, for example:

```
In [96]: x = tf.constant([1.,2.,3.])
2**x
Out[96]:
<tf.Tensor: id=179, shape=(3,), dtype=float32, numpy=array([2.,
4., 8.], dtype=float32)>
```

In particular, for natural exponents e^x, this can be achieved with tf.exp(x), for example:

```
In [97]: tf.exp(1.)
Out[97]:
<tf.Tensor: id=182, shape=(), dtype=float32, numpy=2.7182817>
```

In TensorFlow, natural logarithms x can be implemented with tf.math.log(x), for example:

```
In [98]: x=tf.exp(3.)
tf.math.log(x)
Out[98]:
<tf.Tensor: id=186, shape=(), dtype=float32, numpy=3.0>
```

If you want to calculate the logarithm of other bases, you can use the logarithmic base-changing formula:

$$x = \frac{x}{a}$$

For example, the calculation of $\frac{x}{10}$ can be achieved by

```
In [99]: x = tf.constant([1.,2.])
x = 10**x
tf.math.log(x)/tf.math.log(10.)
Out[99]:
<tf.Tensor: id=6, shape=(2,), dtype=float32, numpy=array([1.,
2.], dtype=float32)>
```

4.9.4 Matrix Multiplication

The neural network contains a large number of matrix multiplication operations. We have previously introduced that the matrix multiplication can be easily implemented by the @ operator and the tf.matmul(a, b) function. It should be noted that the matrix multiplication in TensorFlow can use the batch method, that is, tensors A and B can have dimensions greater than 2. When the dimensions are greater than 2, TensorFlow selects the last two dimensions of A and B to perform matrix multiplication, and all the previous dimensions are considered as batch dimensions.

According to the definition of matrix multiplication, the condition of A being able to multiply a matrix B is that the length of the penultimate dimension (column) of A and the length of the penultimate dimension (row) of B must be equal. For example, tensor a with shape $[4, 3, 28, 32]$ can be multiplied by tensor b with shape $[4, 3, 32, 2]$. The code is as follows:

```
In [100]:
a = tf.random.normal([4,3,28,32])
b = tf.random.normal([4,3,32,2])
a@b
Out[100]:
<tf.Tensor: id=236, shape=(4, 3, 28, 2), dtype=float32, numpy=
array([[[[-1.66706240e+00, -8.32602978e+00],
        [ 9.83304405e+00,  8.15909767e+00],
        [ 6.31014729e+00,  9.26124632e-01],...
```

Matrix multiplication also supports the automatic broadcasting mechanism, for example:

```
In [101]:
a = tf.random.normal([4,28,32])
b = tf.random.normal([32,16])
```

```
tf.matmul(a,b)  # First broadcast b to shape [4, 32, 16] and
then multiply a
Out[101]:
<tf.Tensor: id=264, shape=(4, 28, 16), dtype=float32, numpy=
array([[[-1.11323869e+00, -9.48194981e+00,  6.48123884e+00, ...,
          6.53280640e+00, -3.10894990e+00,  1.53050375e+00],
        [ 4.35898495e+00, -1.03704405e+01,  8.90656471e+00, ...,
```

The preceding operation automatically expands the variable b to a common shape [4,32,16] and then multiplies the variable a in batch form to obtain the results with shape [4,28,16].

4.10 Hands-On Forward Propagation

So far, we have introduced tensor creation, index slicing, dimensional transformations, and common mathematical operations. Finally, we will use the knowledge we have learned to complete the implementation of the three-layer neural network:

$$out = ReLU\{ReLU\{ReLU[X @ W_1 + b_1] @ W_2 + b_2\} @ W_3 + b_3\}$$

The data set we use is the MNIST handwritten digital picture data set. The number of input nodes is 784. The output node numbers of the first, second, and third layers are 256, 128, and 10, respectively. First, let's create the tensor parameters W and b for each nonlinear layer as follows:

```
# Every layer's tensor needs to be optimized. Set initial bias
to be 0s.
# w and b for first layer
w1 = tf.Variable(tf.random.truncated_normal([784, 256],
stddev=0.1))
b1 = tf.Variable(tf.zeros([256]))
```

```
# w and b for second layer
w2 = tf.Variable(tf.random.truncated_normal([256, 128],
stddev=0.1))
b2 = tf.Variable(tf.zeros([128]))
# w and b for third layer
w3 = tf.Variable(tf.random.truncated_normal([128, 10],
stddev=0.1))
b3 = tf.Variable(tf.zeros([10]))
```

In forward calculation, the view of the input tensor with shape $[b, 28, 28]$ is first adjusted to a matrix with shape $[b, 784]$, so that it is suitable for the input format of the network:

```
# Change view[b, 28, 28] => [b, 28*28]
x = tf.reshape(x, [-1, 28*28])
```

Next, finish the calculation of the first layer. We perform the automatic expansion operation here:

```
# First layer calculation, [b, 784]@[784, 256] +
[256] => [b, 256] + [256] => [b, 256] + [b, 256]
h1 = x@w1 + tf.broadcast_to(b1, [x.shape[0], 256])
h1 = tf.nn.relu(h1) # apply activation function
```

Use the same method for the second and third nonlinear function layers. The output layer can use the ReLU activation function:

```
# Second layer calculation, [b, 256] => [b, 128]
h2 = h1@w2 + b2
h2 = tf.nn.relu(h2)
# Output layer calculation, [b, 128] => [b, 10]
out = h2@w3 + b3
```

Transform the real labeled tensor into one-hot encoding and calculate the mean squared error from out as follows:

```
# Calculate mean square error, mse =
mean(sum(y-out)^2)
# [b, 10]
loss = tf.square(y_onehot - out)
# Error metrics, mean: scalar
loss = tf.reduce_mean(loss)
```

The preceding forward calculation process needs to be wrapped in the context of "with tf.GradientTape() as tape," so that the computational graph information can be saved during forward calculation for the automatic differentiation operation.

Use the tape.gradient() function to get the gradient information of the network parameters. The result is stored in the grads list variable as follows:

```
# Calculate gradients for [w1, b1, w2, b2, w3, b3]
grads = tape.gradient(loss, [w1, b1, w2, b2, w3, b3])
```

Then we need to update the parameters by

$$\theta' = \theta - \eta \cdot \frac{\partial L}{\partial \theta}$$

```
# Update parameters using assign_sub (subtract the update
and assign back to the original parameter)
    w1.assign_sub(lr * grads[0])
    b1.assign_sub(lr * grads[1])
    w2.assign_sub(lr * grads[2])
    b2.assign_sub(lr * grads[3])
    w3.assign_sub(lr * grads[4])
    b3.assign_sub(lr * grads[5])
```

Among them, assign_sub() subtracts itself from a given parameter value to implement an in-place update operation. The variation of the network training error is shown in Figure 4-11.

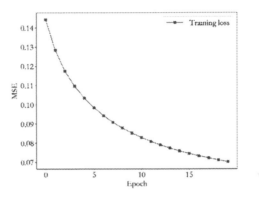

Figure 4-11. *Training error of the forward calculation*

CHAPTER 5

Advanced TensorFlow

Artificial intelligence would be the ultimate version of Google. The ultimate search engine that would understand everything on the Web. It would understand exactly what you wanted, and it would give you the right thing.

—Larry Page

After introducing the basic tensor operations, let's further explore the advanced operations, such as tensor merging and segmentation, norm statistics, tensor filling, and clipping. We will also use the MNIST dataset again to enhance our understanding of tensor operations in TensorFlow.

5.1 Merge and Split

5.1.1 Merge

Merging means combining multiple tensors into one tensor in a certain dimension. Taking the data of a school's gradebooks as an example, tensor A is used to save the gradebooks of classes 1–4. There are 35 students in each class with a total of eight subjects. The shape of tensor A is [4,35,8]. Similarly, tensor B keeps the gradebooks of the other six classes, with a shape of [6,35,8]. By merging these two gradebooks, you can get the gradebooks of all the classes in the school, recorded as tensor C, and the corresponding shape should be [10,35,8], where 10 represents ten classes, 35 represents 35 students, and 8 represents eight subjects.

© Liangqu Long and Xiangming Zeng 2022
L. Long and X. Zeng, *Beginning Deep Learning with TensorFlow*,
https://doi.org/10.1007/978-1-4842-7915-1_5

Tensors can be merged using concatenate and stack operations. The concatenate operation does not generate new dimensions. It only merges along existing dimensions. But the stack operation creates new dimensions. Whether to use the concatenate or stack operation to merge tensors depends on whether a new dimension needs to be created for a specific scene. We will discuss both of them in the following session.

Concatenate. In TensorFlow, tensors can be concatenated using the tf.concat(tensors, axis) function, where the first parameter holds a list of tensors that need to be merged and the second parameter specifies the dimensional index on which to merge. Back to the preceding example, we merge the gradebooks in the class dimension. Here, the index number of the class dimension is 0, that is, axis = 0. The code for merging A and B is as follows:

```
In [1]:
a = tf.random.normal([4,35,8]) # Create gradebook A
b = tf.random.normal([6,35,8]) # Create gradebook B
tf.concat([a,b],axis=0) # Merge gradebooks
Out[1]:
<tf.Tensor: id=13, shape=(10, 35, 8), dtype=float32, numpy=
array([[[ 1.95299834e-01,  6.87859178e-01, -5.80048323e-01, ...,
          1.29430830e+00,  2.56610274e-01, -1.27798581e+00],
        [ 4.29753691e-01,  9.11329567e-01, -4.47975427e-01, ...,
```

In addition to the class dimension, we can also merge tensors in other dimensions. Consider that tensor A saves the first four subjects' scores of all students in all classes, with shape [10,35,4] and tensor B saves the remaining 4 subjects' scores, with shape [10,35,4]. We can get the total gradebook tensor by merging A and B as in the following:

```
In [2]:
a = tf.random.normal([10,35,4])
b = tf.random.normal([10,35,4])
```

```
tf.concat([a,b],axis=2) # Merge along the last dimension
Out[2]:
<tf.Tensor: id=28, shape=(10, 35, 8), dtype=float32, numpy=
array([[[-5.13509691e-01, -1.79707789e+00,  6.50747120e-01, ...,
          2.58447856e-01,  8.47878829e-02,  4.13468748e-01],
        [-1.17108583e+00,  1.93961406e+00,  1.27830813e-02, ...,
```

Syntactically, the concatenate operation can be performed on any dimension. The only constraint is that the length of the non-merging dimension must be the same. For example, the tensors with shape [4,32,8] and shape [6,35,8] cannot be directly merged in the class dimension, because the length of the number of students' dimension is not the same – one is 32 and the other is 35, for example:

```
In [3]:
a = tf.random.normal([4,32,8])
b = tf.random.normal([6,35,8])
tf.concat([a,b],axis=0) # Illegal merge. Second dimension is
different.
Out[3]:
InvalidArgumentError: ConcatOp : Dimensions of inputs
should match: shape[0] = [4,32,8] vs. shape[1] = [6,35,8]
[Op:ConcatV2] name: concat
```

Stack. The concatenate operation merges data directly on existing dimensions and does not create new dimensions. If we want to create a new dimension when merging data, we need to use the tf.stack operation. Consider that tensor *A* saves the gradebook of one class with the shape of [35, 8] and tensor *B* saves the gradebook of another class with the shape of [35, 8]. When merging the data of these two classes, we need to create a new dimension, defined as the class dimension. The new dimension can be placed in any position. Generally, the class dimension is placed before the student dimension, that is, the new shape of the merged tensor should be [2,35,8].

The tf.stack(tensors, axis) function can be used to combine multiple tensors. The first parameter represents the tensor list to be merged, and the second parameter specifies the position where the new dimension is inserted. The usage of axis is the same as that of the tf.expand_dims function. When $axis \geq 0$, a new dimension is inserted before axis. When $axis < 0$, we insert a new dimension after axis. Figure 5-1 shows the new dimension position corresponding to different axis parameter settings for a tensor with shape $[b, c, h, w]$.

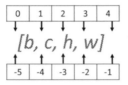

Figure 5-1. *New dimension insertion position for stack operation with different axis values*

Merge the two classes' gradebooks using the stack operation and insert the class dimension at the axis = 0 position. The code is as follows:

```
In [4]:
a = tf.random.normal([35,8])
b = tf.random.normal([35,8])
tf.stack([a,b],axis=0) # Stack a and b and insert new dimension
at axis=0
Out[4]:
<tf.Tensor: id=55, shape=(2, 35, 8), dtype=float32, numpy=
array([[[ 3.68728966e-01, -8.54765773e-01, -4.77824420e-01,
         -3.83714020e-01, -1.73216307e+00,  2.03872994e-02,
          2.63810277e+00, -1.12998331e+00],...
```

We can also choose to insert new dimensions elsewhere. For example, insert the class dimension at the end:

```
In [5]:
a = tf.random.normal([35,8])
b = tf.random.normal([35,8])
tf.stack([a,b],axis=-1) # Insert new dimension at the end
Out[5]:
<tf.Tensor: id=69, shape=(35, 8, 2), dtype=float32, numpy=
array([[[ 0.3456724 , -1.7037214 ],
        [ 0.41140947, -1.1554345 ],
        [ 1.8998919 ,  0.56994915]],...
```

Now the class dimension is on axis = 2, and we need to understand the data according to the view represented by the latest dimension order. If we choose to use tf.concat to merge the preceding transcripts, then it would be

```
In [6]:
a = tf.random.normal([35,8])
b = tf.random.normal([35,8])
tf.concat([a,b],axis=0) # No class dimension
Out[6]:
<tf.Tensor: id=108, shape=(70, 8), dtype=float32, numpy=
array([[-0.5516891 , -1.5031327 , -0.35369992,
   0.31304857,  0.13965549,
        0.6696881 , -0.50115544,  0.15550546],
      [ 0.8622069 ,  1.0188094 ,  0.18977325,  0.6353301 ,
        0.05809061,...
```

It can be seen that tf.concat can also merge data smoothly, but we need to understand the tensor data in the way that the first 35 students come from the first class and the last 35 students come from the second class, which is not very intuitive. For this example, it is obviously more reasonable to create a new dimension through the tf.stack method.

The tf.stack function also needs to meet a certain condition to use. It needs all the tensors to be merged to have the same shape. Let's see what happens when stacking two tensors with different shapes:

```
In [7]:
a = tf.random.normal([35,4])
b = tf.random.normal([35,8])
tf.stack([a,b],axis=-1) # Illegal use of stack function.
Different shapes.
Out[7]:
InvalidArgumentError: Shapes of all inputs must match:
values[0].shape = [35,4] != values[1].shape = [35,8] [Op:Pack]
name: stack
```

The preceding operation attempts to merge two tensors whose shapes are [35, 4] and [35, 8], respectively. Because the shapes of the two tensors are not the same, the merge operation cannot be completed.

5.1.2 Split

The inverse process of the merge operation is split, which splits a tensor into multiple tensors. Let's continue the gradebook example. We get the gradebook tensor of the entire school with shape of [10,35,8]. Now we need to cut the data into ten tensors in the class dimension, and each tensor holds the gradebook data of the corresponding class. tf.split(x, num_or_size_splits, axis) can be used to complete the tensor split operation. The meaning of the parameters in the function is as follows:

- x: The tensor to be split.

- num_or_size_splits: Cutting scheme. When num_or_size_splits is a single value, such as 10, it means that

the tensor x is cut into ten parts with equal length.
When num_or_size_splits is a list, each element of the
list represents the length of each part. For example,
num_or_size_splits=[2, 4, 2, 2] means that the tensor is
cut into four parts, with the length of each part as 2, 4,
2, and 2.

- axis: Specifies the dimension index of the split.

Now we cut the total gradebook tensor into ten pieces as follows:

```
In [8]:
x = tf.random.normal([10,35,8])
# Cut into 10 pieces with equal length
result = tf.split(x, num_or_size_splits=10, axis=0)
len(result)  # Return a list with 10 tensors of equal length
Out[8]: 10
```

We can view the shape of a tensor after cutting, and it should be all
gradebook data of one class with shape of $[1, 35, 8]$:

```
In [9]: result[0] # Check the first class gradebook
Out[9]: <tf.Tensor: id=136, shape=(1, 35, 8),
dtype=float32, numpy=
array([[[-1.7786729 ,  0.2970506 ,  0.02983334,  1.3970423 ,
          1.315918  , -0.79110134, -0.8501629 , -1.5549672 ],
        [ 0.5398711 ,  0.21478991, -0.08685189,  0.7730989 ,...
```

It can be seen that the shape of the first class tensor is [1,35,8],
which still has the class dimension. Let's perform unequal length cutting.
For example, split the data into four parts with each length as $[4, 2, 2, 2]$ for
each part:

```
In [10]: x = tf.random.normal([10,35,8])
# Split tensor into 4 parts
```

```
result = tf.split(x, num_or_size_splits=[4,2,2,2] ,axis=0)
len(result)
Out[10]: 4
```

Check the shape of the first split tensor. According to our splitting scheme, it should contain the gradebooks of four classes. The shape should be [4,35,8]:

```
In [10]: result[0]
Out[10]: <tf.Tensor: id=155, shape=(4, 35, 8),
dtype=float32, numpy=
array([[[-6.95693314e-01,  3.01393479e-01,  1.33964568e-01, ...,
```

In particular, if we want to divide one certain dimension by a length of 1, we can use the tf.unstack(x, axis) function. This method is a special case of tf.split. The splitting length is fixed as 1. We only need to specify the index number of the splitting dimension. For example, unstack the total gradebook tensor in the class dimension:

```
In [11]: x = tf.random.normal([10,35,8])
result = tf.unstack(x,axis=0)
len(result) # Return a list with 10 tensors
Out[11]: 10
```

View the shape of the split tensor:

```
In [12]: result[0] # The first class tensor
Out[12]: <tf.Tensor: id=166, shape=(35, 8),
dtype=float32, numpy=
array([[-0.2034383 ,  1.1851563 ,  0.25327438,
-0.10160723,  2.094969  ,
        -0.8571669 , -0.48985648,  0.55798006],...
```

It can be seen that after splitting through tf.unstack, the split tensor shape becomes [35, 8], that is, the class dimension disappears, which is different from tf.split.

5.2 Common Statistics

During the neural network calculations, various statistical attributes need to be computed, such as maximum, minimum, mean, and norm. Because tensors usually contain a lot of data, it is easier to infer the distribution of tensor values by obtaining the statistical information of these tensors.

5.2.1 Norm

Norm is a measure of the "length" of a vector. It can be generalized to tensors. In neural networks, it is often used to represent the tensor weight and the gradient magnitude. Commonly used norms are:

- L1 norm, defined as the sum of the absolute values of all the elements of the vector:

$$\| x \|_1 = \sum_i |x_i|$$

- L2 norm, defined as the root sum of the squares of all the elements of the vector:

$$\| x \|_2 = \sqrt{\sum_i |x_i|^2}$$

- ∞ norm, defined as the maximum of the absolute values of all elements of a vector:

$$\| x \|_\infty = max_i(|x_i|)$$

For matrices and tensors, the preceding formulas can also be used after flattening the matrices and tensors into a vector. In TensorFlow, the tf.norm(x, ord) function can be used to solve the L1, L2, and ∞norms, where the parameter ord is specified as 1, 1, and np.inf for L1, L2, and ∞ norms, respectively:

```
In [13]: x = tf.ones([2,2])
tf.norm(x,ord=1) # L1 norm
Out[13]: <tf.Tensor: id=183, shape=(), dtype=float32,
numpy=4.0>
In [14]: tf.norm(x,ord=2) # L2 norm
Out[14]: <tf.Tensor: id=189, shape=(), dtype=float32,
numpy=2.0>
In [15]: import numpy as np
tf.norm(x,ord=np.inf) # ∞ norm
Out[15]: <tf.Tensor: id=194, shape=(), dtype=float32,
numpy=1.0>
```

5.2.2 Max, Min, Mean, and Sum

The tf.reduce_max, tf.reduce_min, tf.reduce_mean, and tf.reduce_sum functions can be used to get the maximum, minimum, mean, and sum of tensors in a certain dimension or in all dimensions.

Consider a tensor of shape [4, 10], where the first dimension represents the number of samples and the second dimension represents the probability that the current sample belongs to each of the ten categories. The maximum value of each sample's probability can be obtained through the tf.reduce_max function:

```
In [16]: x = tf.random.normal([4,10])
tf.reduce_max(x,axis=1) # get maximum value at 2nd dimension
```

```
Out[16]:<tf.Tensor: id=203, shape=(4,), dtype=float32,
numpy=array([1.2410722 , 0.88495886, 1.4170984 , 0.9550192 ],
dtype=float32)>
```

The preceding code returns a vector of length 4, which represents the maximum probability value of each sample. Similarly, we can find the minimum value of the probability for each sample as follows:

```
In [17]: tf.reduce_min(x,axis=1) # get the minimum value at 2nd
dimension
Out[17]:<tf.Tensor: id=206, shape=(4,), dtype=float32,
numpy=array([-0.27862206, -2.4480672 , -1.9983795 , -1.5287997 ],
dtype=float32)>
```

Find the mean probabilities of each sample:

```
In [18]: tf.reduce_mean(x,axis=1)
Out[18]:<tf.Tensor: id=209, shape=(4,), dtype=float32,
numpy=array([ 0.39526337, -0.17684573, -0.148988  ,
-0.43544054], dtype=float32)>
```

When the axis parameter is not specified, the tf.reduce_* functions will find the maximum, minimum, mean, and sum of all the data:

```
In [19]:x = tf.random.normal([4,10])
tf.reduce_max(x),tf.reduce_min(x),tf.reduce_mean(x)
Out [19]: (<tf.Tensor: id=218, shape=(), dtype=float32,
numpy=1.8653786>,
 <tf.Tensor: id=220, shape=(), dtype=float32,
numpy=-1.9751656>,
 <tf.Tensor: id=222, shape=(), dtype=float32,
numpy=0.014772797>)
```

When solving the error function, the error of each sample can be obtained through the MSE function, and the average error of the sample needs to be calculated. Here we can use tf.reduce_mean function as follows:

```
In [20]:
out = tf.random.normal([4,10]) # Simulate output
y = tf.constant([1,2,2,0]) # Real labels
y = tf.one_hot(y,depth=10) # One-hot encoding
loss = keras.losses.mse(y,out) # Calculate loss of each sample
loss = tf.reduce_mean(loss) # Calculate mean loss
loss
Out[20]:
<tf.Tensor: id=241, shape=(), dtype=float32, numpy=1.1921183>
```

Similar to the tf.reduce_mean function, the sum function tf.reduce_sum(x, axis) can calculate the sum of all features of the tensor on the corresponding axis:

```
In [21]:out = tf.random.normal([4,10])
tf.reduce_sum(out,axis=-1) # Calculate sum along the last dimension
Out[21]:<tf.Tensor: id=303, shape=(4,), dtype=float32,
numpy=array([-0.588144 ,   2.2382064,  2.1582587,  4.962141 ],
dtype=float32)>
```

In addition, to obtain the maximum or minimum value of the tensor, we sometimes also want to obtain the corresponding position index. For example, for the classification tasks, we need to know the position index of the maximum probability, which is usually used as the prediction category. Considering the classification problem with ten categories, we get the output tensor with shape [2, 10], where 2 represents two samples and 10 indicates the probability of belonging to ten categories. Since the position index of the element represents the probability that the current sample belongs to this category, we often use the index corresponding to the largest probability as the predicted category.

```
In [22]:out = tf.random.normal([2,10])
out = tf.nn.softmax(out, axis=1) # Use softmax to convert to
probability
out
Out[22]:<tf.Tensor: id=257, shape=(2, 10),
dtype=float32, numpy=
array([[0.18773547, 0.1510464 , 0.09431915, 0.13652141, 0.06579739,
        0.02033597, 0.06067333, 0.0666793 , 0.14594753, 0.07094406],
       [0.5092072 , 0.03887136, 0.0390687 , 0.01911005, 0.03850609,
        0.03442522, 0.08060656, 0.10171875, 0.08244187, 0.05604421]],
       dtype=float32)>
```

Taking the first sample as an example, it can be seen that the index with the highest probability (0.1877) is 0. Because the probability on each index represents the probability that the sample belongs to this category, the probability that the first sample belongs to class 0 is the largest. Therefore, the first sample should most likely belong to class 0. This is a typical application where the index number of the maximum needs to be solved.

We can use tf.argmax(x, axis) and tf.argmin(x, axis) to find the index of the maximum and minimum values of x on the axis parameter. For example:

```
In [23]:pred = tf.argmax(out, axis=1)
pred
Out[23]:<tf.Tensor: id=262, shape=(2,), dtype=int64,
numpy=array([0, 0], dtype=int64)>
```

It can be seen that the maximum probability of the two samples appears on index 0, so it is most likely that they both belong to category 0. We can use category 0 as the predicted category for the two samples.

5.3 Tensor Comparison

In order to get the classification metrics such as accuracy, it is generally necessary to compare the prediction result with the real label. Considering the prediction results of 100 samples, the predicted category can be obtained through tf.argmax.

```
In [24]:out = tf.random.normal([100,10])
out = tf.nn.softmax(out, axis=1) # Convert to probability
pred = tf.argmax(out, axis=1) # Find corresponding category
Out[24]:<tf.Tensor: id=272, shape=(100,), dtype=int64, numpy=
array([0, 6, 4, 3, 6, 8, 6, 3, 7, 9, 5, 7, 3, 7, 1, 5, 6, 1, 2,
       9, 0, 6,
       5, 4, 9, 5, 6, 4, 6, 0, 8, 4, 7, 3, 4, 7, 4, 1, 2, 4,
       9, 4,...
```

The pred variable holds the predicted category of the 100 samples. We compare them with the true labels to get a boolean tensor representing whether each sample predicts the correct one. The tf.equal(a, b) (or tf.math.equal(a, b), which is equivalent) function can compare whether the two tensors are equal, for example:

```
In [25]: # Simiulate the true labels
y = tf.random.uniform([100],dtype=tf.int64,maxval=10)
Out[25]:<tf.Tensor: id=281, shape=(100,), dtype=int64, numpy=
array([0, 9, 8, 4, 9, 7, 2, 7, 6, 7, 3, 4, 2, 6, 5, 0, 9, 4, 5,
       8, 4, 2,
       5, 5, 5, 3, 8, 5, 2, 0, 3, 6, 0, 7, 1, 1, 7, 0, 6, 1, 2,
       1, 3, ...
In [26]:out = tf.equal(pred,y) # Compare true and prediction
Out[26]:<tf.Tensor: id=288, shape=(100,), dtype=bool, numpy=
array([False, False, False, False, True, False, False, False, False,
       False, False, False, False, False, True, False,
       False, True,...
```

The tf.equal function returns the comparison result as a boolean tensor. We only need to count the number of True elements to get the correct number of predictions. In order to achieve this, we first convert the boolean type to an integer tensor, that is, True corresponds to 1, and False corresponds to 0, and then sum the number of 1 to get the number of True elements in the comparison result:

```
In [27]:out = tf.cast(out, dtype=tf.float32) # convert to int type
correct = tf.reduce_sum(out) # get the number of True elements
Out[27]:<tf.Tensor: id=293, shape=(), dtype=float32, numpy=12.0>
```

It can be seen that the number of correct predictions in our randomly generated prediction data is 12, so its accuracy is:

$$accuracy = \frac{12}{100} = 12\%$$

This is the normal level of random prediction models.

Except for the tf.equal function, other commonly used comparison functions are shown in Table 5-1.

Table 5-1. *Common comparison functions*

Function	Comparison logic
tf.math.greater	$a > b$
tf.math.less	$a < b$
tf.math.greater_equal	$a \geq b$
tf.math.less_equal	$a \leq b$
tf.math.not_equal	$a \neq b$
tf.math.is_nan	$a = nan$

5.4 Fill and Copy
5.4.1 Fill

The height and width of images and the length of the sequence signals may not be the same. In order to facilitate parallel computing of the network, it is necessary to expand data of different lengths to the same. We previously introduced that the length of data can be increased by copying. However, repeatedly copying data will destroy the original data structure and is not suitable for some situations. A common practice is to fill in a sufficient number of specific values at the beginning or end of the data. These specific values (e.g., 0) generally represent invalid meanings. This operation is called padding.

Consider a two-sentence tensor that each word is represented by a digital code, such as 1 for I, 2 for like, and so on. The first sentence is "I like the weather today." We assume that the sentence number is encoded as $[1, 2, 3, 4, 5, 6]$. The second sentence is "So do I." with encoding as $[7, 8, 1, 6]$. In order to store the two sentences in one tensor, we need to keep the length of these two sentences consistent, that is, we need to expand the length of the second sentence to 6. A common padding scheme is to pad a number of zeros at the end of the second sentence, that is, $[7, 8, 1, 6, 0, 0]$. Now the two sentences can be stacked and combined into a tensor of shape $[2, 6]$.

The padding operation can be implemented by the tf.pad(x, paddings) function. The parameter paddings is a list of multiple nested schemes with the format of [*Left Padding, Right Padding*]. For example, *paddings* = $[[0, 0], [2, 1], [1, 2]]$ indicates that the first dimension is not filled, and the left (the beginning) of the second dimension is filled with two units, and fill one unit on the right (end) of the second dimension, fill one unit on the left of the third dimension, and fill two units on the right of the third dimension. Considering the example of the preceding two sentences, two units need to be filled to the right of the first dimension of the second sentence, and the paddings scheme is $[[0, 2]]$:

```
In [28]:a = tf.constant([1,2,3,4,5,6]) # 1st sentence
b = tf.constant([7,8,1,6]) # 2nd sentence
b = tf.pad(b, [[0,2]]) # Pad two 0's in the end of 2nd sentence
b
Out[28]:<tf.Tensor: id=3, shape=(6,), dtype=int32,
numpy=array([7, 8, 1, 6, 0, 0])>
```

After filling, the shape of the two tensors is consistent, and we can stack them together. The code is as follows:

```
In [29]:tf.stack([a,b],axis=0) # Stack a and b
Out[29]:<tf.Tensor: id=5, shape=(2, 6), dtype=int32, numpy=
array([[1, 2, 3, 4, 5, 6],
       [7, 8, 1, 6, 0, 0]])>
```

In natural language processing, sentences with different lengths need to be loaded. Some sentences are shorter, such as only ten words, and some sentences are longer, such as more than 100 words. In order to be able to save in the same tensor, a threshold that can cover most of the sentence length is generally selected, such as 80 words. For sentences with less than 80 words, we fill with 0s at the end of those sentences. For sentences with more than 80 words, we truncate the sentence to 80 words by removing some words at the end. We will use the IMDB dataset as an example to demonstrate how to transform sentences of unequal length into a structure of equal length. The code is as follows:

```
In [30]:total_words = 10000 # Set word number
max_review_len = 80 # Maximum length for each sentence
embedding_len = 100 # Word vector length
# Load IMDB dataset
```

```
(x_train, y_train), (x_test, y_test) = keras.datasets.imdb.
load_data(num_words=total_words)
# Pad or truncate sentences to the same length with end padding
and truncation
x_train = keras.preprocessing.sequence.pad_sequences(x_train,
maxlen=max_review_len,truncating='post',padding='post')
x_test = keras.preprocessing.sequence.pad_sequences(x_test,
maxlen=max_review_len,truncating='post',padding='post')
print(x_train.shape, x_test.shape)
Out[30]: (25000, 80) (25000, 80)
```

In the preceding code, we set the maximum length of the sentence max_review_len to 80 words. Through the keras.preprocessing.sequence. pad_sequences function, we can quickly complete the padding and truncation implementation. Take one of the sentences as an example, and the transformed vector is like this:

```
[    1  778  128    74   12  630  163   15    4 1766 7982 1051    2   32
    85  156   45    40  148  139  121  664  665   10   10 1361  173    4
   749    2   16 3804    8    4  226   65   12   43  127   24    2   10
    10    0    0    0    0    0    0    0    0    0    0    0    0    0
     0    0    0    0    0    0    0    0    0    0    0    0    0    0
     0    0    0    0    0    0    0    0    0    0]
```

We can see that the final part of the sentence is filled with 0s so that the length of the sentence is exactly 80. In fact, we can also choose to fill the beginning part of the sentence when the length of the sentence is not enough. After processing, all sentence length becomes 80, so that the training set can be uniformly stored in the tensor of shape [25000, 80] and the test set can be stored in the tensor of shape [25000, 80].

Let's introduce an example of filling in multiple dimensions at the same time. Consider padding the height and width dimensions of images. If we have pictures with dimension 28 × 28 and the input layer shape of neural network is 32 × 32, we need to fill the images to get the shape of 32 × 32. We can choose to fill 2 units each in the upper, lower, left, and right of the image matrix as shown in Figure 5-2.

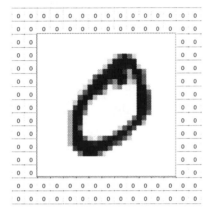

Figure 5-2. *Image padding example*

The preceding padding scheme can be implemented as follows:

```
In [31]:
x = tf.random.normal([4,28,28,1])
# Pad two units at each edge of the image
tf.pad(x,[[0,0],[2,2],[2,2],[0,0]])
Out[31]:
<tf.Tensor: id=16, shape=(4, 32, 32, 1), dtype=float32, numpy=
array([[[[ 0.          ],
         [ 0.          ],
         [ 0.          ],...
```

After the padding operation, the size of the picture becomes 32 × 32, which meets the input requirements of the neural network.

5.4.2 Copy

In the dimensional transformation section, we introduced the tf.tile function of copying the dimension of length 1. Actually, the tf.tile function can be used to repeatedly copy multiple copies of data in any dimension. For example, for image data with shape [4,32,32,3], if the copy scheme is multiples=[2, 3, 3, 1], that means the channel dimension is not copied, three copies in the height and width dimensions, and two copies in the image number dimension. The implementation is as follows:

```
In [32]:x = tf.random.normal([4,32,32,3])
tf.tile(x,[2,3,3,1])
Out[32]:<tf.Tensor: id=25, shape=(8, 96, 96, 3),
dtype=float32, numpy=
array([[[[ 1.20957184e+00,  2.82766962e+00,  1.65782201e+00],
        [ 3.85402292e-01,  2.00732923e+00, -2.79068202e-01],
        [-2.52583921e-01,  7.82584965e-01,  7.56870627e-01],...
```

5.5 Data Limiting

Consider how to implement the nonlinear activation function ReLU. In fact, it can be implemented by simple data limiting operations with the range of elements being limited to $x \in [0, +\infty)$.

In TensorFlow, the lower limit of the data can be set through tf.maximum (x, a), that is, the upper limit of the data can be set through tf.minimum (x, a).

```
In [33]:x = tf.range(9)
tf.maximum(x,2) # Set lower limit of x to 2
Out[33]:<tf.Tensor: id=48, shape=(9,), dtype=int32,
numpy=array([2, 2, 2, 3, 4, 5, 6, 7, 8])>
In [34]:tf.minimum(x,7) # Set x upper limit to 7
```

```
Out[34]:<tf.Tensor: id=41, shape=(9,), dtype=int32,
numpy=array([0, 1, 2, 3, 4, 5, 6, 7, 7])>
```

Based on tf.maximum function, we can implement ReLU as follows:

```
def relu(x): # ReLU function
    return tf.maximum(x,0.) # Set lower limit of x to be 0
```

By combining tf.maximum(x, a) and tf.minimum(x, b), you can limit the upper and lower boundaries of the data at the same time, that is, $x \in [a, b]$.

```
In [35]:x = tf.range(9)
tf.minimum(tf.maximum(x,2),7) # Set x range to be [2, 7]
Out[35]:<tf.Tensor: id=57, shape=(9,), dtype=int32,
numpy=array([2, 2, 2, 3, 4, 5, 6, 7, 7])>
```

More conveniently, we can use the tf.clip_by_value function to achieve upper and lower clipping:

```
In [36]:x = tf.range(9)
tf.clip_by_value(x,2,7) # Set x range to be [2, 7]
Out[36]:<tf.Tensor: id=66, shape=(9,), dtype=int32,
numpy=array([2, 2, 2, 3, 4, 5, 6, 7, 7])>
```

5.6 Advanced Operations

Most of the preceding functions are common and easy to understand. Next, we will introduce some commonly used but slightly more complicated functions.

5.6.1 tf.gather

The tf.gather function can collect data according to the index number. Consider the example of grade books. Assume that there are four classes, 35 students in each class, eight subjects in total, and the tensor shape of the grade books is [4,35,8].

```
x = tf.random.uniform([4,35,8],maxval=100,dtype=tf.int32)
```

Now we need to collect the grade books of the first and second classes. We can give the index number of the class we want to collect (e.g., [0, 1]) and specify the dimension of the class (e.g., axis = 0). And then collect the data through the tf.gather function.

```
In [38]:tf.gather(x,[0,1],axis=0) # Collect data for 1st and
2nd classes
Out[38]:<tf.Tensor: id=83, shape=(2, 35, 8),
dtype=int32, numpy=
array([[[43, 10, 93, 85, 75, 87, 28, 19],
        [52, 17, 44, 88, 82, 54, 16, 65],
        [98, 26,  1, 47, 59,  3, 59, 70],...
```

In fact, the preceding requirements can be more conveniently achieved through slicing. However, for irregular indexing methods, such as the need to spot check the grade data of students 1, 4, 9, 12, 13, and 27, the slicing method is not suitable. The tf.gather function is designed for this situation and is more convenient to use. The implementation is as follows:

```
In [39]: # Collect the grade of students 1,4,9,12,13 and 27
tf.gather(x,[0,3,8,11,12,26],axis=1)
Out[39]:<tf.Tensor: id=87, shape=(4, 6, 8), dtype=int32, numpy=
array([[[43, 10, 93, 85, 75, 87, 28, 19],
        [74, 11, 25, 64, 84, 89, 79, 85],...
```

If we need to collect the grades of the third and fifth subjects of all students, we can specify the subject dimension axis = 2 to achieve the following:

```
# Collect the grades of the 3rd and 5th subjects of all
students
In [40]:tf.gather(x,[2,4],axis=2)
Out[40]:<tf.Tensor: id=91, shape=(4, 35, 2),
dtype=int32, numpy=
array([[[93, 75],
        [44, 82],
        [ 1, 59],...
```

It can be seen that tf.gather is very suitable for situations where the index numbers are not regular. The index numbers can be arranged out of order, and the data collected will also be in the corresponding order. For example:

```
In [41]:a=tf.range(8)
a=tf.reshape(a,[4,2])
Out[41]:<tf.Tensor: id=115, shape=(4, 2), dtype=int32, numpy=
array([[0, 1],
       [2, 3],
       [4, 5],
       [6, 7]])>
In [42]:tf.gather(a,[3,1,0,2],axis=0) # Collect element 4,2,1,3
Out[42]:<tf.Tensor: id=119, shape=(4, 2), dtype=int32, numpy=
array([[6, 7],
       [2, 3],
       [0, 1],
       [4, 5]])>
```

We will make the problem a little more complicated. If we want to check the subject scores of students $[3, 4, 6, 27]$ in class $[2, 3]$, we can do this by combining multiple tf.gather operations. First extract data for class $[2, 3]$:

```
In [43]:
students=tf.gather(x,[1,2],axis=0) # Collect data for
class 2 and 3
Out[43]:<tf.Tensor: id=227, shape=(2, 35, 8),
dtype=int32, numpy=
array([[[ 0, 62, 99,  7, 66, 56, 95, 98],...
```

Then we extract the corresponding data for selected students:

```
In [44]:
tf.gather(students,[2,3,5,26],axis=1) # Collect data for
students 3,4,6,27
Out[44]:<tf.Tensor: id=231, shape=(2, 4, 8),
dtype=int32, numpy=
array([[[[69, 67, 93,  2, 31,  5, 66, 65], ...
```

Now we get the selected tensor with shape $[2, 4, 8]$.

This time we want to spot check all subjects of the second classmate of the second class, all subjects of the third classmate of the third class, and all subjects of the fourth classmate of the fourth class. So how does it work? Data can be manually extracted one by one in a clumsy way. First extract the data of the first sampling point: $x[1, 1]$.

```
In [45]: x[1,1]
Out[45]:<tf.Tensor: id=236, shape=(8,), dtype=int32,
numpy=array([45, 34, 99, 17,  3,  1, 43, 86])>
```

Then extract the data of the second sampling point $x[2, 2]$ and the data of the third sampling point $x[3, 3]$, and finally combine the sampling results together.

```
In [46]: tf.stack([x[1,1],x[2,2],x[3,3]],axis=0)
Out[46]:<tf.Tensor: id=250, shape=(3, 8), dtype=int32, numpy=
array([[45, 34, 99, 17,  3,  1, 43, 86],
       [11, 25, 84, 95, 97, 95, 69, 69],
       [ 0, 89, 52, 29, 76,  7,  2, 98]])>
```

Using the preceding method, we can correctly obtain the result of shape [3, 8], where 3 represents the number of sampling points and 4 represents the data of each sampling point. The biggest problem is that the sampling is performed manually and serially, and the calculation efficiency is extremely low. Is there a better way to achieve this?

5.6.2 tf.gather_nd

With the tf.gather_nd function, we can sample multiple points by specifying the multidimensional coordinates of each sampling point. Going back to the preceding challenge, we want to spot check all the subjects of the second classmate of the second class, all the subjects of the third classmate of the third class, and all the subjects of the fourth classmate of the fourth class. Then the index coordinates of the three sampling points can be recorded as [1, 1], [2, 2], and [3, 3], and we can combine this sampling scheme into a list [[1, 1], [2, 2], [3, 3]].

```
In [47]:
tf.gather_nd(x,[[1,1],[2,2],[3,3]])
Out[47]:<tf.Tensor: id=256, shape=(3, 8), dtype=int32, numpy=
array([[45, 34, 99, 17,  3,  1, 43, 86],
       [11, 25, 84, 95, 97, 95, 69, 69],
       [ 0, 89, 52, 29, 76,  7,  2, 98]])>
```

The result is consistent with the serial sampling method, and the implementation is more concise and efficient.

Generally, when using tf.gather_nd to sample multiple samples, for example, if we want to sample class i, student j, and subject k, we can use the expression $[..., [i, j, k], ...]$. The inner list contains the corresponding index coordinates of each sampling point, for example:

```
In [48]:
tf.gather_nd(x,[[1,1,2],[2,2,3],[3,3,4]])
Out[48]:<tf.Tensor: id=259, shape=(3,), dtype=int32,
numpy=array([99, 95, 76])>
```

In the preceding code, we extracted the grades of subject 1 of class 1 student 2, subject 2 of class 2 student 3, and class 3 of student 3 subject 4. There are a total of three grade data, and the results are summarized into a tensor with shape of [3].

5.6.3 tf.boolean_mask

In addition to sampling by a given index number, sampling can also be performed by a given mask. Continue to take the gradebook tensor with shape [4,35,8] as an example; this time we use the mask method for data extraction.

Consider sampling in the class dimension and set the corresponding mask as:

$$mask = [True, False, False, True]$$

That is, the first and fourth classes are sampled. Using the function tf.boolean_mask(x, mask, axis), the sampling can be performed on the corresponding axis according to the mask scheme, which is realized as:

```
In [49]:
tf.boolean_mask(x,mask=[True, False,False,True],axis=0)
```

```
Out[49]:<tf.Tensor: id=288, shape=(2, 35, 8),
dtype=int32, numpy=
array([[[43, 10, 93, 85, 75, 87, 28, 19],...
```

Note that the length of the mask must be the same as the length of the corresponding dimension. If we are sampling in the class dimension, we must specify the mask with length 4 to specify whether the four classes are sampling.

If mask sampling is performed on eight subjects, we need to set the mask sampling scheme to

$$mask = [True, False, False, True, True, False, False, True]$$

That is, sample the first, fourth, fifth, and eighth subjects:

```
In [50]:
tf.boolean_mask(x,mask=[True,False,False,True,True,False,False,
True],axis=2)
Out[50]:<tf.Tensor: id=318, shape=(4, 35, 4),
dtype=int32, numpy=
array([[[43, 85, 75, 19],...
```

It is not difficult to find that the usage of tf.boolean_mask here is actually very similar to tf.gather, except that one is sampled by the mask method, and the other is directly given the index number.

Now let's consider a multidimensional mask sampling method similar to tf.gather_nd. In order to facilitate the demonstration, we reduced the number of classes to two and the number of students to three. That is, a class has only three students and the tensor shape is [2, 3, 8]. If we want to sample students 1 to 2 of the first class and students 2 to 3 of the second class, we can achieve it using tf.gather_nd:

```
In [51]:x = tf.random.uniform([2,3,8],maxval=100,dtype
=tf.int32)
tf.gather_nd(x,[[0,0],[0,1],[1,1],[1,2]])
Out[51]:<tf.Tensor: id=325, shape=(4, 8), dtype=int32, numpy=
array([[52, 81, 78, 21, 50,  6, 68, 19],
       [53, 70, 62, 12,  7, 68, 36, 84],
       [62, 30, 52, 60, 10, 93, 33,  6],
       [97, 92, 59, 87, 86, 49, 47, 11]])>
```

A total of four students' results were sampled with a shape of $[4, 8]$.

If we use a mask, how do we express it? Table 5-2 expresses the sampling of the corresponding position:

Table 5-2. *Sampling using mask method*

	Student 0	**Student 1**	**Student 2**
Class 0	True	True	False
Class 1	False	True	True

Therefore, through this table, the sampling scheme using the mask method can be well expressed. The code is implemented as follows:

```
In [52]:
tf.boolean_mask(x,[[True,True,False],[False,True,True]])
Out[52]:<tf.Tensor: id=354, shape=(4, 8), dtype=int32, numpy=
array([[52, 81, 78, 21, 50,  6, 68, 19],
       [53, 70, 62, 12,  7, 68, 36, 84],
       [62, 30, 52, 60, 10, 93, 33,  6],
       [97, 92, 59, 87, 86, 49, 47, 11]])>
```

The result is exactly the same as tf.gather_nd method. It can be seen that tf.boolean_mask method can be used for both one- and multidimensional samplings.

The preceding three operations are more commonly used, especially tf.gather and tf.gather_nd. Three additional advanced operations are added in the following.

5.6.4 tf.where

Through the tf.where(cond, a, b) function, we can read data from the parameter a or b according to the true and false conditions of the cond condition. The condition determination rule is as follows:

$$o_i = \{a_i \ cond_i \ \text{为} \ True \ \ b_i \ cond_i \ \text{为} \ False$$

Among them i is the element index of the tensor. The size of the returned tensor is consistent with a and b. When the corresponding position of $cond_i$ is True, the data is copied from a_i to o_i. Otherwise, the data is copied from b_i to o_i. Consider extracting data from two tensors A and B of all 1's and 0's, where the position of True in $cond_i$ extracts element 1 from the corresponding position of A, otherwise extracts 0 from the corresponding position of B. The code is as follows:

```
In [53]:
a = tf.ones([3,3])  # Tensor A
b = tf.zeros([3,3]) # Tensor B
# Create condition matrix
cond = tf.constant([[True,False,False],[False,True,False],[True,
True,False]])
tf.where(cond,a,b)
Out[53]:<tf.Tensor: id=384, shape=(3, 3), dtype=float32, numpy=
array([[1., 0., 0.],
```

```
     [0., 1., 0.],
     [1., 1., 0.]], dtype=float32)>
```

It can be seen that the positions of 1 in the returned tensor are all from tensor *A*, and the positions of 0 in the returned tensor are from tensor *B*.

When the parameter a=b=None, that is, a and b parameters are not specified; tf.where returns the index coordinates of all True elements in the cond tensor. Consider the following cond tensor:

```
In [54]: cond
Out[54]:<tf.Tensor: id=383, shape=(3, 3), dtype=bool, numpy=
array([[ True, False, False],
       [False,  True, False],
       [ True,  True, False]])>
```

True appears four times in total, and the index at the position of each True element is [0, 0], [1, 1], [2, 0], and [2, 1] respectively. The index coordinates of these elements can be obtained directly through the form of tf.where(cond) as follows:

```
In [55]:tf.where(cond)
Out[55]:<tf.Tensor: id=387, shape=(4, 2), dtype=int64, numpy=
array([[0, 0],
       [1, 1],
       [2, 0],
       [2, 1]], dtype=int64)>
```

So what's the use of this? Consider a scenario where we need to extract all the positive data and indexes in a tensor. First construct tensor a and obtain the position masks of all positive numbers through comparison operations:

```
In [56]:x = tf.random.normal([3,3]) # Create tensor a
Out[56]:<tf.Tensor: id=403, shape=(3, 3), dtype=float32, numpy=
```

```
array([[-2.2946844 ,   0.6708417 ,  -0.5222212 ],
       [-0.6919401 ,  -1.9418817 ,   0.3559235 ],
       [-0.8005251 ,   1.0603906 ,  -0.68819374]],
       dtype=float32)>
```

By comparison operation, we get the mask of all positive numbers:

```
In [57]:mask=x>0 # equivalent to tf.math.greater()
mask
Out[57]:<tf.Tensor: id=405, shape=(3, 3), dtype=bool, numpy=
array([[False,  True, False],
       [False, False,  True],
       [False,  True, False]])>
```

Extract the index coordinates of the True element in the mask tensor via tf.where:

```
In [58]:indices=tf.where(mask) # Extract all element
greater than 0
Out[58]:<tf.Tensor: id=407, shape=(3, 2), dtype=int64, numpy=
array([[0, 1],
       [1, 2],
       [2, 1]], dtype=int64)>
```

After getting the index, we can restore all positive elements through tf.gather_nd:

```
In [59]:tf.gather_nd(x,indices) # Extract all positive elements
Out[59]:<tf.Tensor: id=410, shape=(3,), dtype=float32,
numpy=array([0.6708417, 0.3559235, 1.0603906], dtype=float32)>
```

In fact, after we get the mask, we can also get all the positive elements directly through tf.boolean_mask:

```
In [60]:tf.boolean_mask(x,mask) # Extract all positive elements
Out[60]:<tf.Tensor: id=439, shape=(3,), dtype=float32,
numpy=array([0.6708417, 0.3559235, 1.0603906], dtype=float32)>
```

Through the preceding series of comparisons, we can intuitively feel that this function has great practical applications and also get a deep understanding of their nature to be able to achieve our purpose in a more flexible, simple, and efficient way.

5.6.5 tf.scatter_nd

The tf.scatter_nd(indices, updates, shape) function can efficiently refresh part of the tensor data, but this function can only perform refresh operations on all 0 tensors, so it may be necessary to combine other operations to implement the data refresh function for non-zero tensors.

Figure 5-3 shows the refresh calculation principle of the one-dimensional all-zero tensor. The shape of the whiteboard is represented by the shape parameter, the index number of the data to be refreshed is represented by indices, and updates parameter contains the new data. The tf.scatter_nd(indices, updates, shape) function writes the new data to the all-zero tensor according to the index position given by indices and returns the updated result tensor.

Figure 5-3. *scatter_nd function for refreshing data*

We implement a refresh example of the tensor in Figure 5-3 as follows:

```
In [61]: # Create indices for refreshing data
indices = tf.constant([[4], [3], [1], [7]])
# Create data for filling the indices
updates = tf.constant([4.4, 3.3, 1.1, 7.7])
# Refresh data for all 0 vector of length 8
tf.scatter_nd(indices, updates, [8])
Out[61]:<tf.Tensor: id=467, shape=(8,), dtype=float32,
numpy=array([0. , 1.1, 0. , 3.3, 4.4, 0. , 0. , 7.7],
dtype=float32)>
```

It can be seen that on the all-zero tensor of length 8, the data of the corresponding positions are filled in with values from updates.

Consider an example of a three-dimensional tensor. As shown in Figure 5-4, the shape of the all-zero tensor is a feature map with four channels in total, and each channel has a size 4×4. New data updates have a shape $[2, 4, 4]$, which needs to be written in indices $[1, 3]$.

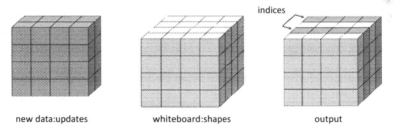

new data:updates   whiteboard:shapes   output

Figure 5-4. *3D tensor data refreshing*

We write the new feature map into the existing tensor as follows:

```
In [62]:
indices = tf.constant([[1],[3]])
updates = tf.constant([
    [[5,5,5,5],[6,6,6,6],[7,7,7,7],[8,8,8,8]],
    [[1,1,1,1],[2,2,2,2],[3,3,3,3],[4,4,4,4]]]
```

```
])
tf.scatter_nd(indices,updates,[4,4,4])
Out[62]:<tf.Tensor: id=477, shape=(4, 4, 4),
dtype=int32, numpy=
array([[[0, 0, 0, 0],
        [0, 0, 0, 0],
        [0, 0, 0, 0],
        [0, 0, 0, 0]],
       [[5, 5, 5, 5], # New data 1
        [6, 6, 6, 6],
        [7, 7, 7, 7],
        [8, 8, 8, 8]],
       [[0, 0, 0, 0],
        [0, 0, 0, 0],
        [0, 0, 0, 0],
        [0, 0, 0, 0]],
       [[1, 1, 1, 1], # New data 2
        [2, 2, 2, 2],
        [3, 3, 3, 3],
        [4, 4, 4, 4]]])>
```

It can be seen that the data is refreshed onto the second and fourth channel feature maps.

5.6.6 tf.meshgrid

The tf.meshgrid function can easily generate the coordinates of the sampling points of the two-dimensional grid, which is convenient for applications such as visualization. Consider the Sinc function with two independent variables x and y as:

$$z = \frac{\sin\sin\left(x^2 + y^2\right)}{x^2 + y^2}$$

If we need to draw a 3D surface of the Sinc function in the interval $x \in [-8, 8]$, $y \in [-8, 8]$, as shown in Figure 5-5, we first need to generate the grid point coordinate set of the x and y axes, so that the output value of the function at each position can be calculated by the expression of the Sinc function z. We can generate 10,000 coordinate sampling points by:

```
points = []
for x in range(-8,8,100): # Loop to generate 100 sampling point
for x-axis
for y in range(-8,8,100): # Loop to generate 100 sampling point
for y-axis
        z = sinc(x,y)
        points.append([x,y,z])
```

Obviously, this serial sampling method is extremely inefficient. Is there a simple and efficient way to generate grid coordinates? The answer is the tf.meshgrid function.

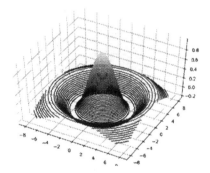

Figure 5-5. *Sinc function*

By sampling 100 data points on the x-axis and y-axis, respectively, the tf.meshgrid(x, y) can be used to generate tensor data of these 10,000 data points and save them in a tensor of shape [100,100,2]. For the convenience of calculation, tf.meshgrid will return two tensors after cutting in the axis = two-dimensional, where tensor *A* contains the x-coordinates of all points and tensor *B* contains the y-coordinates of all points.

```
In [63]:
x = tf.linspace(-8.,8,100) # x-axis
y = tf.linspace(-8.,8,100) # y-axis
x,y = tf.meshgrid(x,y)
x.shape,y.shape
Out[63]: (TensorShape([100, 100]), TensorShape([100, 100]))
```

Using the generated grid point coordinate tensors, the Sinc function is implemented in TensorFlow as follows:

```
z = tf.sqrt(x**2+y**2)
z = tf.sin(z)/z  # sinc function
```

The matplotlib library can be used to draw the 3D surface of the function as shown in Figure 5-5.

```
import matplotlib
from matplotlib import pyplot as plt
from mpl_toolkits.mplot3d import Axes3D

fig = plt.figure()
ax = Axes3D(fig)
# Plot Sinc function
ax.contour3D(x.numpy(), y.numpy(), z.numpy(), 50)
plt.show()
```

5.7 Load Classic Datasets

So far, we have learned the common tensor operations and are ready to implement most of the deep networks. Finally, we will complete this chapter with a classification network model implemented in a tensor format. Before that, we first formally introduce how to use the tools provided by TensorFlow to load datasets conveniently for commonly used classic datasets. For loading custom datasets, we will introduce in the subsequent chapters.

In TensorFlow, the keras.datasets module provides automatic download, management, loading, and conversion functions of commonly used classic datasets, as well as the corresponding Dataset objects, which facilitates multi-threading, preprocessing, shuffling, and batch-training.

Some commonly used classic datasets:

- Boston Housing: the Boston housing price trend dataset, used for training and testing of regression models.

- CIFAR10/100: a real picture dataset for picture classification tasks.

- MNIST/Fashion_MNIST: a handwritten digital picture dataset, used for picture classification tasks.

- IMDB: sentiment classification task dataset, for text classification tasks.

These datasets are used very frequently in machine learning or deep learning. For the newly proposed algorithms, it is generally preferred to test on classic datasets, and then try to migrate to larger and more complex data sets.

We can use the datasets.xxx.load_data() function to automatically load classic datasets, where xxx represents the specific dataset name, such as "CIFAR10" and "MNIST". TensorFlow will cache the data in the .keras/datasets folder in the user directory by default, as shown in Figure 5-6. Users do not need to care about how the dataset is saved. If the current dataset is not in the cache, it will be downloaded, decompressed, and loaded automatically from the network. If it is already in the cache, the load is automatically completed. For example, to automatically load the MNIST dataset:

```
In [66]:
import  tensorflow as tf
from    tensorflow import keras
from    tensorflow.keras import datasets # Load dataset
loading module
# Load MNIST dataset
(x, y), (x_test, y_test) = datasets.mnist.load_data()
print('x:', x.shape, 'y:', y.shape, 'x test:', x_test.shape, 'y
test:', y_test)
Out [66]:
x: (60000, 28, 28) y: (60000,) x test: (10000, 28, 28) y test:
[7 2 1 ... 4 5 6]
```

The load_data() function will return data in the corresponding format. For the image datasets MNIST and CIFAR10, two tuples will be returned. The first tuple holds the training data x and y objects; the second tuple is the test data x_test and y_test objects. All data is stored in a Numpy array container.

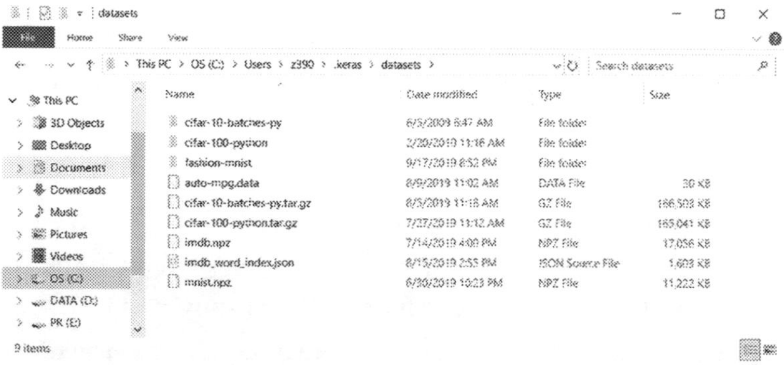

Figure 5-6. *TensorFlow classic dataset saving directory*

After data is loaded into the memory, it needs to be converted into a Dataset object in order to take advantage of the various convenient functions provided by TensorFlow. Dataset.from_tensor_slices can be used to convert the training data image x and label y into Dataset objects:

```
# Convert to Dataset objects
train_db = tf.data.Dataset.from_tensor_slices((x, y))
```

After converting data into a Dataset object, we generally need to add a series of standard processing steps for the dataset, such as random shuffling, preprocessing, and batch loading.

5.7.1 Shuffling

Using the Dataset.shuffle(buffer_size) function, we can randomly shuffle the Dataset objects to prevent the data from being generated in a fixed order during each training, so that the model will not "remember" the label information. The code is implemented as follows:

```
train_db = train_db.shuffle(10000)
```

Here the buffer_size parameter specifies the size of the buffer pool, which is generally set to a larger constant. Calling these utility functions provided by the Dataset will return a new Dataset object.

$$db = db.step1().step2().step3.()$$

This method completes all data processing steps in order, which is very convenient to implement.

5.7.2 Batch Training

In order to take advantage of the parallel computing capabilities of GPUs, multiple samples are generally calculated simultaneously during the network calculation process. We call this training method batch training, and the number of samples in one batch is called batch size. In order to generate batch size samples from the Dataset at one time, the dataset needs to be set to batch training mode. The implementation is as follows:

```
train_db = train_db.batch(128) # batch size is 128
```

Here 128 is the batch size parameter, that is, 128 samples are calculated at one time in parallel. Batch sis generally set according to the user's GPU memory resources. When the GPU memory is insufficient, the batch size can be appropriately reduced.

5.7.3 Preprocessing

The format of the dataset loaded from keras.datasets cannot meet the model input requirements in most cases, so it is necessary to implement the preprocessing step according to the user's logic. The Dataset object can call the user-defined preprocessing logic very conveniently by providing the map(func) utility function, while the preprocessing logic is implemented in the func function. For example, the following code calls a function named preprocess to complete the preprocessing of each sample:

```
# Preprocessing is implemented in the preprocess function
train_db = train_db.map(preprocess)
```

Considering the MNIST handwritten digital picture dataset, image x loaded from keras.datasets after .batch () operation has shape $[b, 28, 28]$, where the pixels are represented by integers from 0 to 255 and the label shape is $[b]$ with digital encoding. The actual neural network input generally needs to normalize the image data to the interval $[0, 1]$ or $[-1, 1]$ around 0. At the same time, according to the network settings, the input view of shape $[28, 28]$ needs to be adjusted to an appropriate format. For label information, we can choose one-hot encoding during preprocessing or the error calculation.

Here we map the MNIST image data to interval $[0, 1]$ and adjust the view to $[b, 28 * 28]$. For label data, we choose to perform one-hot encoding in the preprocessing function. The preprocess function is implemented as follows:

```
def preprocess(x, y): # Customized preprocessing function
    x = tf.cast(x, dtype=tf.float32) / 255.
    x = tf.reshape(x, [-1, 28*28])      # flatten
    y = tf.cast(y, dtype=tf.int32)      # convert to int
    y = tf.one_hot(y, depth=10)     # one-hot encoding
    return x,y
```

5.7.4 Epoch Training

For the Dataset object, we can iterate through the following ways:

```
    for step, (x,y) in enumerate(train_db): # Iterate with step
```
or
```
    for x,y in train_db: # Iterate without step
```

The x and y objects returned each time are batch samples and labels. When one iteration is completed for all samples of train_db, the for loop terminates. Completing a batch of data training is called a Step, and completing an iteration of the entire training set through multiple steps is called an Epoch. In training, it is usually necessary to iterate multiple Epochs on the data set to obtain better training results. For example, fixed training of 20 Epoch is implemented as follows:

```
    for epoch in range(20): # Epoch number
        for step, (x,y) in enumerate(train_db): # Iteration
        step number
            # training...
```

In addition, we can also set a Dataset object so that the dataset will traverse multiple times before exiting such as:

```
train_db = train_db.repeat(20) # Dataset iteration 20 times
```

The preceding code makes the for x, y in train_db iterates 20 Epochs before exiting. No matter which of these methods is used, the same effect can be achieved. Since the previous chapter has completed the actual calculation of forward calculation, we skip it here.

5.8 Hands-On MNIST Dataset

We have already introduced and implemented the forward propagation and dataset. Now let's finish the remaining classification task logic. In the training process, the error data can be effectively monitored by printing out after several steps. The code is as follows:

```
# Print training error every 100 steps
if step % 100 == 0:
    print(step, 'loss:', float(loss))
```

Since loss is a tensor type of TensorFlow, it can be converted to a standard Python floating-point number through the float() function. After several Steps or several Epoch trainings, a test (verification) can be performed to obtain the current performance of the model, for example:

```
if step % 500 == 0: # Do a test every 500 steps
    # evaluate/test
```

Now let's use the tensor operation functions to complete the actual calculation of accuracy. First consider a batch sample x. The network's predicted value can be obtained through forward calculation as follows:

```
for x, y in test_db: # Iterate through test dataset
    h1 = x @ w1 + b1 # 1st layer
    h1 = tf.nn.relu(h1) # Activation function
```

```
h2 = h1 @ w2 + b2 # 2nd layer
h2 = tf.nn.relu(h2) # Activation function
out = h2 @ w3 + b3 # Output layer
```

The shape of the predicted value is $[b, 10]$. It represents the probability that the sample belongs to each category. We select the index number where the maximum probability occurs according to the tf.argmax function, which is the most likely category number of the sample:

```
# Select the max probability category
pred = tf.argmax(out, axis=1)
```

Since y has already been one-hot encoded in preprocessing, we can get the category number for y similarly:

```
y = tf.argmax(y, axis=1)
```

With tf.equal, we can compare whether the two results are equal:

```
correct = tf.equal(pred, y)
```

Sum the number of all True (converted to 1) element in the result, which is the correct number of predictions:

```
total_correct += tf.reduce_sum(tf.cast(correct,
dtype=tf.int32)).numpy()
```

Divide the correct number of predictions by the total number of tests to get the accuracy, and print it out as follows:

```
# Calcualte accuracy
print(step, 'Evaluate Acc:', total_correct/total)
```

After training a simple three-layer neural network with 20 Epochs, we achieved an accuracy of 87.25% on the test set. If we use complex neural network models and fine-tune network hyperparameters, we can get better accuracy. The training error curve is shown in Figure 5-7, and the test accuracy curve is shown in Figure 5-8.

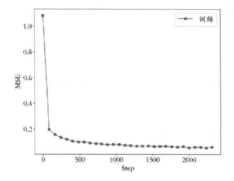

Figure 5-7. *MNIST training loss*

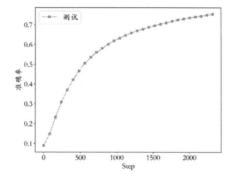

Figure 5-8. *MNIST testing accuracy*

CHAPTER 6

Neural Networks

It is difficult to imagine which big industry will not be changed by artificial intelligence. Artificial intelligence will play a major role in these industries, and this trend is very obvious.

—Andrew Ng

The ultimate goal of machine learning is to find a good set of parameters, so that the model can learn the mapping relationship $f_\theta : x \to y, x, y \in D^{train}$ from the training set and use the trained relationship to predict new samples. Neural networks belong to a branch of research in machine learning. It specifically refers to a model that uses multiple neurons to parameterize the mapping function f_θ.

6.1 Perceptron

In 1943, American neuroscientist Warren Sturgis McCulloch and mathematical logician Walter Pitts were inspired by the structure of biological neurons and proposed a mathematical model of artificial neurons, which was further developed and proposed by American neurophysicist Frank Rosenblatt, which is known as perceptron model. In 1957, Frank Rosenblatt implemented the perceptron model on an IBM-704 computer. This model can complete some simple visual classification tasks, such as distinguishing triangles, circles, and rectangles [1].

The structure of the perceptron model is shown in Figure 6-1. It accepts a one-dimensional vector of length n, $x = [x_1, x_2, ..., x_n]$, and each input node is aggregated as a variable through a connection of weights w_i, $i\epsilon[1, n]$, namely:

$$z = w_1 x_1 + w_2 x_2 + \cdots + w_n x_n + b$$

Among them, b is called the bias of the perceptron, and the one-dimensional vector $w = [w_1, w_2, ..., w_n]$ is called the weight of the perceptron, while z is called the net activation value of the perceptron.

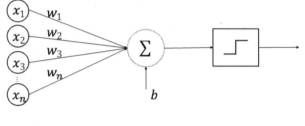

Input x Output a

Figure 6-1. *Perception model*

The preceding formula can be written in vector form:

$$z = w^T x + b$$

Perceptron is a linear model and cannot deal with linear inseparability. The activation value is obtained by adding the activation function after the linear model:

$$a = \sigma(z) = \sigma(w^T x + b)$$

The activation function can be a step function. As shown in Figure 6-2, the output of the step function is only 0/1. When $z < 0$, 0 was then output, representing category 0; when $z \geq 0$, 1 was the output, representing category 1, namely:

$$a = \{1\, w^T x + b \geq 0 \quad 0\, w^T x + b < 0$$

It can also be a sign function as shown in Figure 6-3, and the expression is:

$$a = \{1\; w^T x + b \geq 0 - 1\; w^T x + b < 0$$

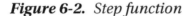

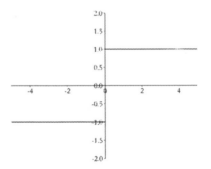

Figure 6-2. *Step function*

Figure 6-3. *Sign function*

After adding the activation function, the perceptron model can be used to complete the binary classification task. The step and the sign functions are discontinuous at $z = 0$, so the gradient descent algorithm cannot be used to optimize the parameters.

In order to enable the perceptron model to automatically learn from the data, Frank Rosenblatt proposed a perceptron learning algorithm, as shown in Algorithm 1.

Algorithm 1: Perceptron training algorithm

Initialize $w = 0, b = 0$

repeat

  Randomly select a sample (x_i, y_i) from training set

  Calculate the output $a = sign(w^T x_i + b)$

  If $a \neq y_i$:

    $w' \leftarrow w + \eta \cdot y_i \cdot x_i$

    $b' \leftarrow b + \eta \cdot y_i$

until you reach the required number of steps

Output:parameters w and b

Here η is learning rate.

Although the perceptron model has been put forward with good development potential, Marvin Lee Minsky and Seymour Papert proved that the linear model represented by the perceptron cannot solve the linear inseparability problem (XOR) in the "Perceptrons" book in 1969, which directly led to the emergence of neural network research to a bottom at the time. Although the perceptron model cannot solve the linear inseparable problem, the book also mentions that it can be solved by nesting multiple layers of neural networks.

6.2 Fully Connected Layer

The underivable nature of the perceptron model severely constrains its potential, making it only capable of solving extremely simple tasks. In fact, modern deep learning models have a parameter scale of millions or even hundreds of millions, but the core structure is not much different from the perceptron model. On the basis of the perceptron model, they replace the discontinuous step activation function with other smooth continuous derivable activation functions and stack multiple network layers to enhance the expressive power of the network.

In this section, we replace the activation function of the perceptron model and stack multiple neurons in parallel to achieve a multi-input and multi-output network layer structure. As shown in Figure 6-4, two neurons are stacked in parallel, that is, two perceptrons with replaced activation functions, forming a network layer of three input nodes and two output nodes. The first output node is:

$$o_1 = \sigma\left(w_{11} \cdot x_1 + w_{21} \cdot x_2 + w_{31} \cdot x_3 + b_1\right)$$

The output of the second node is:

$$o_2 = \sigma\left(w_{12} \cdot x_1 + w_{22} \cdot x_2 + w_{32} \cdot x_3 + b_2\right)$$

Putting them together, the output vector is $o = [o_1, o_2]$. The entire network layer can be expressed by the matrix relationship:

$$\begin{bmatrix} o_1 & o_2 \end{bmatrix} = \begin{bmatrix} x_1 & x_2 & x_3 \end{bmatrix} @ \begin{bmatrix} w_{11} & w_{12} & w_{21} & w_{22} & w_{31} & w_{32} \end{bmatrix} + \begin{bmatrix} b_1 & b_2 \end{bmatrix} \qquad (6\text{-}1)$$

That is:

$$O = X @ W + b$$

The shape of the input matrix X is defined as $[b, d_{in}]$, while the number of samples is b and the number of input nodes is d_{in}. The shape of the weight matrix W is defined as $[d_{in}, d_{out}]$, while the number of output nodes is d_{out}, and the shape of the offset vector b is $[d_{out}]$.

Considering two samples, $x^{(1)} = \left[x_1^{(1)}, x_2^{(1)}, x_3^{(1)}\right]$, $x^{(2)} = \left[x_1^{(2)}, x_2^{(2)}, x_3^{(2)}\right]$, the preceding equation can also be written as:

$$\begin{bmatrix} o_1^{(1)} & o_2^{(1)} & o_1^{(2)} & o_2^{(2)} \end{bmatrix} = \begin{bmatrix} x_1^{(1)} & x_2^{(1)} & x_3^{(1)} & x_1^{(2)} & x_2^{(2)} & x_3^{(2)} \end{bmatrix} @ \begin{bmatrix} w_{11} & w_{12} & w_{21} & w_{22} & w_{31} & w_{32} \end{bmatrix}$$
$$+ \begin{bmatrix} b_1 & b_2 \end{bmatrix}$$

Among it, the output matrix O contains the output of b samples, and the shape is $[b, d_{out}]$. Since each output node is connected to all input nodes, this network layer is called a fully connected layer, or a dense layer, with W as weight matrix and b is the bias vector.

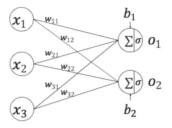

Figure 6-4. *Fully connected layer*

6.2.1 Tensor Mode Implementation

In TensorFlow, to achieve a fully connected layer, you only need to define the weight tensor W and bias tensor b and use the batch matrix multiplication function tf.matmul() provided by TensorFlow to complete the calculation of the network layer. For example, for an input matrix X with two samples and input feature length of each sample $d_{in} = 784$ and the number of output nodes $d_{out} = 256$, the shape of the weight matrix W is [784,256]. The shape of the bias vector b is [256]. After the addition, the shape of the output layer is [2,256], that is, the features of the two samples with each feature length as 256. The code is implemented as follows:

```
In [1]:
x = tf.random.normal([2,784])
w1 = tf.Variable(tf.random.truncated_normal([784, 256],
stddev=0.1))
b1 = tf.Variable(tf.zeros([256]))
o1 = tf.matmul(x,w1) + b1   # linear transformation
o1 = tf.nn.relu(o1)   # activation function
```

```
Out[1]:
    <tf.Tensor: id=31, shape=(2, 256), dtype=float32, numpy=
    array([[ 1.51279330e+00,   2.36286330e+00,   8.16453278e-01,
              1.80338228e+00,   4.58602428e+00,   2.54454136e+00,...
```

In fact, we have used the preceding code many times to implement network layers.

6.2.2 Layer Implementation

The fully connected layer is essentially matrix multiplication and addition operations. But as one of the most commonly used network layers, TensorFlow has a more convenient implementation method: layers. Dense(units, activation). Through the layer.Dense class, you only need to specify the number of output nodes (units) and activation function type (activation). It should be noted that the number of input nodes will be determined according to the input shape during the first operation, and the weight tensor and bias tensor will be automatically created and initialized based on the number of input and output nodes. The weight tensor and bias tensor will not be created immediately due to lazy evaluation. The build function or direct calculation will be required to complete the creation of the network parameters. The activation parameter specifies the activation function of the current layer, which can be a common activation function or a custom activation function, or be specified as none, that is, no activation function.

```
In [2]:
x = tf.random.normal([4,28*28])
from tensorflow.keras import  layers
# Create fully-connected layer with output nodes and activation
function
fc = layers.Dense(512, activation=tf.nn.relu)
```

```
h1 = fc(x)  # calculate and return a new tensor
Out[2]:
<tf.Tensor: id=72, shape=(4, 512), dtype=float32, numpy=
array([[0.63339347, 0.21663809, 0.         , ..., 1.7361937 ,
0.39962345, 2.4346168 ],...
```

We can create a fully connected layer fc with a single line of code in the preceding code with the number of output nodes as 512 and the number of input nodes automatically obtained during calculation. The code creates internal weight tensor and bias tensor automatically as well. We can obtain the weight and bias tensor object through the class member kernel and bias within the class:

```
In [3]: fc.kernel # Get the weight tensor
Out[3]:
<tf.Variable 'dense_1/kernel:0' shape=(784, 512)
dtype=float32, numpy=
array([[-0.04067389,  0.05240148,  0.03931375, ...,
-0.01595572, -0.01075954, -0.06222073],
In [4]: fc.bias # Get the bias tensor
Out[4]:
<tf.Variable 'dense_1/bias:0' shape=(512,)
dtype=float32, numpy=
array([0., 0., 0., 0., 0., 0., 0., 0., 0., 0., 0., 0., 0.,
0., 0., 0.,...])
```

It can be seen that the shape of the weight and the bias tensor are in line with our understanding. When optimizing parameters, we need to obtain a list of all tensor parameters to be optimized in the network, which can be done through the class trainable_variables.

```
In [5]: fc.trainable_variables
Out[5]:  # Return all parameters to be optimized
 [<tf.Variable 'dense_1/kernel:0' shape=(784, 512)
 dtype=float32,...,
 <tf.Variable 'dense_1/bias:0' shape=(512,) dtype=float32,
 numpy=...]
```

In fact, the network layer saves not only the list of trainable_variables to be optimized but also tensors that do not participate in gradient optimization. For example, the batch normalization layer can return all parameter lists that do not need optimization through the non_trainable_variables member. If you want to get a list of all parameters, you can get all internal tensors through the variables member of the class, for example:

```
In [6]: fc.variables # Get all parameters
Out[6]:
[<tf.Variable 'dense_1/kernel:0' shape=(784, 512)
dtype=float32,...,
 <tf.Variable 'dense_1/bias:0' shape=(512,) dtype=float32,
 numpy=...]
```

For fully connected layers, all internal tensors participate in gradient optimization, so the list returned by variables is the same as trainable_variables.

When using the network layer class object for forward calculation, you only need to call the __call__ method of the class, that is, write it in the fc(x) mode, it will automatically call the __call__ method. This setting is automatically completed by the TensorFlow framework. For a fully connected layer class, the operation logic implemented in the call method is very simple.

6.3 Neural Network

By stacking the fully connected layers in Figure 6-4 and ensuring that the number of output nodes of the previous layer matches the number of input nodes of the current layer, a network of any number of layers can be created, which is known as neural networks. As shown in Figure 6-5, by stacking four fully connected layers, a neural network with four layers can be obtained. Since each layer is a fully connected layer, it is called a fully connected network. Among them, the first to third fully connected layers are called hidden layers, and the output of the last fully connected layer is called the output layer of the network. The number of output nodes of the hidden layers is [256,128,64], respectively, and the nodes of the output layer is 10.

When designing a fully connected network, the hyperparameters such as the configuration of the network can be set freely according to the rule of thumb, and only a few constraints need to be followed. For example, the number of input nodes in the first hidden layer needs to match the actual feature length of the data. The number of input layers in each layer matches the number of output nodes in the previous layer. The activation function and number of nodes in the output layer need to be set according to the specific settings of the required output. In general, the design of the neural network models has a greater degree of freedom. As shown in Figure 6-5, the number of output nodes in each layer does not have to be [256,128,64,10] and can be freely matched, such as [256,256,64,10] or [512,64,32,10]. As for which set of hyperparameters is optimal, it requires a lot of field experience and experimentation.

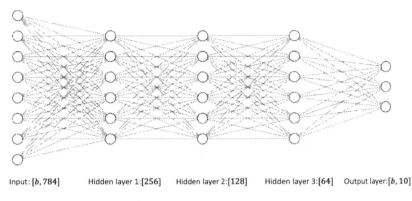

Input: [b, 784]   Hidden layer 1:[256]   Hidden layer 2:[128]   Hidden layer 3:[64]   Output layer:[b, 10]

Figure 6-5. *Four-layer neural network*

6.3.1 Tensor Mode Implementation

For a multi-layer neural network such as Figure 6-5, the weight matrix and bias vector of each layer need to be defined separately. The parameters of each layer can only be used for the corresponding layer and should not be mixed. The network model in Figure 6-5 is implemented as follows:

```
# Hidden layer 1
w1 = tf.Variable(tf.random.truncated_normal([784, 256],
stddev=0.1))
b1 = tf.Variable(tf.zeros([256]))
# Hidden layer 2
w2 = tf.Variable(tf.random.truncated_normal([256, 128],
stddev=0.1))
b2 = tf.Variable(tf.zeros([128]))
# Hidden layer 3
w3 = tf.Variable(tf.random.truncated_normal([128, 64],
stddev=0.1))
b3 = tf.Variable(tf.zeros([64]))
```

```
# Hidden layer 4
w4 = tf.Variable(tf.random.truncated_normal([64, 10],
stddev=0.1))
b4 = tf.Variable(tf.zeros([10]))
```

When calculating, you only need to use the output of the previous layer as the input of the current layer, repeat until the last layer, and use the output of the output layer as the output of the network.

```
with tf.GradientTape() as tape:
    # x: [b, 28*28]
    #  Hidden layer 1 forward calculation, [b, 28*28]
    => [b, 256]
    h1 = x@w1 + tf.broadcast_to(b1, [x.shape[0], 256])
    h1 = tf.nn.relu(h1)
    # Hidden layer 2 forward calculation, [b, 256]
    => [b, 128]
    h2 = h1@w2 + b2
    h2 = tf.nn.relu(h2)
    # Hidden layer 3 forward calculation, [b, 128]
    => [b, 64]
    h3 = h2@w3 + b3
    h3 = tf.nn.relu(h3)
    # Output layer forward calculation, [b, 64]
    => [b, 10]
    h4 = h3@w4 + b4
```

Whether the activation function needs to be added in the last layer usually depends on the specific task.

When using the TensorFlow automatic derivation function to calculate the gradient, the forward calculation process needs to be placed in the tf.GradientTape() environment, so that the gradient() method of the GradientTape object can be used to automatically solve the gradient of the parameter, and the parameter is updated by the optimizers object.

6.3.2 Layer Mode Implementation

For the conventional network layer, it is more concise and efficient to implement through the layer method. First, create new network layer classes and specify the activation function types of each layer:

```
#  Import layers modules
from tensorflow.keras import layers,Sequential

fc1 = layers.Dense(256, activation=tf.nn.relu)
#  Hidden layer 1
fc2 = layers.Dense(128, activation=tf.nn.relu)
#  Hidden layer 2
fc3 = layers.Dense(64, activation=tf.nn.relu) #  Hidden layer 3
fc4 = layers.Dense(10, activation=None) #  Output layer
x = tf.random.normal([4,28*28])
h1 = fc1(x)  #  Get output of hidden layer 1
h2 = fc2(h1) #  Get output of hidden layer 2
h3 = fc3(h2) #  Get output of hidden layer 3
h4 = fc4(h3) #  Get the network output
```

For such a network where data forwards in turn, it can also be encapsulated into a network class object through the sequential container, and the forward calculation function of the class can be called once to complete the forward calculation of all layers. It is more convenient to use and is implemented as follows :

```
from tensorflow.keras import layers,Sequential

#  Encapsulate a neural network through Sequential container
model = Sequential([
    layers.Dense(256, activation=tf.nn.relu) , # Hidden layer 1
    layers.Dense(128, activation=tf.nn.relu) , # Hidden layer 2
```

```
    layers.Dense(64, activation=tf.nn.relu) , # Hidden layer 3
    layers.Dense(10, activation=None) , # Output layer
])
```

In forward calculation, you only need to call the large network objects once to complete the sequential calculation of all layers:

```
out = model(x)
```

6.3.3 Optimization

We call the calculation process of the neural network from input to output as forward propagation. The forward propagation process of the neural network is also the process of the flow of the data tensor from the first layer to the output layer. That is, from the input data, tensors are passed through each hidden layer, until the output is obtained and the error is calculated, which is also the origin of the TensorFlow framework name.

The final step of forward propagation is to complete the error calculation:

$$L = g\big(f_\theta(x),y\big)$$

In the preceding formula, $f_\theta(\cdot)$ represents a neural network model with parameters θ. $g(\cdot)$ called an error function, used to describe the gap between the predicted value of the current network $f_\theta(x)$ and the real label y, such as the commonly used mean square error function. L is called the error or loss of the network, which is generally a scalar. We hope to minimize the training error L by learning a set of parameters on the training set D^{train}:

$$\theta^* = arg\,min_\theta\ g(f_\theta(x),y),\ x \in D^{train}$$

The preceding minimization problem generally uses the backward propagation algorithm to solve and uses the gradient descent algorithm to iteratively update the parameters:

$$\theta' = \theta - \eta \cdot \nabla_{\theta} L$$

where η is the learning rate.

From another perspective to understand the neural network, it completes the function of feature dimension transformation, such as the four-layer MNIST handwritten digital image recognition fully connected network, which in turn completes the feature dimensionality reduction process of $784 \rightarrow 256 \rightarrow 128 \rightarrow 64 \rightarrow 10$. The original features usually have higher dimensions and contain many low-level features and useless information. Through the layer-by-layer feature transformation, higher dimensions are reduced to lower dimensions where high-level abstract feature information highly correlated to the task is generally generated and specific task can be completed through simple logical determination of these features, such as the classification of pictures.

The amount of network parameters is an important indicator to measure the scale of the network. So how to calculate the amount of parameters of the fully connected layer? Consider a network layer with weight matrix W, bias vector b, input feature length d_{in}, and output feature length d_{out}. The number of parameters for W is $d_{in} \cdot d_{out}$. Adding the bias parameter, the total number of parameter is $d_{in} \cdot d_{out} + d_{out}$. For a multilayer fully connected neural network, for example, $784 \rightarrow 256 \rightarrow 128 \rightarrow 64 \rightarrow 10$, the expression of the total parameter amount is:

$$256 \cdot 784 + 256 + 128 \cdot 256 + 128 + 64 \cdot 128 + 64 + 10 \cdot 64 + 10 = 242762$$

The fully connected layer is the most basic type of neural network. It is very important for the research of subsequent neural network models, such as convolutional neural networks and recurrent neural networks. Through learning other network types, we will find that they, more or less, originate from the idea of a fully connected layer network. Because Geoffrey Hinton, Yoshua Bengio, and Yann LeCun have insisted on research in the frontline of neural networks, they have made outstanding contributions to the development of artificial intelligence and won the Turing Award in 2018 (Figure 6-6, from the right are Yann LeCun, Geoffrey Hinton, and Yoshua Bengio).

Figure 6-6. *2018 Turing Award Winners[1]*

6.4 Activation function

In the following, we introduce common activation functions in neural networks. Unlike step and symbolic functions, these functions are smooth and derivable and are suitable for gradient descent algorithms.

6.4.1 Sigmoid

The Sigmoid function is also called the logistic function, which is defined as:

$$Sigmoid(x) \triangleq \frac{1}{1+e^{-x}}$$

---

[1] Image source: www.theverge.com/2019/3/27/18280665/ai-godfathers-turing-award-2018-yoshua-bengio-geoffrey-hinton-yann-lecun

One of its excellent features is the ability to "compress" the input $x \in R$ to an interval $x \in (0, 1)$. The value of this interval is commonly used in machine learning to express the following meanings:

- **Probability distribution** The output of the interval $(0, 1)$ matches the distribution range of probability. The output can be translated into a probability by the Sigmoid function

- **Signal strength** Usually, 0~1 can be understood as the strength of a certain signal, such as the color intensity of the pixel: 1 represents the strongest color of the current channel, and 0 represents the current channel without color. It can also be used to represent the current Gate status, that is, 1 means open and 0 indicates closed.

The Sigmoid function is continuously derivable, as shown in Figure 6-7. The gradient descent algorithm can be directly used to optimize the network parameters.

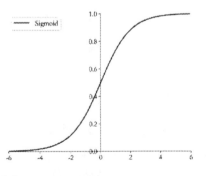

Figure 6-7. *Sigmoid function*

In TensorFlow, the Sigmoid function can be implemented through tf.nn.sigmoid function as follows:

```
In [7]:x = tf.linspace(-6.,6.,10)
x # Create input vector -6~6
Out[7]:
<tf.Tensor: id=5, shape=(10,), dtype=float32, numpy=
array([-6.        , -4.6666665, -3.3333333, -2.        ,
-0.6666665,
        0.666667 , 2.        , 3.333334 , 4.666667 ,
   6.        ]...
In [8]:tf.nn.sigmoid(x) # Pass x to Sigmoid function
Out[8]:
<tf.Tensor: id=7, shape=(10,), dtype=float32, numpy=
array([0.00247264, 0.00931597, 0.03444517, 0.11920291,
0.33924365, 0.6607564 , 0.8807971 , 0.96555483, 0.99068403,
0.9975274 ],
        dtype=float32)>
```

As you can see, the range $[-6, 6]$ of element values in the vector is mapped to the interval $(0, 1)$.

6.4.2 ReLU

Before the ReLU (rectified linear unit), activation function was proposed; the Sigmoid function was usually the first choice for activation functions of neural networks. However, when the input value of Sigmoid function is too large or too small, the gradient value is close to 0, which is known as the gradient dispersion phenomenon. When this phenomenon occurs, the network parameters will not be updated for a long time, which leads to the phenomenon that the training does not converge. The gradient dispersion phenomenon is more likely to occur in deeper network models. The eight-layer AlexNet model proposed in 2012 uses an activation function called ReLU, which makes the number of network layers reach 8. Since then, the ReLU function has become more and more widely used. The ReLU function is defined as:

$$ReLU(x) \triangleq max(0,x)$$

The function curve is shown in Figure 6-8. It can be seen that ReLU suppresses all values less than 0 to 0; for positive numbers, it outputs those directly. This unilateral suppression characteristic comes from biology. In 2001, neuroscientists Dayan and Abott simulated a more accurate model of brain neuron activation, as shown in Figure 6-9. It has characteristics such as unilateral suppression and relatively loose excitation boundaries. The design of the ReLU function is very similar to it [2].

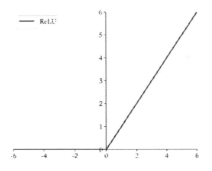

Figure 6-8. *ReLU function*

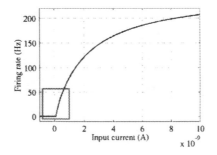

Figure 6-9. *Human brain activation function [2]*

In TensorFlow, the ReLU function can be implemented through tf.nn.relu function as follows:

```
In [9]:tf.nn.relu(x)
Out[9]:
<tf.Tensor: id=11, shape=(10,), dtype=float32, numpy=
array([0.    , 0.    , 0.    , 0.    , 0.    ,
0.666667,    2.   , 3.333334, 4.666667, 6.    ],
dtype=float32)>
```

It can be seen that after the ReLU activation function, the negative numbers are all suppressed to 0, and the positive numbers are retained.

In addition to using the functional interface tf.nn.relu to implement the ReLU function, the ReLU function can also be added to the network as a network layer like the dense layer. The corresponding class is layers. ReLU(). Generally speaking, the activation function class is not the main network computing layer and does not count into the number of network layers.

The design of the ReLU function is derived from neuroscience. The calculation of function values and derivative values is very simple. At the same time, it has excellent gradient characteristics. It has been verified to be very effective in a large number of deep learning applications.

6.4.3 LeakyReLU

The derivative of the ReLU function is always 0 when $x < 0$, which may also cause gradient dispersion. To overcome this problem, the LeakyReLU function (Figure 6-10) is proposed.

$$LeakyReLU \triangleq \{x \quad x \geq 0 \quad px \quad x < 0$$

where p is a small value set by users, such as 0.02. When $p = 0$, the
LeakyReLU function degenerates to the ReLU function. When $p \neq 0$,
a small derivative value can be obtained at $x < 0$, thereby avoiding the
phenomenon of gradient dispersion.

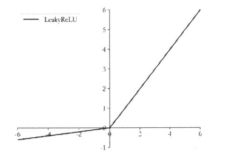

Figure 6-10. *LeakyReLU function*

In TensorFlow, LeakyReLU function can be implemented through
tf.nn.leaky_relu as follows:

```
In [10]:tf.nn.leaky_relu(x, alpha=0.1)
Out[10]:
<tf.Tensor: id=13, shape=(10,), dtype=float32, numpy=
array([-0.6       , -0.46666667, -0.33333334, -0.2       ,
-0.06666666,
       0.666667  ,  2.        ,  3.333334  ,  4.666667  ,
       6.        ],
    dtype=float32)>
```

The alpha parameter represents p. The corresponding class of tf.nn.
leaky_relu is layers.LeakyReLU. You can create a LeakyReLU network layer
through LeakyReLU(alpha) and set the parameter p. Like the Dense layer,
the LeakyReLU layer can be placed in a suitable position on the network.

6.4.4 Tanh

The Tanh function can "compress" the input $x \in R$ to an interval $(-1, 1)$, defined as:

$$tanh\ tanh\ (x) = \frac{\left(e^x - e^{-x}\right)}{\left(e^x + e^{-x}\right)}$$

$$= 2 \cdot sigmoid(2x) - 1$$

It can be seen that the Tanh activation function can be realized after zooming and translated by the Sigmoid function, as shown in Figure 6-11.

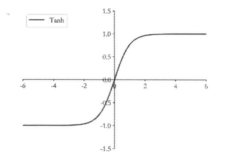

Figure 6-11. *Tanh function*

In Tensorflow, the Tanh function can be implemented using tf.nn.tanh as follows:

```
In [11]:tf.nn.tanh(x)
Out[11]:
<tf.Tensor: id=15, shape=(10,), dtype=float32, numpy=
array([-0.9999877 , -0.99982315, -0.997458  , -0.9640276 ,
-0.58278286, 0.5827831 ,  0.9640276 ,  0.997458  ,  0.99982315,
0.9999877 ],
      dtype=float32)>
```

You can see that the range of vector element values is mapped to $(-1, 1)$.

6.5 Design of Output Layer

Let's discuss the design of the last layer of network in particular. In addition to all hidden layers, it completes the functions of dimensional transformation and feature extraction, and it is also used as an output layer. It is necessary to decide whether to use the activation function and what type of activation function to use according to the specific tasks.

We will classify the discussions based on the range of output values. Common types of output include:

- $o_i \in R^d$ The output belongs to the entire real number space, or a certain part of real number space, such as function value trend prediction and age prediction problems.

- $o_i \in [0, 1]$ The output value falls in the interval $[0, 1]$, such as image generation, and the pixel value of the image is generally expressed by values in interval $[0, 1]$ or the probability of the binary classification problem, such as the probability of the tail or face of a coin.

- $o_i \in [0, 1]$, $\sum_i o_i = 1$ The output value falls within the interval $[0, 1]$, and the sum of all output values is 1. Common problems include multi-classification problems, such as MNIST handwritten digital picture recognition, which the sum of the probability that the picture belongs to ten categories should be 1.

- $o_i \in [-1, 1]$ output value is between -1 and 1.

6.5.1 Common Real Number Space

This type of problem is more common. For example, sine function curve, age prediction, and stock trend prediction all belong to the whole or part of continuous real number space, and the output layer may not have an activation function. The calculation of the error is directly based on the output o of the last layer and the true value y. For example, the mean square error function is used to measure the distance between the output value o and the true value y:

$$L = g(o, y)$$

where g represents a specific error calculation function, such as MSE.

6.5.2 [0, 1] Interval

It is also common for output values to belong to interval $[0, 1]$, such as image generation, and binary classification problems. In machine learning, image pixel values are generally normalized to intervals $[0, 1]$. If the values of the output layer are used directly, the pixel value range will be distributed in the entire real number space. In order to map the pixel values to the effective real number space $[0, 1]$, a suitable activation function needs to be added after the output layer. The Sigmoid function is a good choice here.

Similarly, for binary classification problems, such as the prediction of the face and tail of coins, the output layer can only be one node which is the probability of an event A occurring $P(x)$ giving the network input x. If we use the output scalar o of the network to represent the probability of the occurrence of positive events, then the probability of the occurrence of negative events is $1 - o$. The network structure is shown in Figure 6-12.

$$P(x)=o$$

$$P(x)=1-o$$

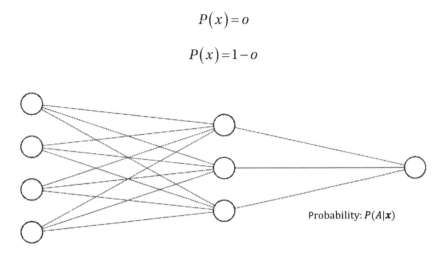

Figure 6-12. *Binary classification network with single output node*

In this case, you only need to add the Sigmoid function after the value of the output layer to translate the output into a probability value. For the binary classification problem, in addition to using a single output node to represent the probability of the occurrence of event A $P(x)$, you can also separately predict $P(x)$ and $P(x)$, and satisfy the constraints:

$$P(x)+P(x)=1$$

where $\underline{A}$ indicates the opposite event of event A. As shown in Figure 6-13, the output layer of the binary classification network is two nodes. The output value of the first node represents the probability of the occurrence of event A $P(x)$, and the output value of the second node represents the probability of the occurrence of the opposite event $P(x)$. The function can only compress a single value to the interval $(0, 1)$ and does not consider the relationship between the two node values. We hope that in addition to satisfy $o_i \in [0,1]$, they can satisfy the constraint that the sum of probabilities is 1:

$$\sum_i o_i = 1$$

This situation is the problem setting to be introduced in the next section.

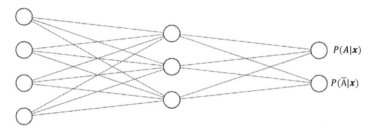

Figure 6-13. *Binary classification network with two outputs*

6.5.3 [0,1] Interval with Sum 1

For cases that the output value $o_i \in [0,1]$, and the sum of all output values is 1, it is the most common problem with multi-classification. As shown in Figure 6-15, each output node of the output layer represents a category. The network structure in the figure is used to handle three classification tasks. The output value distribution of the three nodes represents the probability that the current sample belongs to category A, B, and C: $P(x)$, $P(B|x)$, and $P(C|x)$. Because the sample in the multi-classification problem can only belong to one of the categories, so the sum of the probabilities of all categories should be 1.

How to implement this constraint logic? This can be achieved by adding a Softmax function to the output layer. The Softmax function is defined as:

$$Softmax(z_i) \triangleq \frac{e^{z_i}}{\sum_{j=1}^{d_{out}} e^{z_j}}$$

The Softmax function can not only map the output value to the interval [0, 1] but also satisfy the characteristic that the sum of all output values is 1. As shown in the example in Figure 6-14, the output of the output layer is [2.0,1.0,0.1]. After going through the Softmax function, the output becomes [0.7,0.2,0.1]. Each value represents the probability that the current sample belongs to each category, and the sum of the probability values is 1. The output of the output layer can be translated into category probabilities through the Softmax function, which is used very often in classification problems.

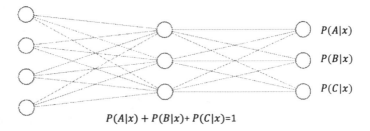

Logits   Softmax   Probability

Figure 6-14. *Softmax function example*

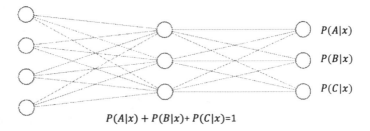

$P(A|x) + P(B|x) + P(C|x) = 1$

Figure 6-15. *Multi-classification network structure*

In TensorFlow, the Softmax function can be implemented through tf.nn.softmax as follows:

```
In [12]: z = tf.constant([2.,1.,0.1])
tf.nn.softmax(z)
Out[12]:
<tf.Tensor: id=19, shape=(3,), dtype=float32,
numpy=array([0.6590012, 0.242433 , 0.0985659], dtype=float32)>
```

Similar to the dense layer, the Softmax function can also be used as a network layer class. It is convenient to add the Softmax layer through the layers.Softmax (axis = -1) class, where the axis parameter specifies the dimension to be calculated.

In the numerical calculation process of the Softmax function, the numerical overflow phenomenon is likely to occur due to the large input value. Similar problem may happen when calculating the cross-entropy. For the stability of numerical calculation, TensorFlow provides a unified interface that implements Softmax and cross-entropy loss function at the same time and also handles the anomalies of numerical instability. It is generally recommended to use these interface functions. The functional interface is tf.keras.losses.categorical_crossentropy(y_true, y_pred, from_logits = False), where y_true represents the one-hot encoded true label and y_pred represents the predicted value of the network. When from_logits is set to True, y_pred represents the variable z that has not went through the Softmax function. When from_logits is set to False, y_pred is expressed as the output of the Softmax function. For numerical calculation stability, generally set from_logits to True, so that tf.keras. losses.categorical_crossentropy will perform Softmax function calculation internally, and there is no need to explicitly call the Softmax function in the model explicitly. For example:

```
In [13]:
z = tf.random.normal([2,10]) # Create output of the
output layer
y_onehot = tf.constant([1,3]) # Create real label
y_onehot = tf.one_hot(y_onehot, depth=10) # one-hot encoding
# The Softmax function is not explicitly used in output
layer, so
# from_logits=True. categorical_cross-entropy function will
use Softmax
# function first in this case.
loss = keras.losses.categorical_crossentropy(y_onehot,z,from_
logits=True)
loss = tf.reduce_mean(loss) # calculate the loss
loss
Out[13]:
<tf.Tensor: id=210, shape=(), dtype=float32, numpy= 2.4201946>
```

In addition to the functional interface, you can also use the losses. CategoricalCrossentropy(from_logits) class method to simultaneously calculate the Softmax and cross-entropy loss functions. For example:

```
In [14]:
criteon = keras.losses.CategoricalCrossentropy(from_
logits=True)
loss = criteon(y_onehot,z)
loss
Out[14]:
<tf.Tensor: id=258, shape=(), dtype=float32, numpy= 2.4201946>
```

6.5.4 (-1, 1) Interval

If you want the range of output values to be distributed in intervals $(-1, 1)$, you can simply use the tanh activation function:

```
In [15]:
x = tf.linspace(-6.,6.,10)
tf.tanh(x)
Out[15]:
<tf.Tensor: id=264, shape=(10,), dtype=float32, numpy=
array([-0.9999877 , -0.99982315, -0.997458  , -0.9640276 ,
-0.58278286, 0.5827831 ,  0.9640276 ,  0.997458  ,  0.99982315,
0.9999877 ],
       dtype=float32)>
```

The design of the output layer has a certain flexibility, which can be designed according to the actual application scenario, and make full use of the characteristics of the existing activation function.

6.6 Error Calculation

After building the model structure, the next step is to select the appropriate error function to calculate the error. Common error functions are mean square error, cross-entropy, KL divergence, and hinge loss. Among them, the mean square error function and cross-entropy function are more common in deep learning. The mean square error function is mainly used for regression problems, and the cross-entropy function is mainly used for classification problem.

6.6.1 Mean Square Error Function

Mean square error (MSE) function maps the output vector and the true vector to two points in the Cartesian coordinate system, by calculating the Euclidean distance between these two points (to be precise, the square of Euclidean distance) to measure the difference between the two vectors:

$$MSE(y,o) \triangleq \frac{1}{d_{out}} \sum_{i=1}^{d_{out}} (y_i - o_i)^2$$

The value of MSE is always greater than or equal to 0. When the MSE function reaches the minimum value of 0, the output is equal to the true label, and the parameters of the neural network reach the optimal state.

The MSE function is widely used in regression problems. In fact, the MSE function can also be used in classification problems. In TensorFlow, MSE calculation can be implemented in a functional or layer manner. For example, implement MSE calculation using a function as follows:

```
In [16]:
o = tf.random.normal([2,10]) # Network output
y_onehot = tf.constant([1,3]) # Real label
y_onehot = tf.one_hot(y_onehot, depth=10)
loss = keras.losses.MSE(y_onehot, o) # Calculate MSE
loss
Out[16]:
<tf.Tensor: id=27, shape=(2,), dtype=float32,
numpy=array([0.779179 , 1.6585705], dtype=float32)>
```

In particular, the MSE function returns the mean square error of each sample. You need to average again in the sample dimension to obtain the mean square error of the average sample. The implementation is as follows:

```
In [17]:
loss = tf.reduce_mean(loss)
loss
Out[17]:
<tf.Tensor: id=30, shape=(), dtype=float32, numpy=1.2188747>
```

It can also be implemented in layer mode. The corresponding class is keras.losses.MeanSquaredError(). Like other classes, the __call__ function can be called to complete the forward calculation. The code is as follows:

```
In [18]:
criteon = keras.losses.MeanSquaredError()
loss = criteon(y_onehot,o)
loss
Out[18]:
<tf.Tensor: id=54, shape=(), dtype=float32, numpy=1.2188747>
```

6.6.2 Cross-Entropy Error Function

Before introducing the cross-entropy loss function, we first introduce the concept of entropy in informatics. In 1948, Claude Shannon introduced the concept of entropy in thermodynamics into information theory to measure the uncertainty of information. Entropy is also called information entropy or Shannon entropy in information science. The greater the entropy, the greater the uncertainty and the greater the amount of information. The entropy of a distribution $P(i)$ is defined as:

$$H(P) \triangleq -\sum_i P(i)P(i)$$

In fact, other base functions can also be used. For example, for the four-category classification problem, if the true label of a sample is category 4, then the one-hot encoding of the label is $[0, 0, 0, 1]$. That is, the classification of this picture is uniquely determined, and it belongs to category 4 with uncertainty 0, and its entropy can be simply calculated as:

$$-0 \cdot 0 - 0 \cdot 0 - 0 \cdot 0 - 1 \cdot 1 = 0$$

That is to say, for a certain distribution, the entropy is 0 and the uncertainty is the lowest.

If the predicted probability distribution is $[0.1, 0.1, 0.1, 0.7]$, its entropy can be calculated as:

$$-0.1 \cdot 0.1 - 0.1 \cdot 0.1 - 0.1 \cdot 0.1 - 0.7 \cdot 0.7 \approx 1.356$$

Considering a random classifier, its prediction probability for each category is equal: $[0.25, 0.25, 0.25, 0.25]$. In the same way, its entropy can be calculated to be about 2, and the uncertainty in this case is slightly larger than the preceding case.

Because, the entropy is always greater than or equal to 0. When the entropy reaches a minimum value of 0, the uncertainty is 0. The distribution of one-hot coding for classification problems is a typical example of entropy of 0. In TensorFlow, we can use tf.math.log to calculate entropy.

After introducing the concept of entropy, we'll derive the definition of cross-entropy based on entropy:

$$H(p \| q) \triangleq -\sum_i p(i) q(i)$$

Through transformation, cross-entropy can be decomposed into the sum of entropy and KL divergence (Kullback-Leibler divergence):

$$H(p \| q) = H(p) + D_{KL}(p \| q)$$

where KL divergence is:

$$D_{KL}(p\|q) = \sum_i p(i) log\left(\frac{p(i)}{q(i)}\right)$$

KL divergence is an indicator used by Solomon Kullback and Richard A. Leibler in 1951 to measure the distance between two distributions. When $p = q$, the minimum value of $D_{KL}(p\|q)$ is 0. The greater the difference between p and q, the greater $D_{KL}(p\|q)$ is. It should be noted that neither the cross-entropy nor the KL divergence is symmetrical, namely:

$$H(p\|q) \neq H(q\|p)$$

$$D_{KL}(p\|q) \neq D_{KL}(q\|p)$$

Cross-entropy is a good measure of the "distance" between two distributions. In particular, when the distribution of y in the classification problem uses one-hot coding, $H(p) = 0$. Then,

$$H(p\|q) = H(p) + D_{KL}(p\|q) = D_{KL}(p\|q)$$

That is, cross-entropy degenerates to the KL divergence between the true label distribution and the output probability distribution.

According to the definition of KL divergence, we derive the calculation expression of cross-entropy in the classification problem:

$$H(p\|q) = D_{KL}(p\|q) = \sum_j y_j log\left(\frac{y_j}{o_j}\right)$$

$$= 1 \cdot log\frac{1}{o_i} + \sum_{j\neq i} 0 \cdot log\left(\frac{0}{o_j}\right)$$

$$= -log o_i$$

where i is the index number of 1 in the one-hot encoding, which is also the real category. It can be seen that the cross-entropy is only related to the probability on the real category o_i, and the larger the corresponding probability o_i, the smaller $H(p\|q)$ is. When the probability on the corresponding category is 1, the cross-entropy achieves the minimum value of 0. At this time, the network output is completely consistent with the real label, and the neural network obtains the optimal state.

Therefore, the process of minimizing the cross-entropy loss function is also the process of maximizing the prediction probability of the correct category. From this perspective, understanding the cross-entropy loss function is very intuitive and easy.

6.7 Types of Neural Networks

The fully connected layer is the most basic type of neural network, and it has made a tremendous contribution to the subsequent research of neural networks. The forward calculation process of the fully connected layer is relatively simple, and the gradient derivation is also relatively simple, but it has one of the biggest defects. When processing data with a large feature length, the parameter amount of the fully connected layer is often large, making the number of parameters of the fully connected network huge and difficult to train. In recent years, the development of social media has produced a large number of digital resources such as pictures, videos, and texts, which has greatly promoted the research of neural networks in the fields of computer vision and natural language processing, and has successively proposed a series of neural network types.

6.7.1 Convolutional Neural Network

How to identify, analyze, and understand data such as pictures and videos is a core problem of computer vision. When the fully connected layer processes high-dimensional pictures and video data, it often has problems such as huge network parameters and very difficult to train. By using the idea of local correlation and weight sharing, Yann Lecun proposed convolutional neural network (CNN) in 1986. With the prosperity of deep learning, the performance of convolutional neural networks in computer vision has greatly surpassed other algorithms, showing a tendency to dominate the field of computer vision. Popular models for image classification include AlexNet, VGG, GoogLeNet, ResNet, and DenseNet. For objective recognition, there are RCNN, Fast RCNN, Faster RCNN, Mask RCNN, YOLO, and SSD. We will introduce the principles of convolutional neural networks in detail in Chapter 10.

6.7.2 Recurrent Neural Network

In addition to data such as pictures and videos with spatial structure, sequence signals are also a very common type of data. One of the most representative sequence signals is text. How to process and understand text data is a core issue of natural language processing. Convolutional neural networks are not good at processing sequence signals due to the lack of memory mechanism and the ability to process signals of indefinite length. Recurrent neural network (RNN), under continuous research by Yoshua Bengio, Jürgen Schmidhuber, and others, is proved to be very good at processing sequence signals. In 1997, Jürgen Schmidhuber proposed the LSTM network. As a variant of RNN, it better overcomes the problems of RNN that lacks long-term memory and is not good at processing long sequences. LSTM has been widely used in natural language processing. Based on the LSTM model, Google has proposed the Seq2Seq model for machine translation, and it has been successfully used in the Google

Neural Machine Translation System (GNMT). Other RNN variants include GRU and bidirectional RNN. We will introduce the principles of recurrent neural networks in detail in Chapter 11.

6.7.3 Attention Mechanism Network

RNN is not the ultimate solution for natural language processing. In recent years, with the attention mechanism proposed, it overcomes the deficiencies of RNN such as training instability and difficulty in parallelization. It has gradually emerged in the fields of natural language processing and image generation. The attention mechanism was originally proposed on the image classification task, but gradually began to become more effective in natural language processing. In 2017, Google proposed the first network model Transformer using a pure attention mechanism, and then based on the Transformer model, a series of attention network models for machine translation, such as GPT, BERT, and GPT-2, were successively proposed. In other fields, the network based on the attention mechanism, especially the self-attention mechanism, has also achieved good results, such as the BigGAN model.

6.7.4 Graph Convolutional Neural Network

Data such as pictures and texts have a regular spatial or temporal structure called Euclidean data. Convolutional neural networks and recurrent neural networks are very good at handling this type of data. For data like a series of irregular spatial topologies, social networks, communication networks, and protein molecular structures, those networks seem to be powerless. In 2016, Thomas Kipf et al. proposed a graph convolution network (GCN) model based on the first-order approximate spectral convolution algorithm. The GCN algorithm is simple to implement and can be intuitively understood from the perspective of spatial first-order neighbor information aggregation and therefore has achieved good results on semi-supervised tasks. Subsequently, a series of network models have been proposed, such as GAT, EdgeConv, and DeepGCN.

6.8 Hands-On of Automobile Fuel Consumption Prediction

In this section, we will use the fully connected network model to complete the prediction of MPG (mile per gallon) of the car.

6.8.1 Dataset

We use the auto MPG dataset, which include the real data of various vehicle performance indicators and other factors such as the number of cylinders, weight, and horsepower. The first five items of the dataset is shown in Table 6-1. In addition, the numeric field of origin indicates the category, the other fields are numeric types. For the place of origin, 1 indicates the USA, 2 indicates Europe, and 3 indicates Japan.

Table 6-1. *First five items of the auto MPG dataset*

MPG	Cylinders	Displacement	Horsepower	Weight	Acceleration	Model Year	Origin
18.0	8	307.0	130.0	3504.0	12.0	70	1
15.0	8	350.0	165.0	3693.0	11.5	70	1
18.0	8	318.0	150.0	3436.0	11.0	70	1
16.0	8	304.0	150.0	3433.0	12.0	70	1
17.0	8	302.0	140.0	3449.0	10.5	70	1

The auto MPG dataset includes a total of 398 records. We download and read the dataset from the UCI server to a DataFrame object. The code is as follows:

```
# Download the dataset online
dataset_path = keras.utils.get_file("auto-mpg.data", "http://
archive.ics.uci.edu/ml/machine-learning-databases/auto-mpg/
auto-mpg.data")
# Use Pandas library to read the dataset
column_names = ['MPG','Cylinders','Displacement','Horsepower','
Weight', 'Acceleration', 'Model Year', 'Origin']
raw_dataset = pd.read_csv(dataset_path, names=column_names,
                      na_values = "?", comment='\t',
                      sep=" ", skipinitialspace=True)
dataset = raw_dataset.copy()
# Show some data
dataset.head()
```

The data in the original table may contain missing values. These record items need to be cleared:

```
dataset.isna().sum() # Calculate the number of missing values
dataset = dataset.dropna() # Drop missing value records
dataset.isna().sum() # Calculate the number of missing
values again
```

After clearing, the dataset record items were reduced to 392 items.

Since the origin field is categorical data, we first remove it and then convert it into three new fields, USA, Europe, and Japan, which represent whether they are from this origin:

```
origin = dataset.pop('Origin')
dataset['USA'] = (origin == 1)*1.0
dataset['Europe'] = (origin == 2)*1.0
```

```
dataset['Japan'] = (origin == 3)*1.0
dataset.tail()
```

Split the data into training (80%) and testing (20%) datasets:

```
train_dataset = dataset.sample(frac=0.8,random_state=0)
test_dataset = dataset.drop(train_dataset.index)
```

Move MPG out and use its real label:

```
train_labels = train_dataset.pop('MPG')
test_labels = test_dataset.pop('MPG')
```

Calculate the mean and standard deviation of each field value of the training set and complete the standardization of the data, through the norm() function; the code is as follows:

```
train_stats = train_dataset.describe()
train_stats.pop("MPG")
train_stats = train_stats.transpose()
# Normalize the data
def norm(x): # minus mean and divide by std
  return (x - train_stats['mean']) / train_stats['std']
normed_train_data = norm(train_dataset)
normed_test_data = norm(test_dataset)
```

Print the shape of training and testing datasets:

```
print(normed_train_data.shape,train_labels.shape)
print(normed_test_data.shape, test_labels.shape)
(314, 9) (314,) # 314 records in training dataset with 9
features.
(78, 9) (78,) # 78 records in training dataset with 9 features.
```

Create TensorFlow dataset:

```
train_db = tf.data.Dataset.from_tensor_slices((normed_train_
data.values, train_labels.values))
train_db = train_db.shuffle(100).batch(32) # Shuffle and batch
```

We can observe the influence of each field on MPG by simply observing the distribution between each field in the dataset, as shown in Figure 6-16. It can be roughly observed that the relationship between car displacement, weight, and MPG is relatively simple. As the displacement or weight increases, the MPG of the car decreases and the energy consumption increases; the smaller the number of cylinders, the better MPG can be, which is in line with our life experience.

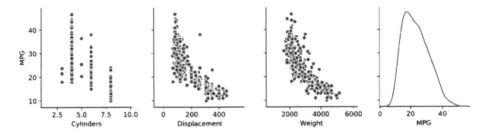

Figure 6-16. *Relations between features*

6.8.2 Create a Network

Considering the small size of the auto MPG dataset, we only create a three-layer fully connected network to complete the MPG prediction task. There are nine input features, so the number of input nodes in the first layer is 9. The number of output nodes of the first layer and the second layer is designed as 64 and 64. Since there is only one kind of prediction value, the output node of the output layer is designed as 1. Because MPG belong to the real number space, the activation function of the output layer may not be added.

We implement the network as a custom network class. We only need to create each sub-network layer in the initialization function and implement the calculation logic of the custom network class in the forward calculation function. The custom network class inherits from the keras.Model class, which is also the standard writing method of the custom network class, in order to conveniently use the various convenient functions such as trainable_variables and save_weights provided by the keras.Model class. The network model class is implemented as follows:

```python
class Network(keras.Model):
    # regression network
    def __init__(self):
        super(Network, self).__init__()
        # create 3 fully-connected layers
        self.fc1 = layers.Dense(64, activation='relu')
        self.fc2 = layers.Dense(64, activation='relu')
        self.fc3 = layers.Dense(1)

    def call(self, inputs, training=None, mask=None):
        # pass through the 3 layers sequentially
        x = self.fc1(inputs)
        x = self.fc2(x)
        x = self.fc3(x)

        return x
```

6.8.3 Training and Testing

After the creation of the main network model class, let's instantiate the network object and create the optimizer as follows:

```python
model = Network() # Instantiate the network
# Build the model with 4 batch and 9 features
model.build(input_shape=(4, 9))
```

```
model.summary() # Print the network
# Create the optimizer with learning rate 0.001
optimizer = tf.keras.optimizers.RMSprop(0.001)
```

Next, implement the network training part. Through the double-layer loop training network composed of Epoch and Step, a total of 200 Epochs are trained.

```
for epoch in range(200): # 200 Epoch
    for step, (x,y) in enumerate(train_db): # Loop through
                                        training set once

        # Set gradient tape
        with tf.GradientTape() as tape:
            out = model(x) # Get network output
            loss = tf.reduce_mean(losses.MSE(y, out))
            # Calculate MSE
            mae_loss = tf.reduce_mean(losses.MAE(y, out))
            # Calculate MAE

        if step % 10 == 0: # Print training loss every 10 steps
            print(epoch, step, float(loss))
        # Calculate and update gradients
        grads = tape.gradient(loss, model.trainable_variables)
        optimizer.apply_gradients(zip(grads, model.trainable_
        variables))
```

For regression problems, in addition to the mean square error (MSE), the mean absolute error (MAE) can also be used to measure the performance of the model.

$$mae \triangleq \frac{1}{d_{out}} \sum_i |y_i - o_i|$$

We can record the MAE at the end of each Epoch for the training and testing dataset and draw the change curve as shown in Figure 6-17.

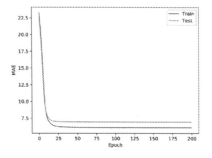

Figure 6-17. *MAE curve*

It can be seen that when training reaches about the 25th Epoch, the decline of MAE becomes slower, in which the MAE of the training set continues to decline slowly, but the MAE of the test set remains almost unchanged, so we can end the training around the 25th Epoch and use the network parameters at this time to predict new input.

6.9 References

[1]. Nick, 2017. A brief history of artificial intelligence.

[2]. X. Glorot, A. Bordes and Y. Bengio, "Deep Sparse Rectifier Neural Networks,"*Proceedings of the Fourteenth International Conference on Artificial Intelligence and Statistics*, Fort Lauderdale, FL, USA, 2011.

[3]. J. Mizera-Pietraszko and P. Pichappan, Lecture Notes in Real-Time Intelligent Systems, Springer International Publishing, 2017.

CHAPTER 7

Backward Propagation Algorithm

The longer you look back, the farther you can look forward.

—Winston S. Churchill

In Chapter 6, we have systematically introduced the basic neural network algorithm: starting from the representation of inputs and outputs; introducing the perceptron model, multi-input and multi-output fully connected layers; and then expanding to multilayer neural networks. We also introduced the design of the output layer under different scenarios and the commonly used loss functions and their implementation.

In this chapter, we will learn one of the core algorithms in the neural network from the theoretical level: error back propagation (BP). In fact, the back propagation algorithm has been proposed in the early 1960s, but it has not attracted the attention of the industry. In 1970, Seppo Linnainmaa proposed an automatic chain derivation method in his master's thesis and implemented the back propagation algorithm. In 1974, Paul Werbos first proposed the possibility of applying the back propagation algorithm to neural networks in his doctoral thesis, but unfortunately, Paul Werbos did

© Liangqu Long and Xiangming Zeng 2022
L. Long and X. Zeng, *Beginning Deep Learning with TensorFlow*,
https://doi.org/10.1007/978-1-4842-7915-1_7

not publish subsequent related research. In fact, Paul Werbos believes that this research idea is meaningful for solving perceptron problems, but due to the cold winter of artificial intelligence, the community has generally lost its belief in solving those problems. Until about 10 years later in 1986, Geoffrey Hinton et al. applied the back propagation algorithm to neural networks[1], making the back propagation algorithm vigorous in the neural network community.

With the functions of automatic derivation and automatic parameter updating of deep learning frameworks, algorithm designers can build complex models and networks with little need for in-depth knowledge of back propagation algorithms and can easily train network models by calling optimization tools. However, the back propagation algorithm and gradient descent algorithm are the core of the neural network, and it is very important to deeply understand its principle. We first review the mathematical concepts such as derivatives and gradients, and then derive the gradient forms of commonly used activation and loss functions, and begin to gradually derive the gradient propagation methods of perceptron and multilayer neural networks. If you want to refresh your memory or learn more about linear algebra and calculus, [2] and [3] have more details.

7.1 Derivatives and Gradients

In high school, we came into contact with the concept of derivative, which is defined as the limit of the ratio of the increment Δy of the function output value to the increment Δx of the independent variable x when the independent variable x produces a slight disturbance Δx as Δx approaches to zero:

$$a = \frac{\Delta y}{\Delta x} = \frac{f(x + \Delta x) - f(x)}{\Delta x}$$

The derivative of the function $f(x)$ can be written as $f(x)$ or $\dfrac{dy}{dx}$. From a geometric point of view, the derivative of a univariate function is the slope of the tangent of the function here, that is, the rate of change of the function value along the direction of x. Consider an example in physics, for example, the expression of the displacement function of free-fall motion $y = \dfrac{1}{2}gt^2$. The derivative with respect to time is $\dfrac{dy}{dt} = \dfrac{d\frac{1}{2}gt^2}{dt} = gt$. Considering that velocity v is defined as the rate of change of displacement, so the derivative of displacement with respect to time is velocity, that is, $v = gt$.

In fact, the derivative is a very broad concept. Because most of the functions we have encountered before are univariate functions, the independent variable has only two directions: x^+ and x^-. When the number of independent variables of a function is greater than one, the concept of the derivative of the function is extended to the rate of change of the function value in any direction. The derivative itself is a scalar and has no direction, but the derivative characterizes the rate of change of the function value in a certain direction. Among these arbitrary directions, several directions along the coordinate axis are relatively special, which is also called partial derivative. For univariate functions, the derivative is written as $\dfrac{dy}{dx}$. For the partial derivative of the multivariate function, it is recorded as $\dfrac{\partial y}{\partial x_1}, \dfrac{\partial y}{\partial x_2}, \cdots$. Partial derivatives are special cases of derivatives and have no direction.

Consider a neural network model that is essentially a multivariate function, such as a weight matrix W of shape $[784, 256]$, which contains a connection weight of 784×256, and we need to ask for a partial derivative of 784×256. It should be noted that in mathematical expression habits, the independent variables to be discussed are generally recorded as x, but in neural networks, they are generally used to represent inputs, such as pictures, text, and voice data. The independent variables of the network are network parameter sets $\theta = \{w_1, b_1, w_2, b_2, \cdots\}$. When the gradient descent algorithm is used to optimize the network, all partial derivatives of the network need to be requested. Therefore, we are also concerned about the derivative of the error function L output along the direction of the independent variable θ_i, that is, $\dfrac{\partial L}{\partial w_1}, \dfrac{\partial L}{\partial b_1}, \cdots$. Write all partial derivatives of the function in vector form:

$$\nabla_\theta L = \left(\frac{\partial L}{\partial \theta_1}, \frac{\partial L}{\partial \theta_2}, \frac{\partial L}{\partial \theta_3}, \cdots, \frac{\partial L}{\partial \theta_n} \right)$$

The gradient descent algorithm can be updated in the form of a vector:

$$\theta' = \theta - \eta \cdot \nabla_\theta L$$

η *is* learning rate. The gradient descent algorithm is generally to find the minimum value of the function L, and sometimes it is also desirable to solve the maximum value of the function, which need to update the gradient in the following way:

$$\theta' = \theta + \eta \cdot \nabla_\theta L$$

This update method is called the gradient ascent algorithm. The gradient descent algorithm and the gradient ascent algorithm are the same in principle. One is to update in the opposite direction of the gradient, and the other is to update in the direction of the gradient. Both need to solve partial derivatives. Here, the vector $\left(\dfrac{\partial L}{\partial \theta_1}, \dfrac{\partial L}{\partial \theta_2}, \dfrac{\partial L}{\partial \theta_3}, \cdots, \dfrac{\partial L}{\partial \theta_n} \right)$ is called the gradient of the function, which is composed of all partial derivatives and represents the direction. The direction of the gradient indicates the direction in which the function value rises fastest, and the reverse of the gradient indicates the direction in which the function value decreases fastest.

The gradient descent algorithm does not guarantee the global optimal solution, which is mainly caused by the non-convexity of the objective function. Consider the non-convex function in Figure 7-1. The dark blue area is the minimum area. Different optimization trajectories may obtain different optimal numerical solutions. These numerical solutions are not necessarily global optimal solutions.

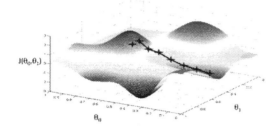

Figure 7-1. *Non-convex function example*

Neural network model expressions are usually very complex, and the model parameters can reach tens or hundreds of millions of levels. Almost all neural network optimization problems rely on deep learning frameworks to automatically calculate the gradient of network parameters and then use gradient descent to iteratively optimize the network parameters until the performance meets the requirement. The main algorithms implemented in deep learning frameworks are back propagation and gradient descent algorithms. So understanding the principles of these two algorithms is helpful to understand the role of deep learning frameworks.

Before introducing the back propagation algorithm of the multilayer neural network, we first introduce the common attributes of the derivative, the gradient derivation of the common activation function, and the loss function and then derive the gradient propagation law of the multilayer neural network.

7.2 Common Properties of Derivatives

This section introduces the derivation rules and sample explanations of common functions, which paves the way for the derivation of neural network related functions.

7.2.1 Common Derivatives

- The derivative of constant function c is 0. For example, the derivative of $y = 2$ is $\dfrac{dy}{dx} = 0$.

- The derivative of linear function $y = ax + c$ is a. For example, the derivative of $y = 2x + 1$ is $\dfrac{dy}{dx} = 2$.

- The derivative of function x^a is ax^{a-1}. For example, the derivative of $y = x^2$ is $\dfrac{dy}{dx} = 2x$.

- The derivative of exponential function a^x is $a^x \ln \ln a$. For example, the derivative of $y = e^x$ is $\dfrac{dy}{dx} = e^x \ln \ln e = e^x$

- The derivative of log function x is $\dfrac{1}{x \ln a}$. For example, the derivative of $y = \ln\ln x$ is $\dfrac{dy}{dx} = \dfrac{1}{x \ln e} = \dfrac{1}{x}$

7.2.2 Common Property of Derivatives

- $(f + g)' = f' + g'$

- $(fg)' = f' \cdot g + f \cdot g'$

- $\left(\dfrac{f}{g} \right)' = \dfrac{f'g - fg'}{g^2}, g \neq 0$

- Consider function of function $f(g(x))$, let $u = g(x)$, the derivative is:

$$\frac{df(g(x))}{dx} = \frac{df(u)}{du} \frac{dg(x)}{dx} = f'(u) \cdot g'(x)$$

7.2.3 Hands-On Derivative Finding

Considering objective function $L = x \cdot w^2 + b^2$, its derivative is:

$$\frac{\partial L}{\partial w} = \frac{\partial x \cdot w^2}{\partial w} = x \cdot 2w$$

$$\frac{\partial L}{\partial b} = \frac{\partial b^2}{\partial b} = 2b$$

Considering objective function $L = x \cdot e^w + e^b$, its derivative is:

$$\frac{\partial L}{\partial w} = \frac{\partial x \cdot e^w}{\partial w} = x \cdot e^w$$

$$\frac{\partial L}{\partial b} = \frac{\partial e^b}{\partial b} = e^b$$

Considering objective function $L = [y - (xw + b)]^2 = [(xw + b) - y]^2$, let $g = xw + b - y$, and the derivative is:

$$\frac{\partial L}{\partial w} = 2g \cdot \frac{\partial g}{\partial w} = 2g \cdot x = 2(xw + b - y) \cdot x$$

$$\frac{\partial L}{\partial b} = 2g \cdot \frac{\partial g}{\partial b} = 2g \cdot 1 = 2(xw + b - y)$$

Considering objective function $L = a \ln(xw + b)$, let $g = xw + b$, and the derivative is:

$$\frac{\partial L}{\partial w} = a \cdot \frac{\partial \ln(g)}{\partial g} \cdot \frac{\partial g}{\partial w} = a \cdot \frac{1}{g} \cdot \frac{\partial g}{\partial w} = \frac{a}{xw + b} \cdot x$$

$$\frac{\partial L}{\partial b} = a \cdot \frac{\partial \ln(g)}{\partial g} \cdot \frac{\partial g}{b} = a \cdot \frac{1}{g} \cdot \frac{\partial g}{\partial b} = \frac{a}{xw + b}$$

7.3 Derivative of Activation Function

Here we introduce the derivation of the activation function commonly used in neural networks.

7.3.1 Derivative of Sigmoid Function

The expression of Sigmoid function is:

$$\sigma(x) = \frac{1}{1 + e^{-x}}$$

Let's derive the derivative expression of the Sigmoid function:

$$\frac{d}{dx}\sigma(x) = \frac{d}{dx}\left(\frac{1}{1 + e^{-x}}\right)$$

$$= \frac{e^{-x}}{\left(1 + e^{-x}\right)^2}$$

$$= \frac{\left(1 + e^{-x}\right) - 1}{\left(1 + e^{-x}\right)^2}$$

$$= \frac{1+e^{-x}}{\left(1+e^{-x}\right)^2} - \left(\frac{1}{1+e^{-x}}\right)^2$$

$$= \sigma(x) - \sigma(x)^2$$

$$= \sigma(1-\sigma)$$

It can be seen that the derivative expression of the Sigmoid function can finally be expressed as a simple operation of the output value of the activation function. Using this property, we can calculate its derivate by caching the output value of the Sigmoid function of each layer in the gradient calculation of the neural network. The derivative function of the Sigmoid function is shown in Figure 7-2.

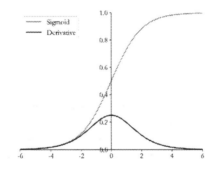

Figure 7-2. *Sigmoid function and its derivative*

In order to help understand the implementation details of the backpropagation algorithm, this chapter chooses not to use TensorFlow's automatic derivation function. This chapter uses Numpy to implement a multilayer neural network optimized by back propagation algorithm. Here the derivative of the Sigmoid function is implemented by Numpy:

```
import numpy as np # import numpy library
def sigmoid(x): # implement sigmoid function
    return 1 / (1 + np.exp(-x))
```

```
def derivative(x):  # calculate derivative of sigmoid
    # Using the derived expression of the derivatives
    return sigmoid(x)*(1-sigmoid(x))
```

7.3.2 Derivative of ReLU Function

Recall the expression of the ReLU function:

$$ReLU(x) = max(0,x)$$

The derivation of its derivative is very simple:

$$\frac{d}{dx} ReLU = \{1 \; x \geq 0 \; 0 \; x < 0$$

It can be seen that the derivative calculation of the ReLU function is simple. When x is greater than or equal to zero, the derivative value is always 1. In the process of back propagation, it will neither amplify the gradient, causing gradient exploding, nor shrink the gradient, causing gradient vanishing phenomenon. The derivative curve of the ReLU function is shown in Figure 7-3.

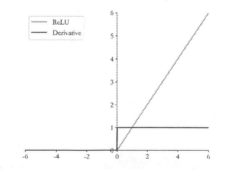

Figure 7-3. *ReLU function and its derivative*

Before the ReLU function was widely used, the activation function in neural networks was mostly Sigmoid. However, the Sigmoid function was prone to gradient dispersion. When the number of layers of the network increased, because the gradient values become very small, the parameters of the network cannot be effectively updated. As a result, deeper neural networks cannot be trained, resulting in the research of neural networks staying at the shallow level. With the introduction of the ReLU function, the phenomenon of gradient dispersion is well alleviated, and the number of layers of the neural network can reach deeper layers. For example, the ReLU activation function is used in AlexNet, and the number of layers reaches eight. Some convolutional neural networks with over 100 layers also mostly uses the ReLU activation function.

Through Numpy, we can easily achieve the derivative of the ReLU function, the code is as follows:

```
def derivative(x):  # Derivative of ReLU
    d = np.array(x, copy=True)
    d[x < 0] = 0
    d[x >= 0] = 1
    return d
```

7.3.3 Derivative of LeakyReLU Function

Recall the expression of LeakyReLU function:

$$LeakyReLU = \{x \; x \geq 0 \;\; px \; x < 0$$

Its derivative can be derived as:

$$\frac{d}{dx} LeakyReLU = \{1 \; x \geq 0 \;\; p \; x < 0$$

It's different from the ReLU function because when x is less than zero, the derivative value of the LeakyReLU function is not 0, but a constant p, which is generally set to a smaller value, such as 0.01 or 0.02. The derivative curve of the LeakyReLU function is shown in Figure 7-4.

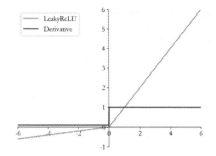

Figure 7-4. *LeakyReLU function and its derivative*

The LeakyReLU function effectively overcomes the defects of the ReLU function and is also widely used. We can implement the derivative of LeakyReLU function through Numpy as follows:

```
def derivative(x, p): # p is the slope of negative part of
LeakyReLU
    dx = np.ones_like(x)  # initialize a vector with 1
    dx[x < 0] = p  # set negative part to p
    return dx
```

7.3.4 Derivative of Tanh Function

Recall the expression of the Tanh function:

$$tanh(x) = \frac{\left(e^x - e^{-x}\right)}{\left(e^x + e^{-x}\right)}$$

$$= 2 \cdot sigmoid(2x) - 1$$

Its derivative expression is:

$$\frac{d}{dx}tanh\,tanh(x) = \frac{\left(e^x + e^{-x}\right)\left(e^x + e^{-x}\right) - \left(e^x - e^{-x}\right)\left(e^x - e^{-x}\right)}{\left(e^x + e^{-x}\right)^2}$$

$$= 1 - \frac{\left(e^x - e^{-x}\right)^2}{\left(e^x + e^{-x}\right)^2} = 1 - (x)$$

The Tanh function and its derivative curve are shown in Figure 7-5.

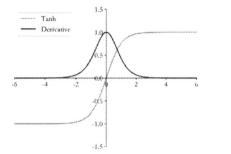

Figure 7-5. *Tanh function and its derivative*

In Numpy, the derivative of the Tanh function is implemented through the Sigmoid function as follows:

```
def sigmoid(x):  # sigmoid function
    return 1 / (1 + np.exp(-x))

def tanh(x):  # tanh function
    return 2*sigmoid(2*x) - 1

def derivative(x):  # derivative of tanh
    return 1-tanh(x)**2
```

7.4 Gradient of Loss Function

The common loss functions have been introduced previously. Here we mainly derive the gradient expressions of the mean square error loss function and the cross-entropy loss function.

7.4.1 Gradient of Mean Square Error Function

The mean square error loss function expression is:

$$L = \frac{1}{2}\sum_{k=1}^{K}\left(y_k - o_k\right)^2$$

The terms $\frac{1}{2}$ in the preceding formula are used to simplify the calculation, and $\frac{1}{K}$ can also be used for averaging instead. None of these scaling operations will change the gradient direction. Then its partial derivative $\frac{\partial L}{\partial o_i}$ can be expanded to:

$$\frac{\partial L}{\partial o_i} = \frac{1}{2}\sum_{k=1}^{K}\frac{\partial}{\partial o_i}\left(y_k - o_k\right)^2$$

Decomposition by the law of derivative of composite function:

$$\frac{\partial L}{\partial o_i} = \frac{1}{2}\sum_{k=1}^{K}2\cdot\left(y_k - o_k\right)\cdot\frac{\partial\left(y_k - o_k\right)}{\partial o_i}$$

That is:

$$\frac{\partial L}{\partial o_i} = \sum_{k=1}^{K}\left(y_k - o_k\right)\cdot -1\cdot\frac{\partial o_k}{\partial o_i}$$

$$= \sum_{k=1}^{K}\left(o_k - y_k\right)\cdot\frac{\partial o_k}{\partial o_i}$$

Considering that $\frac{\partial o_k}{\partial o_i}$ is 1 when $k = i$ and $\frac{\partial o_k}{\partial o_i}$ is 0 for other cases, that is, the partial derivative $\frac{\partial L}{\partial o_i}$ is only related to the ith node, so the summation symbol in the preceding formula can be removed. The derivative of the mean square error function can be expressed as:

$$\frac{\partial L}{\partial o_i} = \left(o_i - y_i \right)$$

7.4.2 Gradient of Cross-Entropy Function

When calculating the cross-entropy loss function, the Softmax function and the cross-entropy function are generally implemented in a unified manner. We first derive the gradient of the Softmax function, and then derive the gradient of the cross-entropy function.

Gradient of Softmax Recall of the expression of Softmax:

$$p_i = \frac{e^{z_i}}{\displaystyle\sum_{k=1}^{K} e^{z_k}}$$

Its function is to convert the values of the output nodes into probabilities and ensure that the sum of probabilities is 1, as shown in Figure 7-6.

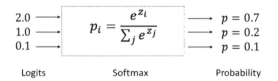

Figure 7-6. *Softmax illustration*

Recall:

$$f(x) = \frac{g(x)}{h(x)}$$

The derivative of the function is:

$$f'(x) = \frac{g'(x)h(x) - h'(x)g(x)}{h(x)^2}$$

For Softmax function, $g(x) = e^{z_i}$, $h(x) = \sum_{k=1}^{K} e^{z_k}$. We'll derive its gradient at two conditions: $i = j$ and $i \neq j$.

- $i = j$. The derivative of Softmax $\dfrac{\partial p_i}{\partial z_j}$ is:

$$\frac{\partial p_i}{\partial z_j} = \frac{\partial \dfrac{e^{z_i}}{\sum_{k=1}^{K} e^{z_k}}}{\partial z_j} = \frac{e^{z_i} \sum_{k=1}^{K} e^{z_k} - e^{z_j} e^{z_i}}{\left(\sum_{k=1}^{K} e^{z_k}\right)^2}$$

$$= \frac{e^{z_i} \left(\sum_{k=1}^{K} e^{z_k} - e^{z_j}\right)}{\left(\sum_{k=1}^{K} e^{z_k}\right)^2}$$

$$= \frac{e^{z_i}}{\sum_{k=1}^{K} e^{z_k}} \times \frac{\left(\sum_{k=1}^{K} e^{z_k} - e^{z_j}\right)}{\sum_{k=1}^{K} e^{z_k}}$$

The preceding expression is the multiplication of p_i and $1 - p_j$, and $p_i = p_j$. So when $i = j$, the derivative of Softmax $\dfrac{\partial p_i}{\partial z_j}$ is:

$$\frac{\partial p_i}{\partial z_j} = p_i(1 - p_j), i = j$$

- $i \neq j$. Extend the Softmax function:

$$\frac{\partial p_i}{\partial z_j} = \frac{\partial \dfrac{e^{z_i}}{\sum_{k=1}^{K} e^{z_k}}}{\partial z_j} = \frac{0 - e^{z_j} e^{z_i}}{\left(\sum_{k=1}^{K} e^{z_k}\right)^2}$$

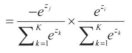

That is:

$$\frac{\partial p_i}{\partial z_j} = -p_j \cdot p_i$$

It can be seen that although the gradient derivation process of the Softmax function is slightly complicated, the final expression is still very concise. The partial derivative expression is as follows:

$$\frac{\partial p_i}{\partial z_j} = \{p_i\left(1-p_j\right) when\ i = j \ - p_i\cdot p_j \qquad when\ i \neq j$$

Gradient of cross-entropy function Consider the expression of the cross-entropy loss function:

$$L = -\sum_k y_k \log\log\left(p_k\right)$$

Here we directly derive the partial derivative of the final loss value L to the logits variable z_i of the network output, which expands to:

$$\frac{\partial L}{\partial z_i} = -\sum_k y_k \frac{\partial \log\log\left(p_k\right)}{\partial z_i}$$

Decompose the composite function $\log\log h$ into:

$$= -\sum_k y_k \frac{\partial \log\log\left(p_k\right)}{\partial p_k} \cdot \frac{\partial p_k}{\partial z_i}$$

That is:

$$= -\sum_k y_k \frac{1}{p_k} \cdot \frac{\partial p_k}{\partial z_i}$$

where $\dfrac{\partial p_k}{\partial z_i}$ is the partial derivative of the Softmax function that we have derived.

Split the summation symbol into the two cases: $k = i$ and $k \neq i$, and substitute the expression of $\dfrac{\partial p_k}{\partial z_i}$, we can get:

$$\frac{\partial L}{\partial z_i} = -y_i(1 - p_i) - \sum_{k \neq i} y_k \frac{1}{p_k}(-p_k \cdot p_i)$$

$$= -y_i(1 - p_i) + \sum_{k \neq i} y_k \cdot p_i$$

$$= -y_i + y_i p_i + \sum_{k \neq i} y_k \cdot p_i$$

That is:

$$\frac{\partial L}{\partial z_i} = p_i \left(y_i + \sum_{k \neq i} y_k \right) - y_i$$

In particular, the one-hot encoding method for the label in the classification problem has the following relationship:

$$\sum_k y_k = 1$$

$$y_i + \sum_{k \neq i} y_k = 1$$

Therefore, the partial derivative of cross-entropy can be further simplified to:

$$\frac{\partial L}{\partial z_i} = p_i - y_i$$

7.5 Gradient of Fully Connected Layer

After introducing the basic knowledge of gradients, we formally entered the derivation of the neural network's back propagation algorithm. The structure of the neural network is diverse, and it is impossible to analyze the gradient expressions one by one. We will use a neural network with a fully connected layer network, a Sigmoid function as the activation function, and a softmax + MSE loss function as the error function to derive the gradient propagation law.

7.5.1 Gradient of a Single Neuron

For a neuron model using Sigmoid activation function, its mathematical model can be written as:

$$o^{(1)} = \sigma\left(w^{(1)T} x + b^{(1)} \right)$$

The superscript of the variable represents the number of layers. For example, $o^{(1)}$ represents the output of the first layer and x is the input of the network. We take the partial derivative derivation $\dfrac{\partial L}{\partial w_{j1}}$ of the weight parameter w_{j1} as an example. For the convenience of demonstration, we draw the neuron model as shown in Figure 7-7. Bias b is not shown in the figure, and the number of input nodes is J. The weight connection from the input of the jth node to the output $o^{(1)}$ is denoted as $w_{j1}^{(1)}$, where the superscript indicates the number of layers to which the weight parameter belongs, and the subscript indicates the starting node number and the ending node number of the current connection. For example, the subscript $j1$ indicates the jth node of the previous layer to the first node of the current layer. The variable before the activation function σ is called $z_1^{(1)}$, and the variable after the activation function σ is called $o_1^{(1)}$. Because there is only one output node, so $o_1^{(1)} = o^{(1)} = o$. The error value L is calculated by the error function between the output and the real label.

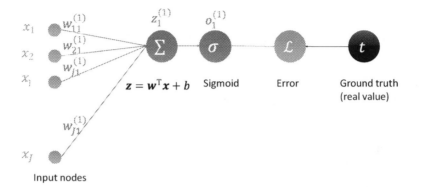

Input nodes

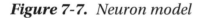

Figure 7-7. *Neuron model*

If we use the mean square error function, considering that a single neuron has only one output $o_1^{(1)}$, then the loss can be expressed as:

$$L = \frac{1}{2}\left(o_1^{(1)} - t\right)^2 = \frac{1}{2}\left(o_1 - t\right)^2$$

Among them, t is the real label value. Adding $\frac{1}{2}$ does not affect the direction of the gradient, and the calculation is simpler. We take the weight variable w_{j1} of the jth ($j \in [1, J]$) node as an example and consider the partial derivative $\dfrac{\partial L}{\partial w_{j1}}$ of the loss function L:

$$\frac{\partial L}{\partial w_{j1}} = \left(o_1 - t\right)\frac{\partial o_1}{\partial w_{j1}}$$

Considering $o_1 = \sigma(z_1)$ and the derivative of the Sigmoid function is $\sigma' = \sigma(1 - \sigma)$, we have:

$$\frac{\partial L}{\partial w_{j1}} = (o_1 - t)\frac{\partial \sigma(z_1)}{\partial w_{j1}}$$

$$= (o_1 - t)\sigma(z_1)(1 - \sigma(z_1))\frac{\partial z_1^{(1)}}{\partial w_{j1}}$$

Write $\sigma(z_1)$ as o_1:

$$\frac{\partial L}{\partial w_{j1}} = (o_1 - t)o_1(1 - o_1)\frac{\partial z_1^{(1)}}{\partial w_{j1}}$$

Consider $\dfrac{\partial z_1^{(1)}}{\partial w_{j1}} = x_j$, we have:

$$\frac{\partial L}{\partial w_{j1}} = (o_1 - t)o_1(1 - o_1)x_j$$

It can be seen from the preceding formula that the partial derivative of the error to the weight w_{j1} is only related to the output value o_1, the true value t, and the input x_j connected to the current weight.

7.5.2 Gradient of Fully Connected Layer

We generalize the single neuron model to a single-layer network of fully connected layers, as shown in Figure 7-8. The input layer obtains the output vector $o^{(1)}$ through a fully connected layer and calculates the mean square error with the real label vector t. The number of input nodes is J, and the number of output nodes is K.

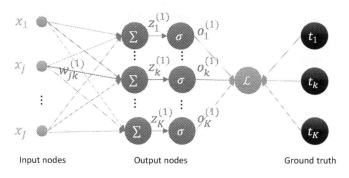

Figure 7-8. *Fully connected layer*

The multi-output fully connected network layer model differs from the single neuron model in that it has many more output nodes $o_1^{(1)}, o_2^{(1)}, o_3^{(1)}, \cdots, o_K^{(1)}$, and each output node corresponds to a real label t_1, t_2, ..., t_K. w_{jk} is the connection weight of the jth input node and the kth output node. The mean square error can be expressed as:

$$L = \frac{1}{2}\sum_{i=1}^{K}\left(o_i^{(1)} - t_i\right)^2$$

Since $\dfrac{\partial L}{\partial w_{jk}}$ is only associated with node $o_k^{(1)}$, the summation symbol in the preceding formula can be removed, that is, $i = k$:

$$\frac{\partial L}{\partial w_{jk}} = \left(o_k - t_k\right)\frac{\partial o_k}{\partial w_{jk}}$$

Substitute $o_k = \sigma(z_k)$:

$$\frac{\partial L}{\partial w_{jk}} = \left(o_k - t_k\right)\frac{\partial \sigma(z_k)}{\partial w_{jk}}$$

Consider the derivative of the Sigmoid function $\sigma' = \sigma(1 - \sigma)$:

$$\frac{\partial L}{\partial w_{jk}} = (o_k - t_k)\sigma(z_k)(1 - \sigma(z_k))\frac{\partial z_k^{(1)}}{\partial w_{jk}}$$

Write $\sigma(z_k)$ as o_k:

$$\frac{\partial L}{\partial w_{jk}} = (o_k - t_k)o_k(1 - o_k)\frac{\partial z_k^{(1)}}{\partial w_{jk}}$$

Consider $\dfrac{\partial z_k^{(1)}}{\partial w_{jk}} = x_j$:

$$\frac{\partial L}{\partial w_{jk}} = (o_k - t_k)o_k(1 - o_k)x_j$$

It can be seen that the partial derivative of w_{jk} is only related to the output node $o_k^{(1)}$ of the current connection, the label $t_k^{(1)}$ of the corresponding true, and the corresponding input node x_j.

Let $\delta_k = (o_k - t_k)o_k(1 - o_k)$, $\dfrac{\partial L}{\partial w_{jk}}$ becomes:

$$\frac{\partial L}{\partial w_{jk}} = \delta_k x_j$$

The variable δ_k characterizes a certain characteristic of the error gradient propagation of the end node of the connection line. After using the representation δ_k, the partial derivative $\dfrac{\partial L}{\partial w_{jk}}$ is only related to the start node x_j and the end node δ_k of the current connection. Later we will see the role of δ_k in cyclically deriving gradients.

Now that the gradient propagation method of the single-layer neural network (i.e., the output layer) has been derived, next we try to derive the gradient propagation method of the penultimate layer. After completing the propagation derivation of the penultimate layer, similarly, the gradient propagation mode of all hidden layers can be derived cyclically to obtain gradient calculation expressions of all layer parameters.

Before introducing the back propagation algorithm, we first learn a core rule of derivative propagation – the chain rule.

7.6 Chain Rule

Earlier, we introduced the gradient calculation method of the output layer. We now introduce the chain rule, which is a core formula that can derive the gradient layer by layer without explicitly deducing the mathematical expression of the neural network.

In fact, the chain rule has been used more or less in the process of deriving the gradient. Considering the compound function $y = f(u)$, $u = g(x)$, we can derive $\dfrac{dy}{dx}$ from $\dfrac{dy}{du}$ and $\dfrac{du}{dx}$:

$$\frac{dy}{dx} = \frac{dy}{du} \cdot \frac{du}{dx} = f'(g(x)) \cdot g'(x)$$

Consider the compound function with two variables $z = f(x, y)$, where $x = g(t)$, $y = h(t)$, then the derivative $\dfrac{dz}{dt}$ can be derived from $\dfrac{\partial z}{\partial x}$ and $\dfrac{\partial z}{\partial y}$:

$$\frac{dz}{dt} = \frac{\partial z}{\partial x}\frac{dx}{dt} + \frac{\partial z}{\partial y}\frac{dy}{dt}$$

For example, $z = (2t+1)^2 + e^{t^2}$, let $x = 2t + 1$, $y = t^2$, then $z = x^2 + e^y$. Using preceding formula, we have:

$$\frac{dz}{dt} = \frac{\partial z}{\partial x}\frac{dx}{dt} + \frac{\partial z}{\partial y}\frac{dy}{dt} = 2x \cdot 2 + e^y \cdot 2t$$

Let $x = 2t + 1$, $y = t^2$:

$$\frac{dz}{dt} = 2(2t+1) \cdot 2 + e^{t^2} \cdot 2t$$

That is:

$$\frac{dz}{dt} = 4(2t+1) + 2te^{t^2}$$

The loss function L of the neural network comes from each output node $o_k^{(K)}$, as shown in Figure 7-9, where the output node $o_k^{(K)}$ is associated with the output node $o_j^{(J)}$ of the hidden layer, so the chain rule is very suitable for the gradient derivation of the neural network. Let us consider how to apply the chain rule to the loss function.

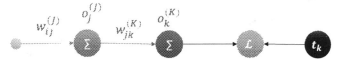

Figure 7-9. *Gradient propagation illustration*

In forward propagation, the data goes through $w_{ij}^{(J)}$ to the node $o_j^{(J)}$ in the penultimate layer and then propagates to the node $o_k^{(K)}$ in the output layer. When there is only one node per layer, the chain rule can be used to decompose $\dfrac{\partial L}{\partial w_{ij}^{(J)}}$ layer by layer into:

$$\frac{\partial L}{\partial w_{ij}^{(J)}} = \frac{\partial L}{\partial o_j^{(J)}}\frac{\partial o_j^{(J)}}{\partial w_{ij}^{(J)}} = \frac{\partial L}{\partial o_k^{(K)}}\frac{\partial o_k^{(K)}}{\partial o_j^{(J)}}\frac{\partial o_j^{(J)}}{\partial w_{ij}^{(J)}}$$

where $\dfrac{\partial L}{\partial o_k^{(K)}}$ can be directly derived from the error function and $\dfrac{\partial o_k^{(K)}}{\partial o_j^{(J)}}$ can be derived from the fully connected layer formula. The derivative $\dfrac{\partial o_j^{(J)}}{\partial w_{ij}^{(J)}}$ is the input $x_i^{(I)}$. It can be seen that through the chain rule, we do not need specific mathematical expressions for the derivative of $L = f\left(w_{ij}^{(J)}\right)$; instead, we can directly decompose the partial derivatives and iteratively derive the derivatives layer by layer.

Here we simply use TensorFlow automatic derivation function to experience the charm of the chain rule.

```
import tensorflow as tf
# Create vectors
x = tf.constant(1.)
w1 = tf.constant(2.)
b1 = tf.constant(1.)
w2 = tf.constant(2.)
b2 = tf.constant(1.)
# Create gradient recorder
with tf.GradientTape(persistent=True) as tape:
    # Manually record gradient info for non-tf.Variable
      variables
    tape.watch([w1, b1, w2, b2])
    # Create two layer neural network
    y1 = x * w1 + b1
    y2 = y1 * w2 + b2

# Solve partial derivatives
dy2_dy1 = tape.gradient(y2, [y1])[0]
dy1_dw1 = tape.gradient(y1, [w1])[0]
dy2_dw1 = tape.gradient(y2, [w1])[0]

# Valdiate chain rule
print(dy2_dy1 * dy1_dw1)
print(dy2_dw1)
```

In the preceding code, we calculated $\frac{\partial y_2}{\partial y_1}$, $\frac{\partial y_1}{\partial w_1}$, and $\frac{\partial y_2}{\partial w_1}$ through auto-gradient calculation in Tensorflow and through chain rule we know $\frac{\partial y_2}{\partial y_1} \cdot \frac{\partial y_1}{\partial w_1}$ and $\frac{\partial y_2}{\partial w_1}$ should be equal. Their results are as follows:

```
tf.Tensor(2.0, shape=(), dtype=float32)
tf.Tensor(2.0, shape=(), dtype=float32)
```

7.7 Back Propagation Algorithm

Now let's derive the gradient propagation law of the hidden layer. Briefly review the partial derivative formula of the output layer:

$$\frac{\partial L}{\partial w_{jk}} = \left(o_k - t_k \right) o_k \left(1 - o_k \right) x_j = \delta_k x_j$$

Consider the partial derivative of the penultimate layer $\dfrac{\partial L}{\partial w_{ij}}$, as shown in Figure 7-10. The number of output layer nodes is K, and the output is $o^{(K)} = \left[o_1^{(K)}, o_2^{(K)}, \cdots, o_K^{(K)} \right]$. The penultimate layer has J nodes, and output is $o^{(J)} = \left[o_1^{(J)}, o_2^{(J)}, \cdots, o_J^{(J)} \right]$. The antepenultimate layer has I nodes, and the output is $o^{(I)} = \left[o_1^{(I)}, o_2^{(I)}, \cdots, o_I^{(I)} \right]$.

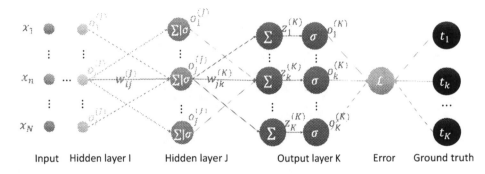

Figure 7-10. *Back propagation algorithm*

In order to express conciseness, the superscripts of some variables are sometimes omitted. First, expand the mean square error function:

$$\frac{\partial L}{\partial w_{ij}} = \frac{\partial}{\partial w_{ij}} \frac{1}{2} \sum_k \left(o_k - t_k \right)^2$$

Because L is associated with w_{ij} through each output node o_k, the summation sign cannot be removed here, and the mean square error function can be disassembled using the chain rule:

$$\frac{\partial L}{\partial w_{ij}} = \sum_{k} \left(o_k - t_k \right) \frac{\partial}{\partial w_{ij}} o_k$$

Substitute $o_k = \sigma(z_k)$:

$$\frac{\partial L}{\partial w_{ij}} = \sum_{k} \left(o_k - t_k \right) \frac{\partial}{\partial w_{ij}} \sigma \left(z_k \right)$$

The derivative of the Sigmoid function is $\sigma' = \sigma(1 - \sigma)$, so:

$$\frac{\partial L}{\partial w_{ij}} = \sum_{k} \left(o_k - t_k \right) \sigma \left(z_k \right) \left(1 - \sigma \left(z_k \right) \right) \frac{\partial z_k}{\partial w_{ij}}$$

Write $\sigma(z_k)$ as o_k, and consider chain rule, we have:

$$\frac{\partial L}{\partial w_{ij}} = \sum_{k} \left(o_k - t_k \right) o_k \left(1 - o_k \right) \frac{\partial z_k}{\partial o_j} \cdot \frac{\partial o_j}{\partial w_{ij}}$$

where $\dfrac{\partial z_k}{\partial o_j} = w_{jk}$, so:

$$\frac{\partial L}{\partial w_{ij}} = \sum_{k} \left(o_k - t_k \right) o_k \left(1 - o_k \right) w_{jk} \frac{\partial o_j}{\partial w_{ij}}$$

Because $\dfrac{\partial o_j}{\partial w_{ij}}$ is not associated with k, we have:

$$\frac{\partial L}{\partial w_{ij}} = \frac{\partial o_j}{\partial w_{ij}} \sum_{k} \left(o_k - t_k \right) o_k \left(1 - o_k \right) w_{jk}$$

Because $o_j = \sigma(z_j)$ and $\sigma' = \sigma(1 - \sigma)$, we have:

$$\frac{\partial L}{\partial w_{ij}} = o_j\left(1 - o_j\right)\frac{\partial z_j}{\partial w_{ij}}\sum_k\left(o_k - t_k\right)o_k\left(1 - o_k\right)w_{jk}$$

where $\dfrac{\partial z_j}{\partial w_{ij}}$ is o_i, so:

$$\frac{\partial L}{\partial w_{ij}} = o_j\left(1 - o_j\right)o_i\sum_k\underbrace{\left(o_k - t_k\right)o_k\left(1 - o_k\right)}_{\delta_k^{(K)}}\cdot w_{jk}$$

where $\delta_k^{(K)} = \left(o_k - t_k\right)o_k\left(1 - o_k\right)$, so:

$$\frac{\partial L}{\partial w_{ij}} = o_j\left(1 - o_j\right)o_i\sum_k\delta_k^{(K)}w_{jk}$$

Similarly as the format of $\dfrac{\partial L}{\partial w_{jk}} = \delta_k^{(K)}x_j$, define δ_j^J as:

$$\delta_j^J \triangleq o_j\left(1 - o_j\right)\sum_k\delta_k^{(K)}w_{jk}$$

At this time, $\dfrac{\partial L}{\partial w_{ij}}$ can be written as a simple multiplication of the output value o_i of the currently connected start node and the gradient variable information $\delta_j^{(J)}$ of the end node:

$$\frac{\partial L}{\partial w_{ij}} = \delta_j^{(J)}o_i^{(I)}$$

It can be seen that by defining variable δ, the gradient expression of each layer becomes more clear and concise, where δ can be simply understood as the contribution value of the current weight w_{ij} to the error function.

Let's summarize the propagation law of the partial derivative of each layer.

Output layer:

$$\frac{\partial L}{\partial w_{jk}} = \delta_k^{(K)} o_j$$

$$\delta_k^{(K)} = o_k \left(1 - o_k\right)\left(o_k - t_k\right)$$

Penultimate layer:

$$\frac{\partial L}{\partial w_{ij}} = \delta_j^{(J)} o_i$$

$$\delta_j^{(J)} = o_j \left(1 - o_j\right) \sum_k \delta_k^{(K)} w_{jk}$$

Antepenultimate layer:

$$\frac{\partial L}{\partial w_{ni}} = \delta_i^{(I)} o_n$$

$$\delta_i^{(I)} = o_i \left(1 - o_i\right) \sum_j \delta_j^{(J)} w_{ij}$$

where o_n is the input of the antepenultimate layer.

According to this law, the partial derivative of the current layer can be obtained only by calculating the values $\delta_k^{(K)}$, $\delta_j^{(J)}$, and $\delta_i^{(I)}$ of each node of each layer iteratively, so as to obtain the gradient of the weight matrix W of each layer, and then iteratively optimize the network parameters through the gradient descent algorithm.

So far, the back propagation algorithm is fully introduced.

Next, we will conduct two hands-on cases: the first case is to use the automatic derivation provided by TensorFlow to optimize the extreme value of the Himmelblau function. The second case is to implement the back propagation algorithm based on Numpy and complete the multi-layer neural network training for binary classification problem.

7.8 Hands-On Optimization of Himmelblau

The Himmelblau function is one of the commonly used sample functions for testing optimization algorithms. It contains two independent variables x and y, and the mathematical expression is:

$$f(x,y) = (x^2 + y - 11)^2 + (x + y^2 - 7)^2$$

First, we implement the expression of the Himmelblau function through the following code:

```
def himmelblau(x):
    # Himmelblau function implementation. Input x is a list
    with 2 elements.
    return (x[0] ** 2 + x[1] - 11) ** 2 + (x[0] + x[1] **
    2 - 7) ** 2
```

Then we complete the visualization of the Himmelblau function. Use np.meshgrid function (meshgrid function is also available in TensorFlow) to generate two-dimensional plane grid point coordinates as follows:

```
x = np.arange(-6, 6, 0.1) # x-axis
y = np.arange(-6, 6, 0.1) # y-axis
print('x,y range:', x.shape, y.shape)
X, Y = np.meshgrid(x, y)
print('X,Y maps:', X.shape, Y.shape)
Z = himmelblau([X, Y])
```

Use the Matplotlib library to visualize the Himmelblau function, as shown in Figure 7-11:

```
# Plot the Himmelblau function
fig = plt.figure('himmelblau')
ax = fig.gca(projection='3d')
ax.plot_surface(X, Y, Z)
ax.view_init(60, -30)
ax.set_xlabel('x')
ax.set_ylabel('y')
plt.show()
```

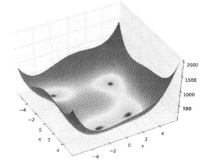

Figure 7-11. *Himmelblau function*

Figure 7-12 is a contour map of the Himmelblau function. It can be roughly seen that it has four local minimum points, and the local minimum values are all 0, so these four local minimum values are also global minimum values. We can calculate the precise coordinates of the local minimum by analytical methods; they are:

$$(3,2),(-2.805,3.131),(-3.779,-3.283),(3.584,-1.848)$$

Knowing the analytical solution of the extreme value, we now use the gradient descent algorithm to optimize the minimum numerical solution of the Himmelblau function.

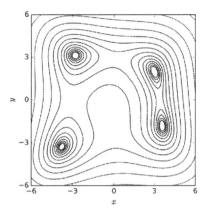

Figure 7-12. *Himmelblau function contour plot*

We can use TensorFlow automatic derivation to find the partial derivative of the sum of the function and iteratively update the sum value as follows:

```
# The influence of the initialization value of the parameter
on the optimization cannot be ignored, you can try different
initialization values # Test the minimum value of function
optimization
# [1., 0.], [-4, 0.], [4, 0.]
x = tf.constant([4., 0.]) # Initialization

for step in range(200):# Loop 200 times
    with tf.GradientTape() as tape: #record gradient
        tape.watch([x]) # Add to the gradient recording list
        y = himmelblau(x) # forward propagation
    # backward propagation
    grads = tape.gradient(y, [x])[0]
    # update paramaters with learning rate of 0.01
    x -= 0.01*grads
    # print info
```

```
if step % 20 == 19:
    print ('step {}: x = {}, f(x) = {}'
            .format(step, x.numpy(), y.numpy()))
```

After 200 iterations of updating, the program can find a minimum solution, at which point the function value is close to zero. The numerical solution is

```
step 199: x = [ 3.584428  -1.8481264], f(x) =
1.1368684856363775e-12
```

This is almost the same as one of the analytical solutions $(3.584, -1.848)$.

In fact, by changing the initialization state of the network parameters, the program can obtain a variety of minimum numerical solutions. The initialization state of the parameters may affect the search trajectory of the gradient descent algorithm, and it may even search out completely different numerical solutions, as shown in Table 7-1. This example explains the effect of different initial states on the gradient descent algorithm.

Table 7-1. *The effect of initial values on optimization results*

Initial value of x	Numerical solution	Analytical solution
(4,0)	(3.58,-1.84)	(3.58,-1.84)
(1,0)	(3,1.99)	(3,2)
(-4,0)	(-3.77,-3.28)	(-3.77,-3.28)
(-2,2)	(-2.80,3.13)	(-2.80,3.13)

7.9 Hands-On Back Propagation Algorithm

In this section, we will use the gradient derivation results of the multi-layer fully connected network introduced earlier, and directly use Python to calculate the gradient of each layer, and manually update according to the gradient descent algorithm. Since TensorFlow has an automatic derivation function, we choose Numpy without the automatic derivation functionality to implement the network, and use Numpy to manually calculate the gradient, and manually update the network parameters.

It should be noted that the gradient propagation formula derived in this chapter is for multiple fully connected layers with only Sigmoid function, and the loss function is a network type of mean square error function. For other types of networks, such as networks with ReLU activation function and cross-entropy loss function, the gradient propagation expression needs to be derived again, but the method is similar. It is precisely because the method of manually deriving the gradient is more limited, it is rarely used in practice.

We will implement a four-layer fully connected network to complete the binary classification task. The number of network input nodes is 2, and the number of nodes in the hidden layer is designed as 20, 50, and 25. The two nodes in the output layer represent the probability of belonging to categories 1 and 2, respectively, as shown in Figure 7-13. Here, the Softmax function is not used to constrain the sum of the network output probability values. Instead, the mean square error function is directly used to calculate the error between prediction and the one-hot encoded real label. All activation functions are Sigmoid. This design is to directly use our gradient propagation formula.

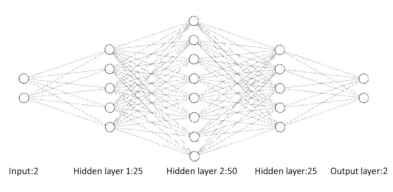

Input:2   Hidden layer 1:25   Hidden layer 2:50   Hidden layer:25   Output layer:2

Figure 7-13. *Network structure*

7.9.1 Dataset

Through the convenient tool provided by the scikit-learn library, 2000 linear inseparable 2-class datasets are generated. The feature length of the data is 2. The sampled data distribution is shown in Figure 7-14. The red points are in one category, and the blue points belong to the other category. The distribution of each category is crescent-shaped and is linearly inseparable, which means a linear network cannot be used to obtain good results. In order to test the performance of the network, we divide the training set and the test set according to the ratio 7:3. Two thousand $\cdot$ 0·s3 = 600 sample points are used for testing and do not participate in the training. The remaining 1400 points are used for network training.

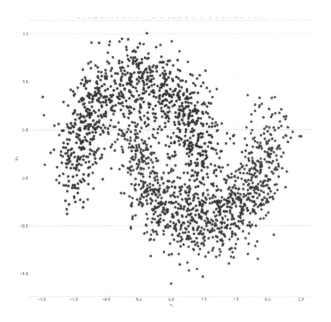

Figure 7-14. *Dataset distribution*

The collection of the data set is directly generated using the make_ moons function provided by scikit-learn, and the number of sampling points and testing ratio are set as follows:

```
N_SAMPLES = 2000 # number of sampling points
TEST_SIZE = 0.3 # testing ratio
# Use make_moons function to generate data set
X, y = make_moons(n_samples = N_SAMPLES, noise=0.2, random_
state=100)
# Split traning and testing data set
X_train, X_test, y_train, y_test = train_test_split(X, y, test_
size=TEST_SIZE, random_state=42)
print(X.shape, y.shape)
```

The distribution of the dataset can be drawn by the following visualization code, as shown in Figure 7-14.

```
# Make a plot
def make_plot(X, y, plot_name, file_name=None, XX=None,
YY=None, preds=None, dark=False):
    if (dark):
        plt.style.use('dark_background')
    else:
        sns.set_style("whitegrid")
    plt.figure(figsize=(16,12))
    axes = plt.gca()
    axes.set(xlabel="$x_1$", ylabel="$x_2$")
    plt.title(plot_name, fontsize=30)
    plt.subplots_adjust(left=0.20)
    plt.subplots_adjust(right=0.80)
    if(XX is not None and YY is not None and preds is
    not None):
        plt.contourf(XX, YY, preds.reshape(XX.shape), 25,
        alpha = 1, cmap=cm.Spectral)
        plt.contour(XX, YY, preds.reshape(XX.shape),
        levels=[.5], cmap="Greys", vmin=0, vmax=.6)
    # Use color to distinguish labels
    plt.scatter(X[:, 0], X[:, 1], c=y.ravel(), s=40, cmap=plt.
    cm.Spectral, edgecolors='none')

    plt.savefig('dataset.svg')
    plt.close()
# Make distribution plot
make_plot(X, y, "Classification Dataset Visualization ")
plt.show()
```

7.9.2 Network Layer

A new layer class is used to implement a network layer. Parameters such as the number of input nodes, the number of output nodes, and the type of activation function are passed into the network layer. The weights and bias tensor bias are automatically generated based on the number of input and output nodes during initialization as in the following:

```python
class Layer:
    # Fully connected layer
    def __init__(self, n_input, n_neurons, activation=None,
    weights=None, bias=None):
        """
        :param int n_input: input nodes
        :param int n_neurons: output nodes
        :param str activation: activation function
        :param weights: weight vectors
        :param bias: bias vectors
        """
        # Initialize weights through Normal distribution
        self.weights = weights if weights is not None else
        np.random.randn(n_input, n_neurons) * np.sqrt(1 / n_
        neurons)
        self.bias = bias if bias is not None else np.random.
        rand(n_neurons) * 0.1
        self.activation = activation # activation function,
        e.g. 'sigmoid'
        self.last_activation = None # output of activation
        function o
        self.error = None
        self.delta = None
```

The forward propagation function of the network layer is implemented as follows, where the last_activation variable is used to save the output value of the current layer:

```
def activate(self, x):
    # Forward propagation function
    r = np.dot(x, self.weights) + self.bias  # X@W+b
    # Get output through activation function
    self.last_activation = self._apply_activation(r)
    return self.last_activation
```

The self._apply_activation function in the preceding code implements the forward calculation process of different types of activation functions, although here we only use the Sigmoid activation function.

```
def _apply_activation(self, r):
    # Calculate output of activation function
    if self.activation is None:
        return r # No activation function
    # ReLU
    elif self.activation == 'relu':
        return np.maximum(r, 0)
    # tanh
    elif self.activation == 'tanh':
        return np.tanh(r)
    # sigmoid
    elif self.activation == 'sigmoid':
        return 1 / (1 + np.exp(-r))

    return r
```

For different types of activation functions, their derivatives are calculated as follows:

```python
def apply_activation_derivative(self, r):
    # Calculate the derivative of activation functions
    # If no activation function, derivative is 1
    if self.activation is None:
        return np.ones_like(r)
    # ReLU
    elif self.activation == 'relu':
        grad = np.array(r, copy=True)
        grad[r > 0] = 1.
        grad[r <= 0] = 0.
        return grad
    # tanh
    elif self.activation == 'tanh':
        return 1 - r ** 2
    # Sigmoid
    elif self.activation == 'sigmoid':
        return r * (1 - r)

    return r
```

It can be seen that the derivative of the Sigmoid function is implemented as $r(1 - r)$, where r is $\sigma(z)$.

7.9.3 Network model

After creating a single-layer network class, we implement the NeuralNetwork class of the network model, which internally maintains the network layer object of each layer. You can add the network layer through the add_layer function to achieve the purpose of creating a network model with different structures as in the following:

```
class NeuralNetwork:
    # Neural Network Class
    def __init__(self):
        self._layers = []  # list of network class

    def add_layer(self, layer):
        # Add layers
        self._layers.append(layer)
```

The forward propagation of the network only needs to cyclically adjust the forward calculation function of each network layer object. The code is as follows:

```
    def feed_forward(self, X):
        # Forward calculation
        for layer in self._layers:
            # Loop through every layer
            X = layer.activate(X)
        return X
```

According to the network structure configuration in Figure 7-13, we use the NeuralNetwork class to create a network object and add a four-layer fully connected network. The code is as follows:

```
nn = NeuralNetwork()
nn.add_layer(Layer(2, 25, 'sigmoid'))  # Hidden layer 1, 2=>25
nn.add_layer(Layer(25, 50, 'sigmoid')) # Hidden layer 2, 25=>50
```

```
nn.add_layer(Layer(50, 25, 'sigmoid')) # Hidden layer 3, 50=>25
nn.add_layer(Layer(25, 2, 'sigmoid'))  # Hidden layer, 25=>2
```

The back propagation of the network model is slightly more complicated. We need to start from the last layer and calculate the variable δ of each layer, and then store the calculated variable δ in the delta variable of the Layer class according to the derived gradient formula as in the following:

```
def backpropagation(self, X, y, learning_rate):
    # Back propagation
    # Get result of forward calculation
    output = self.feed_forward(X)
    for i in reversed(range(len(self._layers))):
    # reverse loop
        layer = self._layers[i]  # get current layer
        # If it's output layer
        if layer == self._layers[-1]:  # output layer
            layer.error = y - output
            # calculate delta
            layer.delta = layer.error * layer.apply_
            activation_derivative(output)
        else:  # For hidden layer
            next_layer = self._layers[i + 1]
            layer.error = np.dot(next_layer.weights,
            next_layer.delta)
            # Calculate delta
            layer.delta = layer.error * layer.apply_
            activation_derivative(layer.last_activation)
            ... # See following code
```

After the reverse calculation of the variable δ of each layer, it is only necessary to calculate the gradient of the parameters of each layer according to the formula $\dfrac{\partial L}{\partial w_{ij}} = o_i \delta_j^{(J)}$ and update the network parameters. Because the delta in the code is actually calculated as $-\delta$, the plus sign is used when updating. The code is as follows:

```python
def backpropagation(self, X, y, learning_rate):
    ... # Continue above code
    # Update weights
    for i in range(len(self._layers)):
        layer = self._layers[i]
        # o_i is output of previous layer
        o_i = np.atleast_2d(X if i == 0 else self._
        layers[i - 1].last_activation)
        # Gradient descent
        layer.weights += layer.delta * o_i.T *
        learning_rate
```

Therefore, in the back propagation function, the variable δ of each layer is reversely calculated, and the gradient values of the parameters of each layer are calculated according to the gradient formula, and the parameter update is completed according to the gradient descent algorithm.

7.9.4 Network Training

The binary classification network here is designed with two output nodes, so the real label needs to be one-hot encoded. The code is as follows:

```python
def train(self, X_train, X_test, y_train, y_test, learning_
rate, max_epochs):
    # Train network
    # one-hot encoding
```

```
y_onehot = np.zeros((y_train.shape[0], 2))
y_onehot[np.arange(y_train.shape[0]), y_train] = 1
```

Calculate the mean square error of the one-hot encoded real label and the output of the network, and call the back propagation function to update the network parameters, and iterate the training set 1000 times as in the following:

```
mses = []
for i in range(max_epochs):  # Train 1000 epoches
    for j in range(len(X_train)):  # Train one sample
                                   per time
        self.backpropagation(X_train[j], y_onehot[j],
        learning_rate)
    if i % 10 == 0:
        # Print MSE Loss
        mse = np.mean(np.square(y_onehot - self.feed_
        forward(X_train)))
        mses.append(mse)
        print('Epoch: #%s, MSE: %f' % (i, float(mse)))

        # Print accuracy
        print('Accuracy: %.2f%%' % (self.accuracy(self.
        predict(X_test), y_test.flatten()) * 100))

return mses
```

7.9.5 Network Performance

We record the training loss value L of each Epoch and draw it as a curve, as shown in Figure 7-15.

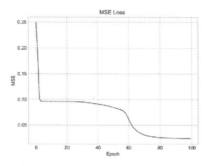

Figure 7-15. *Training error plot*

After training 1000 Epochs, the accuracy rate obtained on 600 samples in the test set is:

```
Epoch: #990, MSE: 0.024335
Accuracy: 97.67%
```

It can be seen that by manually calculating the gradient formula and manually updating the network parameters, we can also obtain a lower error rate for simple binary classification tasks. Through fine-tuning network hyperparameters and other techniques, you can also get better network performance.

In each Epoch, we complete an accuracy test on the test set and draw it into a curve, as shown in Figure 7-16. It can be seen that with the progress of Epoch, the accuracy of the model has been steadily improved, the initial stage is faster, and the subsequent improvement is relatively smooth.

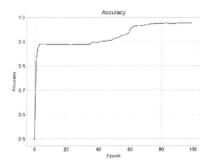

Figure 7-16. *Testing accuracy*

Through this binary classification of fully connected network based on Numpy's manual calculation of gradients, I believe readers can more deeply appreciate the role of deep learning frameworks in algorithm implementation. Without frameworks such as TensorFlow, we can also implement complex neural networks, but flexibility, stability, development efficiency, and computational efficiency are poor. Algorithm design and training based on these deep learning frameworks will greatly improve the work of algorithm developers' effectiveness. At the same time, we can also realize that the framework is just a tool. More importantly, our understanding of the algorithm itself is the most important ability of algorithm developers.

7.10 References

[1].   D. E. Rumelhart, G. E. Hinton and R. J. Williams, "Learning Representations by Back-propagating Errors", *Nature,* 323, 6088, pp. 533-536, 1986.

[2].   Singh, Kuldeep. *Linear Algebra: Step by Step.* 1st edition. Oxford, United Kingdom: Oxford University Press, 2013.

[3].   Stewart, James. *Calculus: Early Transcendentals.* 8th edition. Boston, MA, USA: Cengage Learning, 2015.

CHAPTER 8

Keras Advanced API

> The problem of artificial intelligence is not only a problem of computer science, but also a problem of mathematics, cognitive science and philosophy.
>
> —François Chollet

Keras is an open-source neural network computing library mainly developed in the Python language. It was originally written by François Chollet. It is designed as a highly modular and extensible high-level neural network interface, so that users can quickly complete model building and training without excessive professional knowledge. The Keras library is divided into a frontend and a backend. The backend generally calls the existing deep learning framework to implement the underlying operations, such as Theano, CNTK, and TensorFlow. The frontend interface is a set of unified interface functions abstracted by Keras. Users can easily switch between different backend operations through Keras. Because of Keras's high abstraction and ease of use, according to KDnuggets, Keras market share reached 26.6% as of 2019, an increase of 19.7%, second only to TensorFlow in deep learning frameworks.

There is a staggered relationship between TensorFlow and Keras that is both competitive and cooperative. Even the founder of Keras works at Google. As early as November 2015, TensorFlow was added to Keras backend support. Since 2017, most components of Keras have been

© Liangqu Long and Xiangming Zeng 2022
L. Long and X. Zeng, *Beginning Deep Learning with TensorFlow*,
https://doi.org/10.1007/978-1-4842-7915-1_8

integrated into the TensorFlow framework. In 2019, Keras was officially identified as the only high-level interface API for TensorFlow 2, replacing the high-level interfaces such as tf.layers included in the TensorFlow 1. In other words, now you can only use the Keras interface to complete TensorFlow layer model building and training. In TensorFlow 2, Keras is implemented in the tf.keras submodule.

What is the difference and connection between Keras and tf.keras? In fact, Keras can be understood as a set of high-level API protocols for building and training neural networks. Keras itself has already implemented this protocol. Installing the standard Keras library can easily call TensorFlow, CNTK, and other backends to complete accelerated calculations. In TensorFlow, a set of Keras protocol is also implemented through tf.keras, which is deeply integrated with TensorFlow, and is only based on TensorFlow backend operations, and supports TensorFlow more perfectly. For developers using TensorFlow, tf.keras can be understood as an ordinary submodule, which is no different from other submodules such as tf.math and tf.data. Unless otherwise specified, Keras refers to tf.keras instead of the standard Keras library in the following chapters.

8.1 Common Functional Modules

Keras provides a series of high-level neural network-related classes and functions, such as classic dataset loading function, network layer class, model container, loss function class, optimizer class, and classic model class.

For classic datasets, one line of code can download, manage, and load datasets. These datasets include Boston house price prediction dataset, CIFAR picture dataset, MNIST/FashionMNIST handwritten digital picture dataset, and IMDB text dataset. We have already introduced some of them in previous chapters.

8.1.1 Common Network Layer Classes

For the common neural network layer, we can use the tensor mode of the underlying interface functions to achieve, which are generally included in the tf.nn module. For common network layers, we generally use the layer method to complete the model construction. A large number of common network layers are provided in the tf.keras.layers namespace (hereinafter using layers to refer to tf.keras.layers), such as fully connected layers, activation function layers, pooling layers, convolutional layers, and recurrent neural network layers. For these network layer classes, you only need to specify the relevant parameters of the network layer at the time of creation and use the __call__ method to complete the forward calculation. When using the __call__ method, Keras will automatically call the forward propagation logic of each layer, which is generally implemented in the call function of the class.

Taking the Softmax layer as an example, it can use the tf.nn.softmax function to complete the Softmax operation in the forward propagation, or it can build the Softmax network layer through the layers. Softmax(axis) class, where the axis parameter specifies the dimension for Softmax operation. First, import the relevant sub-modules as follows:

```
import tensorflow as tf
# Do not use "import keras" which will import the standard
Keras, not the one in Tensorflow
from tensorflow import keras
from tensorflow.keras import layers # import common layer class
```

Then create a Softmax layer and use the __call__ method to complete the forward calculation:

```
In [1]:
x = tf.constant([2.,1.,0.1])  # create input tensor
layer = layers.Softmax(axis=-1)  # create Softmax layer
out = layer(x)  # forward propagation
```

After passing through the Softmax network layer, the probability distribution output is:

```
Out[1]:
<tf.Tensor: id=2, shape=(3,), dtype=float32,
numpy=array([0.6590012, 0.242433 , 0.0985659], dtype=float32)>
```

Of course, we can also directly complete the calculation through the tf.nn.softmax() function as follows:

```
out = tf.nn.softmax(x)
```

8.1.2 Network Container

For common networks, we need to manually call the class instance of each layer to complete the forward propagation operation. When the network layer becomes deeper, this part of the code appears very bloated. Multiple network layers can be encapsulated into a large network model through the network container Sequential provided by Keras. Only the instance of the network model needs to be called once to complete the sequential propagation operation of the data from the first layer to the last layer.

For example, the two-layer fully connected network with a separate activation function layer can be encapsulated as a network through the Sequential container.

```
from tensorflow.keras import layers, Sequential
network = Sequential([
    layers.Dense(3, activation=None), # Fully-connected layer
    without activation function
    layers.ReLU(),# activation function layer
    layers.Dense(2, activation=None), # Fully-connected layer
    without activation function
    layers.ReLU() # activation function layer
])
```

```
x = tf.random.normal([4,3])
out = network(x)
```

The Sequential container can also continue to add a new network layer through the add() method to dynamically create a network:

```
In [2]:
layers_num = 2
network = Sequential([]) # Create an empty container
for _ in range(layers_num):
    network.add(layers.Dense(3)) # add fully-connected layer
    network.add(layers.ReLU())# add activation layer
network.build(input_shape=(4, 4))
network.summary()
```

The preceding code can create a network structure with the number of layers specified by the layers_num parameter. When the network creation is completed, the network layer class does not create member variables such as internal weight tensors. Using the build method, you can specify the input size which will automatically create internal tensors for all layers. Through the summary() function, you can easily print out the network structure and parameters. The results are as follows:

```
Out[2]:
Model: "sequential_2"
```

Layer (type)	Output Shape	Param Number
dense_2 (Dense)	multiple	15
re_lu_2 (ReLU)	multiple	0
dense_3 (Dense)	multiple	12

re_lu_3 (ReLU)	multiple	0

```
=================================================================
Total params: 27
Trainable params: 27
Non-trainable params: 0
```

The layer column includes the name of each layer which is maintained internally by TensorFlow and is not the same as the object name of Python. The output shape column indicates the output shape of each layer. Note that the values for the output shape column are all "multiple" because we only built or compiled the network at this point and haven't really trained or executed the network. After we call the network with real inputs, the real out shape of each layer will be reflected in the output shape column. Param number column is the number of parameters of each layer. Total params counts the total number of parameters. Trainable params is the total number of parameters to be optimized. Non-trainable params is the total number of parameters that do not need to be optimized.

When we encapsulate multiple network layers through Sequential container, the parameter list of each layer will be automatically incorporated into the Sequential container. The trainable_variables and variables of the Sequential object contain the list of tensors to be optimized and tensors of all layers, for example:

```
In [3]: # print name and shape of trainable variables
for p in network.trainable_variables:
    print(p.name, p.shape)
Out[3]:
dense_2/kernel:0 (4, 3)
dense_2/bias:0 (3,)
dense_3/kernel:0 (3, 3)
dense_3/bias:0 (3,)
```

The Sequential container is one of the most commonly used classes. It is very useful for quickly building multi-layer neural networks. It should be used as much as possible to simplify the implementation of network models.

8.2 Model Configuration, Training, and Testing

When training the network, the general process is to obtain the output value of the network through forward calculation, then calculate the network error through the loss function, and then calculate and update the gradients through automatic differentiation tool, and test the network performance occasionally. For this commonly used training logic, it can be directly implemented through high-level interfaces provided by Keras.

8.2.1 Model Configuration

In Keras, there are two special classes: keras.Model and keras.layers.Layer. The Layer class is the parent class of the network layer, and it defines some common functions of the network layer, such as adding weights and managing weight lists. The model class is the parent class of the network. In addition to the functions of the layer class, convenient functions such as saving model, loading model, and training and testing model are added. Sequential is also a subclass of model, so it has all the functions of the model class.

Let's introduce the model configuration and training functions of the model class and its subclasses. Taking the network encapsulated by the Sequential container as an example, we first create a five-layer fully connected network for MNIST handwritten digital picture recognition. The code is as follows:

```
# Create a 5-layer fully connected network
network = Sequential([layers.Dense(256, activation='relu'),
                      layers.Dense(128, activation='relu'),
                      layers.Dense(64, activation='relu'),
                      layers.Dense(32, activation='relu'),
                      layers.Dense(10)])
network.build(input_shape=(4, 28*28))
network.summary()
```

After the network is created, the normal process is to iterate over multiple Epochs in the dataset, generate training data in batches, do forward propagation calculation, then calculate the error value through the loss function, and automatically calculate the gradient and update the network parameters by back propagation. Since this part of the logic is very general, the compile() and fit() functions are provided in Keras to facilitate the logic. We can specify the optimizer, loss function, evaluation metrics, and other settings used by the network through the compile function directly. This step is called configuration.

```
# Import optimizer, loss function module
from tensorflow.keras import optimizers,losses
# Use Adam optimizer with learning rate of 0.01
# Use cross-entropy loss function with Softmax
network.compile(optimizer=optimizers.Adam(lr=0.01),
        loss=losses.CategoricalCrossentropy(from_logits=True),
        metrics=['accuracy'] # Set accuracy as
        evaluation metric
)
```

The optimizer, loss function, and other parameters specified in the compile() function are also the parameters that we need to set during our own training. Keras implements this part of the common logic internally to improve development efficiency.

8.2.2 Model Training

After the model is configured, the datasets for training and validation can be sent through the fit() function. This step is called model training.

```
# Training dataset is train_db, and validation dataset is val_db
# Train 5 epochs and validate every 2 epoch
# Training record and history is saved in history variable
history = network.fit(train_db, epochs=5, validation_data=val_db,
validation_freq=2)
```

train_db can be a tf.data.Dataset object or a Numpy array. The Epochs parameter specifies the number of Epochs for training iterations. The validation_data parameter specifies the dataset used for validation, and the validation frequency is controlled by validation_freq.

The preceding code can achieve the network training and validation functions. The fit function will return the history of the training process data records, where history.history is a dictionary object, including the loss of the training process, evaluation metrics, and other records, such as:

```
In [4]: history.history # print training record
Out[4]:
{'loss': [0.31980024444262184,  # training loss
  0.1123824894875288,
  0.07620834542314212,
  0.05487803366283576,
  0.041726120284820596],  # training accuracy
```

```
'accuracy': [0.904, 0.96638334, 0.97678334, 0.9830833,
0.9870667],
 'val_loss': [0.09901347314302303, 0.09504951824009701],
# validation loss
 'val_accuracy': [0.9688, 0.9703]}  # validation accuracy
```

The operation of the fit() function represents the training process of the network, so it will consume considerable training time and return after the training is completed. The historical data generated during the training can be obtained through the return value object. It can be seen that the code implemented through the Compile&Fit method is very concise and efficient, which greatly reduces the development time. However, because the interface is very high level, the flexibility is also reduced, and it is up to the user to decide whether to use it.

8.2.3 Model Testing

The model class can not only easily complete the network configuration, training, and validation, but also is very convenient for prediction and testing. We will elaborate on the difference between validation and testing in the chapter of overfitting. Here, validation and testing can be understood as a way of model evaluation.

The Model.predict(x) method can complete the model prediction, for example:

```
# Load one batch of test dataset
x,y = next(iter(db_test))
print('predict x:', x.shape) # print the batch shape
out = network.predict(x) # prediction
print(out)
```

where the out is the output of the network. Through the preceding code, the trained model can be used to predict the label information of new samples.

If you simply need to test the performance of the model, you can use Model.evaluate(db) to test all the samples on the db dataset and print out the performance indicators, for example:

```
network.evaluate(db_test)
```

8.3 Model Saving and Loading

After the model training is completed, the model needs to be saved to a file system to facilitate subsequent model testing and deployment. In fact, it is also a good habit to save the model state during training, which is especially important for training large-scale networks. Generally, a large-scale network requires several days or even weeks of training. Once the training process is interrupted or an accident occurs, the previous training progress will be lost. If the model state can be saved to the file system intermittently, even if an accident such as a downtime occurs, it can be recovered from the latest network state file, thereby avoiding wasting a lot of training time and computing resources. Therefore, the preservation and loading of the model is very important.

In Keras, there are three commonly used methods for saving and loading models.

8.3.1 Tensor Method

The state of the network is mainly reflected in the structure of the network and tensor data within the network layer. Therefore, under the condition of having the source file of the network structure, it is the most lightweight way to directly save the network tensor parameters to the file system.

Taking the MNIST handwritten digital picture recognition model as an example, the current network parameters can be saved by calling the Model.save_weights(path) method. The code is as follows:

```
network.save_weights('weights.ckpt') # Save tensor data of
the model
```

The preceding code saves the network model to the weights.ckpt file. When needed, first we create a network object, and then call the load_weights(path) method of the network object to load the tensor value saved in the specified model file to the current network parameters, for example:

```
# Save tensor data of the model
network.save_weights('weights.ckpt')
print('saved weights.')
del network # delete network object
# Create similar network
network = Sequential([layers.Dense(256, activation='relu'),
                      layers.Dense(128, activation='relu'),
                      layers.Dense(64, activation='relu'),
                      layers.Dense(32, activation='relu'),
                      layers.Dense(10)])
network.compile(optimizer=optimizers.Adam(lr=0.01),
        loss=tf.losses.CategoricalCrossentropy(from_
        logits=True),
        metrics=['accuracy']
    )
# Load weights from file
network.load_weights('weights.ckpt')
print('loaded weights!')
```

This method of saving and loading the network is the most lightweight. The file only saves the values of the tensor parameters, and there are no other additional structural parameters. But it needs to use the same network structure to be able to restore the network state correctly, so it is generally used in the case of having network source files.

8.3.2 Network Method

Let's introduce a method that does not require network source files and only needs model parameter files to recover the network model. The model structure and model parameters can be saved to the path file through the Model.save(path) function, and the network structure and network parameters can be restored through keras.models.load_model(path) without the need for network source files .

First, save the MNIST handwritten digital picture recognition model to a file, and delete the network object:

```
# Save model and parameters to a file
network.save('model.h5')
print('saved total model.')
del network # Delete the network
```

The structure and state of the network can be recovered through the model.h5 file, and there is no need to create network objects in advance. The code is as follows:

```
# Recover the model and parameters from a file
network = keras.models.load_model('model.h5')
```

As you can see, in addition to storing model parameters, the model. h5 file should also save network structure information. You can directly recover the network object from the file without creating a model in advance.

8.3.3 SavedModel method

TensorFlow is favored by the industry, not only because of the excellent neural network layer API support, but also because it has powerful ecosystem, including mobile and web support. When the model needs to be deployed to other platforms, the SavedModel method proposed by TensorFlow is platform-independent.

By tf.saved_model.save(network, path), the model can be saved to the path directory as follows:

```
#  Save model and parameters to a file
tf.saved_model.save(network, 'model-savedmodel')
print('saving savedmodel.')
del network # Delete network object
```

The following network files appear in the file system model-savedmodel directory, as shown in Figure 8-1:

Name	Date modified	Type	Size
assets	8/13/2019 7:53 PM	File folder	
variables	8/13/2019 7:53 PM	File folder	
saved_model.pb	8/13/2019 7:53 PM	PB File	240 KB

Figure 8-1. *SavedModel method directory*

Users don't need to care about the file saving format, they only need to restore the model object through the tf.saved_model.load function. After recovering the model instance, we complete the calculation of the test accuracy rate and achieve the following:

```
print('load savedmodel from file.')
# Recover network and parameter from files
network =  tf.saved_model.load('model-savedmodel')
# Accuracy metrics
acc_meter = metrics.CategoricalAccuracy()
```

```
for x,y in ds_val:    # Loop through test dataset
    pred = network(x) # Forward calculation
    acc_meter.update_state(y_true=y, y_pred=pred)
    # Update stats
# Print accuracy
print("Test Accuracy:%f" % acc_meter.result())
```

8.4 Custom Network

Although Keras provides many common network layer classes, the network used for deep learning are far more than that. Researchers generally implement relatively new network layers on their own. Therefore, it is very important to master the custom network layer and the implementation of the network.

For the network layer that needs to create customized logic, it can be implemented through a custom class. When creating a customized network layer class, you need to inherit from the layers.Layer base class. When creating a custom network class, you need to inherit from the keras. Model base class, so the custom class created in this way can easily use the Layer/Model base class. The parameter management and other functions provided by the class can also be used interactively with other standard network layer classes.

8.4.1 Custom Network Layer

For a custom network layer, we at least need to implement the initialization (__init__) method and the forward propagation logic. Let's take a specific custom network layer as an example, assuming that a fully connected layer without bias vectors is needed, that is, bias is 0, and the fixed activation function is ReLU. Although this can be created through the standard dense layer, we still explain how to implement a custom network layer by implementing this "special" network layer class.

First, create a class and inherit from the base layer class. Create an initialization method, and call the initialization function of the parent class. Because it is a fully connected layer, two parameters need to be set: the length of the input feature inp_dim and the length of the output feature outp_dim, and the shape size is created by self.add_variable(name, shape). The name tensor W is set to be optimized.

```
class MyDense(layers.Layer):
    # Custom layer
    def __init__(self, inp_dim, outp_dim):
        super(MyDense, self).__init__()
        # Create weight tensor and set to be trainable
        self.kernel = self.add_variable('w', [inp_dim,
        outp_dim], trainable=True)
```

It should be noted that self.add_variable will return a Python reference to the tensor W, and the variable name is maintained internally by TensorFlow and is used less often. We instantiate the MyDense class and view its parameter list, for example:

```
In [5]: net = MyDense(4,3) # Input dimension is 4 and output
dimension is 3.
net.variables,net.trainable_variables  # Check the trainable
parameters
Out[5]:
# All parameters
([<tf.Variable 'w:0' shape=(4, 3) dtype=float32, numpy=...
# Trainable parameters
 [<tf.Variable 'w:0' shape=(4, 3) dtype=float32, numpy=...
```

You can see that the tensor W is automatically included in the parameter list.

By modifying to self.kernel = self.add_variable('w', [inp_dim, outp_dim], trainable = False), we can set the tensor *W* not to be trainable and then observe the management state of the tensor:

```
([<tf.Variable 'w:0' shape=(4, 3) dtype=float32, numpy=...],
# All parameters
[])# Trainable parameters
```

As you can see, the tensor is not managed by trainable_variables at this time. In addition, class member variables created as tf.Variable in class initialization are also automatically included in tensor management, for example:

```
self.kernel = tf.Variable(tf.random.normal([inp_dim,
outp_dim]), trainable=False)
```

The list of managed tensors is printed out as follows:

```
# All parameters
([<tf.Variable 'Variable:0' shape=(4, 3) dtype=float32, numpy=...],
[])# Trainable parameters
```

After the initialization of the custom class, we will design the forward calculation logic. For this example, only the matrix operation $O = X @ W$ needs to be completed and the fixed ReLU activation function can be used. The code is as follows:

```
def call(self, inputs, training=None):
    # Forward calculation
    # X@W
    out = inputs @ self.kernel
    # Run activation function
    out = tf.nn.relu(out)
    return out
```

As aforementioned, the forward calculation logic is implemented in the call(inputs, training = None) function, where inputs parameter represents input and is passed in by the user. The training parameter is used to specify the state of the model: True means training mode and False indicates testing mode, and default value is None, which is the test mode. Since the training and test modes of the fully connected layer are logically consistent, no additional processing is required here. For the network layer whose test and training modes are inconsistent, the logic to be executed needs to be designed according to the training parameters.

8.4.2 Customized Network

After completing the custom fully connected layer class implementation, we created the MNIST handwritten digital picture model based on the "unbiased fully connected layer" described previously.

The custom network class can be easily encapsulated into a network model through the Sequential container like other standard classes:

```
network = Sequential([MyDense(784, 256), # Use custom layer
                    MyDense(256, 128),
                    MyDense(128, 64),
                    MyDense(64, 32),
                    MyDense(32, 10)])
network.build(input_shape=(None, 28*28))
network.summary()
```

It can be seen that by stacking our custom network layer classes, a five-layer fully connected layer network can also be realized. Each layer of the fully connected layer has no bias tensor, and the activation function uses the ReLU function.

The Sequential container is suitable for a network model in which data propagates in order from the first layer to the second layer, and then from the second layer to the third layer, and propagates in this manner. For complex network structures, for example, the input of the third layer is not only the output of the second layer, but also the output of the first layer. At this time, it is more flexible to use a customized network. First, create a class that inherits from the model base class, and then respectively create the corresponding network layer object as follows:

```
class MyModel(keras.Model):
    # Custom network class
    def __init__(self):
        super(MyModel, self).__init__()
        # Create the network
        self.fc1 = MyDense(28*28, 256)
        self.fc2 = MyDense(256, 128)
        self.fc3 = MyDense(128, 64)
        self.fc4 = MyDense(64, 32)
        self.fc5 = MyDense(32, 10)
```

Then implement the forward operation logic of the custom network as follows:

```
    def call(self, inputs, training=None):
        # Forward calculation
        x = self.fc1(inputs)
        x = self.fc2(x)
        x = self.fc3(x)
        x = self.fc4(x)
        x = self.fc5(x)
        return x
```

This example can be implemented directly using the Sequential container method. But the forward calculation logic of the customized network can be freely defined and more general. We will see the superiority of the customized network in the chapter of convolutional neural networks.

8.5 Model Zoo

For commonly used network models, such as ResNet and VGG, you do not need to manually create them. They can be implemented directly with the keras.applications submodule with a line of code. At the same time, you can also load pre-trained models by setting the weights parameters.

8.5.1 Load Model

Taking the ResNet50 network model as an example, the network after removing the last layer of ResNet50 is generally used as the feature extraction subnetwork for the new task, that is, using the pre-trained network parameters on the ImageNet dataset to initialize and appending one fully connected layer corresponding to the number of data categories according to the category of the task, so that new tasks can be learned quickly and efficiently on the basis of the pre-trained network.

First, use the Keras model zoo to load the pre-trained ResNet50 network by ImageNet. The code is as follows:

```python
# Load ImageNet pre-trained network. Exclude the last layer.
resnet = keras.applications.ResNet50(weights='imagenet',inclu
de_top=False)
resnet.summary()
# test the output
x = tf.random.normal([4,224,224,3])
out = resnet(x) # get output
out.shape
```

The preceding code automatically downloads the model structure and pre-trained network parameters on the ImageNet dataset from the server. By setting the include_top parameter to False, we choose to remove the last layer of ResNet50. The size of the output feature map of the network is $[b, 7, 7, 2048]$. For a specific task, we need to set a custom number of output nodes. Taking 100 classification tasks as an example, we rebuild a new network based on ResNet50. Create a new pooling layer (the pooling layer here can be understood as a function of downsampling in high and wide dimensions) and reduce the features dimension from $[b, 7, 7, 2048]$ to $[b, 2048]$ as in the following.

```
In [6]:
# New pooling layer
global_average_layer = layers.GlobalAveragePooling2D()
# Use last layer's output as this layer's input
x = tf.random.normal([4,7,7,2048])
# Use pooling layer to reduce dimension from [4,7,7,2048] to
[4,1,1,2048],and squeeze to [4,2048]
out = global_average_layer(x)
print(out.shape)
Out[6]: (4, 2048)
```

Finally, create a new fully connected layer and set the number of output nodes to 100. The code is as follows:

```
In [7]:
# New fully connected layer
fc = layers.Dense(100)
# Use last layer's output as this layer's input
x = tf.random.normal([4,2048])
out = fc(x)
print(out.shape)
Out[7]: (4, 100)
```

After creating a pre-trained ResNet50 feature sub-network, a new pooling layer, and a fully connected layer, we re-use the Sequential container to encapsulate a new network:

```
# Build a new network using previous layers
mynet = Sequential([resnet, global_average_layer, fc])
mynet.summary()
```

You can see the structure information of the new network model is:

```
Layer (type)                    Output
Shape                 Param Number
=================================================================
resnet50 (Model)                (None, None, None, 2048)  23587712
_____
global_average_pooling2d (Gl (None, 2048)                        0
_____
dense_4 (Dense)                 (None, 100)                  204900
=================================================================
Total params: 23,792,612
Trainable params: 23,739,492
Non-trainable params: 53,120
```

By setting resnet.trainable = False, you can choose to freeze the network parameters of the ResNet part and only train the newly created network layer, so that the network model training can be completed quickly and efficiently. Of course, you can also update all the parameters of the network.

8.6 Metrics

In the training process of the network, metrics such as accuracy and recall rate are often required. Keras provides some commonly used metrics in the keras.metrics module.

There are four main steps in the use of Keras metrics: creating a new metrics container, writing data, reading statistical data, and clearing the measuring container.

8.6.1 Create a Metrics Container

In the keras.metrics module, it provides many commonly used metrics classes, such as mean, accuracy, and cosine similarity. In the following, we take the mean error as an example.

```
loss_meter = metrics.Mean()
```

8.6.2 Write Data

New data can be written through the update_state function, and the metric will record and process the sampled data according to its own logic. For example, the loss value is collected once at the end of each step:

```
# Record the sampled data, and convert the tensor to an
ordinary value through the float() function
    loss_meter.update_state(float(loss))
```

After the preceding sampling code is placed at the end of each batch operation, the meter will automatically calculate the average value based on the sampled data.

8.6.3 Read Statistical Data

After sampling multiple times of data, you can choose to call the measurer's result() function to obtain statistical values. For example, the interval statistical loss average is as follows:

```
# Print the average loss during the statistical period
print(step, 'loss:', loss_meter.result())
```

8.6.4 Clear the Container

Since the metric container will record all historical data, it is necessary to clear the historical status when starting a new round of statistics. It can be realized by reset_states() function. For example, after reading the average error every time, clear the statistical information to start the next round of statistics as follows:

```
if step % 100 == 0:
    # Print the average loss
    print(step, 'loss:', loss_meter.result())
    loss_meter.reset_states() # reset the state
```

8.6.5 Hands-On Accuracy Metric

According to the method of using the metric tool, we use the accuracy metric to count the accuracy rate during the training process. First, create a new accuracy measuring container as follows:

```
acc_meter = metrics.Accuracy()
```

After each forward calculation is completed, record the training accuracy rate. It should be noted that the parameters of the update_state function of the accuracy class are the predicted value and the true value, not the accuracy rate of the current batch. We write the label and prediction result of the current batch sample into the metric as follows:

```
# [b, 784] => [b, 10, network output
out = network(x)
# [b, 10] => [b], feed into argmax()
pred = tf.argmax(out, axis=1)
pred = tf.cast(pred, dtype=tf.int32)
# record the accuracy
acc_meter.update_state(y, pred)
```

After counting the predicted values of all batches in the test set, print the average accuracy of the statistics and clear the metric container. The code is as follows:

```
print(step, 'Evaluate Acc:', acc_meter.result().
numpy())
acc_meter.reset_states() # reset metric
```

8.7 Visualization

In the process of network training, it is very important to improve the development efficiency and monitor the training progress of the network through the web terminal and visualize the training results. TensorFlow provides a special visualization tool called TensorBoard, which writes monitoring data to the file system through TensorFlow and uses the web backend to monitor the corresponding file directory, thus allowing users to view network monitoring data.

The use of TensorBoard requires cooperation between the model code and the browser. Before using TensorBoard, you need to install the TensorBoard library. The installation command is as follows:

```
# Install TensorBoard
pip install tensorboard
```

Next, we introduce how to use the TensorBoard tool to monitor the progress of network training in the model side and the browser side.

8.7.1 Model Side

On the model side, you need to create a summary class that writes monitoring data when needed. First, create an instance of the monitoring object class through tf.summary.create_file_writer, and specify the directory where the monitoring data is written. The code is as follows:

```
# Create a monitoring class, the monitoring data will be
written to the log_dir directory
summary_writer = tf.summary.create_file_writer(log_dir)
```

We take monitoring error and visual image data as examples to introduce how to write monitoring data. After the forward calculation is completed, for the scalar data such as error, we record the monitoring data through the tf.summary.scalar function and specify the time stamp step parameter. The step parameter here is similar to the time scale information corresponding to each data and can also be understood as the coordinates of the data curve, so it should not be repeated. Each type of data is distinguished by the name of the string, and similar data needs to be written to the database with the same name. For example:

```
with summary_writer.as_default():
    # write the current loss to train-loss database
    tf.summary.scalar('train-loss', float(loss),
    step=step)
```

TensorBoard distinguishes different types of monitoring data by string ID, so for error data, we named it "train-loss"; other types of data cannot be written to prevent data pollution.

For picture-type data, you can write monitoring picture data through the tf.summary.image function. For example, during training, the sample image can be visualized by the tf.summary.image function. Since the tensor in TensorFlow generally contains multiple samples, the tf.summary. image function accepts tensor data of multiple pictures and sets the max_ outputs parameter to select the maximum number of displayed pictures. The code is as follows:

```
with summary_writer.as_default():
    # log accuracy
    tf.summary.scalar('test-acc', float(total_correct/
    total), step=step)
    # log images
    tf.summary.image("val-onebyone-images:",
    val_images, max_outputs=9, step=step)
```

Run the model program, and the corresponding data will be written to the specified file directory in real time.

8.7.2 Browser Side

When running the program, the monitoring data is written to the specified file directory. If you want to remotely view and visualize these data in real time, you also need to use a browser and a web backend. The first step is to open the web backend. Run command "tensorboard --logdir path" in terminal and specify the file directory path monitored by the web backend, then you can open the web backend monitoring process, as shown in Figure 8-2:

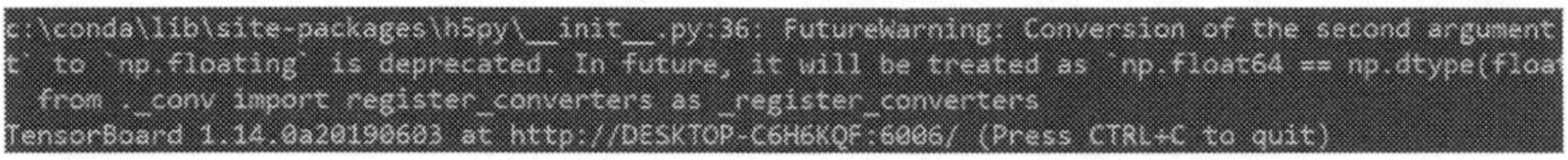

```
c:\conda\lib\site-packages\h5py\__init__.py:36: FutureWarning: Conversion of the second argument
t to `np.floating` is deprecated. In future, it will be treated as `np.float64 == np.dtype(floa
from ._conv import register_converters as _register_converters
TensorBoard 1.14.0a20190603 at http://DESKTOP-C6H6KQF:6006/ (Press CTRL+C to quit)
```

Figure 8-2. *Open web server*

Open a browser and enter the URL http://localhost: 6006 (you can
also remotely access through the IP address, the specific port number
may change depending on the command prompt) to monitor the progress
of the network training. TensorBoard can display multiple monitoring
records at the same time. On the left side of the monitoring page, you can
select monitoring records, as shown in Figure 8-3:

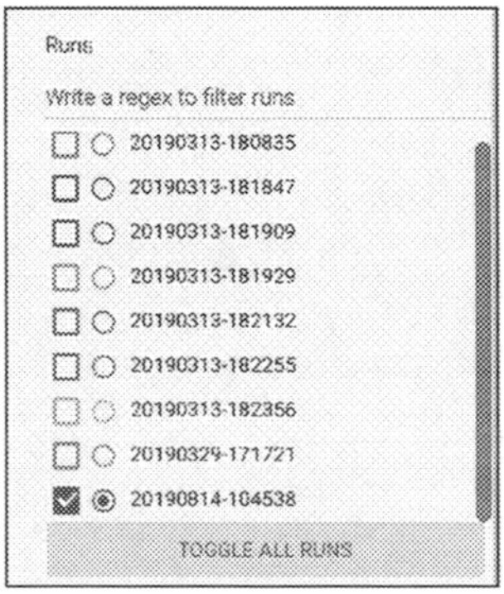

Figure 8-3. *Snapshot of TensorBoard*

On the upper end of the monitoring page, you can choose different
types of data monitoring pages, such as scalar monitoring page SCALARS
and picture visualization page IMAGES. For this example, we need to
monitor the training error and test accuracy rate for scalar data, and its
curve can be viewed on the SCALARS page, as shown in Figure 8-4 and
Figure 8-5.

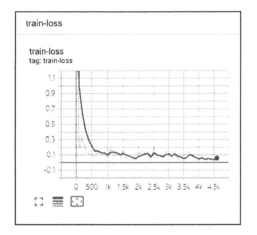

Figure 8-4. *Training loss curve*

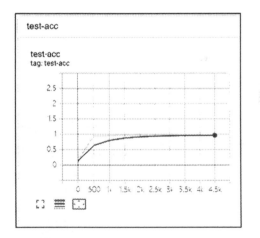

Figure 8-5. *Training accuracy curve*

On the IMAGES page, you can view images at each step as shown in Figure 8-6.

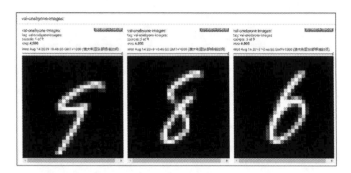

Figure 8-6. *Pictures from each step*

In addition to monitoring scalar data and image data, TensorBoard also supports functions such as viewing histogram distribution of tensor data through tf.summary.histogram, and printing text information through tf.summary.text. For example:

```
with summary_writer.as_default():
    tf.summary.scalar('train-loss', float(loss),
    step=step)
tf.summary.histogram('y-hist',y, step=step)
    tf.summary.text('loss-text',str(float(loss)))
```

You can view the histogram of the tensor on the HISTOGRAMS page, as shown in Figure 8-7, and you can view the text information on the TEXT page, as shown in Figure 8-8.

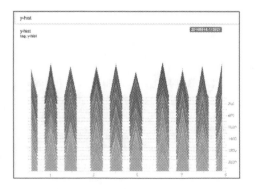

Figure 8-7. *TensorBoard histogram*

Figure 8-8. *TensorBoard text visualization*

In fact, in addition to TensorBoard, the Visdom tool developed by Facebook can also facilitate the visualization of data, and supports a variety of visualization methods in real time, and is more convenient to use. Figure 8-9 shows the visualization of Visdom data. Visdom can directly accept PyTorch's tensor-type data but cannot directly accept TensorFlow's tensor-type data. It needs to be converted to a Numpy array. For readers pursuing rich visualization methods and real-time monitoring, Visdom may be a better choice.

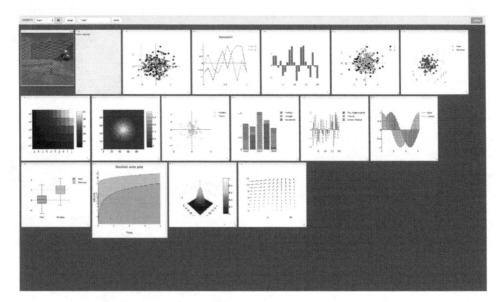

Figure 8-9. *Visdom snapshot[1]*

8.8 Summary

In this chapter, we introduced the usage of Keras advanced API which can save us a lot of time during network development. We can use the container method to construct networks easily. Training and testing a neural network can be quickly implemented using the Keras built-in functions. After the network is trained and tested, we can also save the trained model and reload the model in the future using Keras. Besides common network layers, Keras also provides functionalities to build customized network layers for different use cases. We also discussed how to load popular network models using Keras as well as setting up evaluation metrics and visualizing model performance using TensorBoard. The tools we learned through this chapter can help us increase network development efficiency significantly.

---

[1] Image source: https://github.com/facebookresearch/visdom

CHAPTER 9

Overfitting

> Everything should be made as simple as possible, but not simpler.
>
> —Albert Einstein

The main purpose of machine learning is to learn the real model of the data from the training set, so that it can perform well on the unseen test set. We call this the generalization ability. Generally speaking, the training set and the test set are sampled from the same data distribution. The sampled samples are independent of each other, but come from the same distribution. We call this assumption the independent identical distribution (i.i.d.) assumption.

The expressive power of the model has been mentioned earlier, also known as the capacity of the model. When the model's expressive power is weak, such as a single linear layer, it can only learn a linear model and cannot approximate the nonlinear model well. When the model's expressive power is too strong, it may be possible to reduce the noise modalities of the training set, but leads to poor performance on the test set (generalization ability is weak). Therefore, for different tasks, designing a model with appropriate capacity can achieve better generalization performance.

© Liangqu Long and Xiangming Zeng 2022
L. Long and X. Zeng, *Beginning Deep Learning with TensorFlow*,
https://doi.org/10.1007/978-1-4842-7915-1_9

9.1 Model Capacity

In layman's terms, the capacity or expressive capacity of a model refers to the model's ability to fit complex functions. An indicator reflecting the capacity of the model is the size of the hypothesis space of the model, that is, the size of the set of functions that the model can represent. The larger and more complete the hypothesis space, the more likely it is to search from the hypothesis space for a function that approximates the real model. Conversely, if the hypothesis space is very limited, it is difficult to find a function that approximates the real model.

Consider sampling from real distribution:

$$p_{data} = \{(x,y)|y = sin(x), x \in [-5,5]\}$$

A small number of points are sampled from the real distribution to form the training set, which contains the observation error ϵ, as shown by the small dots in Figure 9-1. If we only search the model space of all first-degree polynomials and set the bias to 0, that is, $y = ax$, as shown by the straight line of the first-degree polynomial in Figure 9-1. Then it is difficult to find a straight line that closely approximates the distribution of real data. Slightly increase the hypothesis space so that the hypothesis space is all third-degree polynomial functions, that is, $y = ax^3 + bx^2 + cx$, it is obvious that this hypothesis space is obviously larger than the hypothesis space of the first-degree polynomial, we can find a curve (as shown in Figure 9-1) that reflects the relationship of the data better than the first-order polynomial model, but it is still not good enough. Increase the hypothesis space again so that the searchable function is a polynomial of degree 5, that is, $y = ax^5 + bx^4 + cx^3 + dx^2 + ex$. In this hypothesis space, a better function can be searched, as shown by the polynomial of degree 5 in Figure 9-1. After increasing the hypothesis space again, as shown in the polynomial curves of 7, 9, 11, 13, 15, and 17 in Figure 9-1, the larger the hypothesis space of the function, the more likely it is to find a function model that better approximates the real distribution.

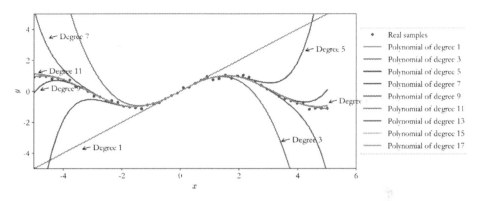

Figure 9-1. *Polynomial capability*

However, an excessively large hypothesis space will undoubtedly increase the search difficulty and computational cost. In fact, under the constraints of limited computing resources, a larger hypothesis space may not necessarily be able to search for a better model. Due to the existence of observation errors, a larger hypothesis space may contain a larger number of functions with too strong expression ability, which can also learn the observation errors of the training samples, thus hurting the generalization ability of the model. Choosing the right model capacity is a difficult problem.

9.2 Overfitting and Underfitting

Because the distribution of real data is often unknown and complicated, it is impossible to deduce the type of distribution function and related parameters. Therefore, when choosing the capacity of the learning model, people often choose a slightly larger model capacity based on empirical values. However, when the capacity of the model is too large, it may appear to perform better on the training set, but perform worse on the test set, as shown in Figure 9-2. When the capacity of the model is too small, it may have poor performance in both the training set and the testing set as shown in the area to the left of the red vertical line in Figure 9-2.

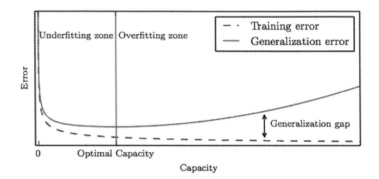

Figure 9-2. *The relation between model capacity and error [1]*

When the capacity of the model is too large, in addition to learning the modalities of the training set data, the network model also learns additional observation errors, resulting in the learned model performing better on the training set, but poor in unseen samples, that is, the generalization ability of the model is weak. We call this phenomenon overfitting. When the capacity of the model is too small, the model cannot learn the modalities of the training set data well, resulting in poor performance on both the training set and the unseen samples. We call this phenomenon underfitting.

Here is a simple example to explain the relationship between the model's capacity and the data distribution. Figure 9-3 plots the distribution of certain data. It can be roughly speculated that the data may belong to a certain degree 2 polynomial distribution. If we use a simple linear function to learn, we will find it difficult to learn a better function, resulting in the underfitting phenomenon that the training set and the test set do not perform well, as shown in Figure 9-3 (a). However, if you use a more complex function model to learn, it is possible that the learned function will excessively "fit" the training set samples, but resulting in poor performance on the test set, that is, overfitting, as shown in Figure 9-3 (c). Only when the capacity of the learned model and the real model roughly match, the model can have a good generalization ability, as shown in Figure 9-3 (b).

318

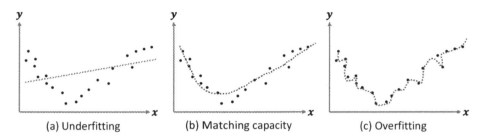

(a) Underfitting   (b) Matching capacity   (c) Overfitting

Figure 9-3. *Overfitting and underfitting*

Consider the distribution p_{data} of data points (x, y), where

$$y = sin(1.2 \cdot \pi \cdot x)$$

During sampling, random Gaussian noise is added to obtain a dataset of 120 points, as shown in Figure 9-4. The curve in the figure is the real model function, the black round points are the training samples, and the green matrix points are the test samples.

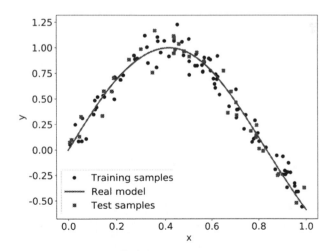

Figure 9-4. *Dataset and the real function*

319

In the case of a known real model, it is natural to design a function space with appropriate capacity to obtain a good learning model. As shown in Figure 9-5, we assume that the model is a second-degree polynomial model, and the learned function curve is approximating the real model. However, in actual scenarios, the real model is often unknown, so if the design hypothesis space is too small, it will be impossible to search for a suitable learning model. If the design hypothesis space is too large, it will result in poor model generalization ability.

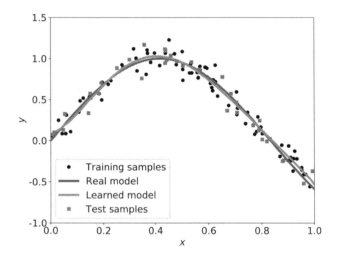

Figure 9-5. *Appropriate model capability*

So how to choose the capacity of the model? Statistical learning theory provides us with some ideas. The VC dimension (Vapnik-Chervonenkis dimension) is a widely used method to measure the capacity of functions. Although these methods provide a certain degree of theoretical guarantee for machine learning, these methods are rarely applied to deep learning. Part of the reason is that the neural network is too complicated to determine the VC dimension of the mathematical model behind the network structure.

Although statistical learning theory is difficult to give the minimum capacity required by a neural network, it can be used to guide the design and training of a neural network based on Occam's razor. Occam's razor principle was a rule of solution proposed by William of Occam, a fourteenth-century logician and Franciscan monk of the Franciscans. He stated in his book that "Don't waste more things and do things that you can do well with less." In other words, if the two-layer neural network structure can express the real model well, then the three-layer neural network can also express well, but we should prefer to use the simpler two-layer neural network because its parameters' amount is smaller, it is easier to train, and it is easier to get a good generalization error through fewer training samples.

9.2.1 Underfitting

Let us consider the phenomenon of underfitting. As shown in Figure 9-6, black dots and green rectangles are independently sampled from the distribution of a parabolic function. Because we already know the real model, if we use a linear function with lower capacity than the real model to fit the data, it is difficult for the model to perform well. The specific performance is that the learned linear model has a larger error (such as the mean square error) on the training set, and the error on the test set is also larger.

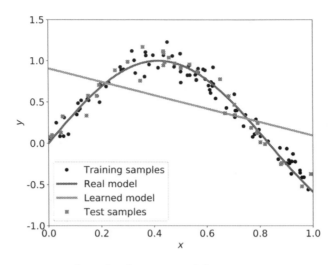

Figure 9-6. *A typical underfitting model*

When we find that the current model has maintained a high error on the training set, it is difficult to optimize and reduce the error, and it also performs poorly on the test set, we can consider whether there is a phenomenon of underfitting. The problem of underfitting can be solved by increasing the number of layers of the neural network or increasing the size of the intermediate dimension. However, because modern deep neural network models can easily reach deeper layers, the capacity of the model used for learning is generally sufficient. In real applications, more overfitting phenomena occur.

9.2.2 Overfitting

Consider the same problem, the black dots of the training set and the green rectangles of the test machine are independently sampled from a parabolic model with the same distribution. When we set the hypothesis space of the model to 25th polynomial, it is much larger than the functional capacity of the real model. It is found that the learned model is likely to overfit the training sample, resulting in the error of the learning model on the training sample is very small, even smaller than the error

322

of the real model on the training set. But for the test sample, the model performance drops sharply, and the generalization ability is very poor, as shown in Figure 9-7.

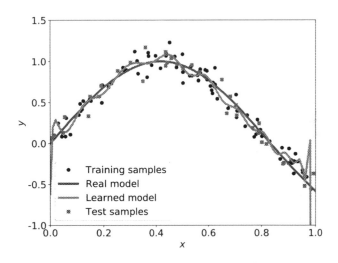

Figure 9-7. *A typical overfitting model*

The phenomenon of overfitting in modern deep neural networks is very easy to occur, mainly because the neural network has a very strong expressive ability and the number of samples in the training set is not enough, it is easy to appear that the capacity of the neural network is too large. So how to effectively detect and reduce overfitting?

Next, we will introduce a series of methods to help detect and suppress overfitting.

9.3 Dataset Division

Earlier we introduced that the dataset needs to be divided into a training set and a test set. In order to select model hyperparameters and detect overfitting, it is generally necessary to split the original training set into a new training set and a validation set, that is, the dataset needs to be divided into three subsets: training set, validation set, and test set.

9.3.1 Validation Set and Hyperparameters

The difference between the training set and the test set has been introduced earlier. The training set D^{train} is used to train model parameters, and the test set D^{test} is used to test the generalization ability of the model. The samples in the test set cannot participate in the model training, preventing the model from "memorizing" the characteristics of the data and damaging the generalization ability of the model. Both the training set and the test set are sampled from the same data distribution. For example, the MNIST handwritten digital picture set has a total of 70,000 sample pictures, of which 60,000 pictures are used as the training set, and the remaining 10,000 pictures are used for the test set. The separation ratio of the training set and the test set can be defined by the user. For example, 80% of the data is used for training, and the remaining 20% is used for testing. When the size of the data set is small, in order to test the generalization ability of the model more accurately, the proportion of the test set can be increased appropriately. Figure 9-8 demonstrates the division of the MNIST handwritten digital picture collection: 80% is used for training, and the remaining 20% is used for testing.

Figure 9-8. *Training and testing dataset division*

However, it is not enough to divide the dataset into only the training set and the test set. Because the performance of the test set cannot be used as feedback for model training, we need to be able to pick out more suitable model hyperparameters during model training to determine whether the model is overfitting. Therefore, we need to divide the training set into training set and validation set, as shown in Figure 9-9. The divided training set has the same function as the original training set and is used to train the parameters of the model, while the validation set is used to select the hyperparameters of the model. Its functions include:

- Adjust the learning rate, weight decay coefficient, training times, etc. according to the performance of the validation set.

- Readjust the network topology according to the performance of the validation set.

- According to the performance of the validation set, determine whether it is overfitting or underfitting.

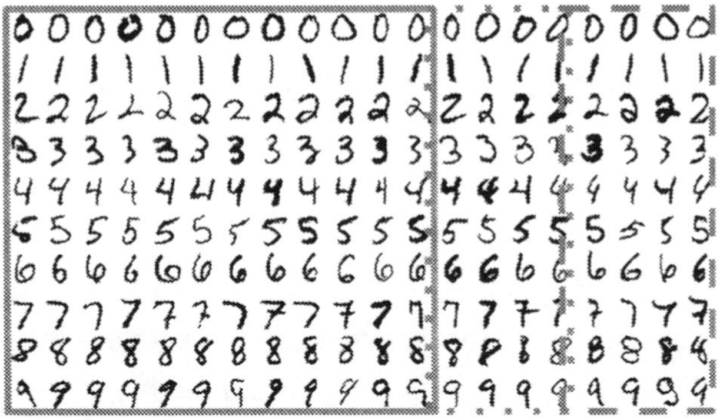

Figure 9-9. *Training, validation, and test dataset*

Similar to the division of the training set-test set, the training set, validation set, and test set can be divided according to a custom ratio, such as the common 60%-20%-20% division. Figure 9-9 shows the MNIST handwriting dataset schematic diagram of the division.

The difference between the validation set and the test set is that the algorithm designer can adjust the settings of various hyperparameters of the model according to the performance of the validation set to improve the generalization ability of the model, but the performance of the test set cannot be used to adjust the model. Otherwise, the functions of the test set and the validation set will overlap, so the performance on the test set will not represent the generalization ability of the model.

In fact, some developers will incorrectly use the test set to select the best model, and then use it as a model generalization performance report. For those cases, the test set is actually the validation set, so the "generalization performance" reported is essentially the performance on the validation set, not the real generalization performance. In order to prevent this kind of "cheating," you can choose to generate multiple test sets, so that even if the developer uses one of the test sets to select the model, we can also use other test sets to evaluate the model, which is also commonly used in Kaggle competitions.

9.3.2 Early Stopping

Generally, we call one batch updating in the training set one Step, and iterating through all the samples in the training set once is called an Epoch. The validation set can be used after several Steps or Epochs to calculate the validation performance of the model. If the validation steps are too frequent, it can accurately observe the training status of the model, but it also introduces additional computation costs. It is generally recommended to perform a validation operation after several Epochs.

Taking the classification task as an example, the training performance indicators include training error, training accuracy, etc. Correspondingly, there are also validation error and validation accuracy during validation process, and test error and test accuracy during testing process. The training accuracy and validation accuracy can roughly infer whether the model is overfitting and underfitting. If the training error of the model is low and the training accuracy is high, but the validation error is high and the validation accuracy rate is low, overfitting may occur. If the errors on both the training set and the validation set are high and the accuracy is low, then underfitting may occur.

When overfitting is observed, the capacity of the network model can be redesigned, such as reducing the number of layers of the network, reducing the number of parameters of the network, adding regularization methods, and adding constraints on the hypothesis space, so that the actual capacity of the model reduces to solve the overfitting phenomenon. When the underfitting phenomenon is observed, you can try to increase the capacity of the network, such as deepening the number of layers of the network, increasing the number of network parameters, and trying more complicated network structures.

In fact, since the actual capacity of the network can change as the training progresses, even with the same network settings, different overfitting and underfitting conditions may be observed. Figure 9-10 shows a typical training curve for classification problems. The red curve is the training accuracy, and the blue curve is the test accuracy. As we can see from the figure, as the training progresses in the early stage of training, the training accuracy and test accuracy of the model show an increasing trend, and there is no overfitting phenomenon at this time. In the later stage of training, even with the same network structure, due to the change in the actual capacity of the model, we observed the phenomenon of overfitting. That is, the training accuracy continues to improve, but the generalization ability becomes weaker (the test accuracy decreases).

This means that for neural networks, even if the network hyperparameters amount remains unchanged (i.e., the maximum capacity of the network is fixed), the model may still appear to be overfitting, because the effective capacity of the neural network is closely related to the state of the network parameters. The effective capacity of the neural network can be very large, and the effective capacity can also be reduced by means of sparse parameters and regularization. In the early and middle stages of training, the phenomenon of overfitting did not appear. As the number of training Epochs increased, the overfitting became more and more serious. In Figure 9-10, the vertical dotted line is in the best state of the network, there is no obvious overfitting phenomenon, and the generalization ability of the network is the best.

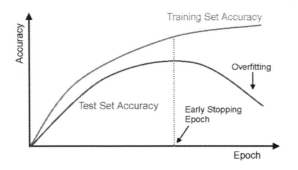

Figure 9-10. *Training process diagram*

So how to choose the right Epoch to stop training early (early stopping) to avoid overfitting? We can predict the possible position of the most suitable Epoch by observing the change of the validation metric. Specifically, for the classification problem, we can record the validation accuracy of the model and monitor its change. When it is found that the validation accuracy has not decreased for successive Epochs, we can predict that the most suitable Epoch may have been reached, so we can stop training. Figure 9-11 plots the curve of training and validation accuracy with training Epoch during a specific training process. It can be observed that when Epoch is around 30, the model reaches its optimal state and we can stop training in advance.

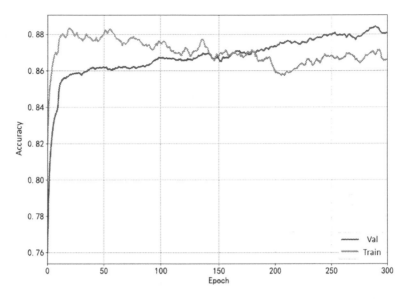

Figure 9-11. *Training curve example*

Algorithm 1 is a pseudo-code that uses an early stop model training algorithm.

Algorithm 1:Network training with early stopping

Initialize parameter θ

repeat

  for *step* = 1, …,*N* **do**

    random select Batch $\{(x, y)\} \sim D^{train}$

    $\theta \leftarrow \theta - \eta \nabla_\theta L(f(x), y)$

  end

  if every *n***th Epoch do**

  Calculate validation set $\{(x, y)\} \sim D^{val}$ **performance**

    if validation performance doesn't increase for certain successive steps do

      save the network and stop training

    end

---

do
until training reaches maximum Epoch
Use the saved network to calculate test set $\{(x, y)\} \sim D^{test}$ performance
Output:Network parameter θ and testing accuracy

9.4 Model Design

The validation set can determine whether the network model is overfitting or underfitting, which provides a basis for adjusting the capacity of the network model. For neural networks, the numbers of layers and parameters of the network are very important reference indicators for network capacity. By reducing the number of layers and reducing the size of the network parameters in each layer, the network capacity can be effectively reduced. Conversely, if the model is found to be underfitting, we can increase the capacity of the network by increasing the number of layers and the amount of parameters in each layer.

To demonstrate the effect of the number of network layers on network capacity, we visualized the decision boundary of a classification task. Figure 9-12, Figure 9-13, Figure 9-14, and Figure 9-15, respectively, demonstrate the decision boundary map for training two-category classification task under different network layers, where the red rectangular block and the blue circular block, respectively, represent the two types of samples on the training set. Only adjust the number of layers of the network while keeping other hyperparameters consistent. As shown in the figure, you can see that as the number of network layers increases, the learned model decision boundary is more and more close to training samples, indicating overfitting. For this task, the two-layer neural network can obtain good generalization ability. The deeper layer of the network does not improve the overall model performance. Instead, it can lead to overfitting, and the generalization ability becomes worse, and the computation cost is also higher.

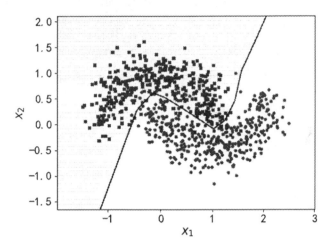

Figure 9-12. *Two layers*

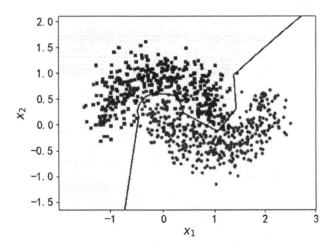

Figure 9-13. *Three layers*

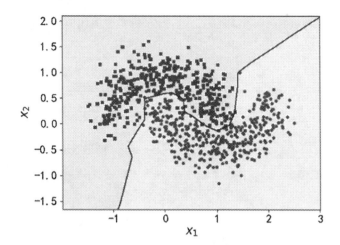

Figure 9-14. *Four layers*

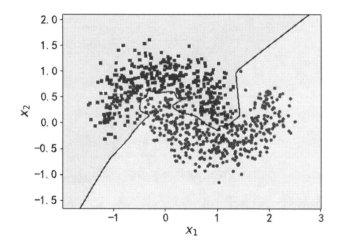

Figure 9-15. *Six layers*

9.5 Regularization

By designing network models with different layers and sizes, the initial function hypothesis space can be provided for the optimization algorithm, but the actual capacity of the model can change as the network parameters are optimized and updated. Take the polynomial function model as an example:

$$y = \beta_0 + \beta_1 x + \beta_2 x^2 + \beta_3 x^3 + \cdots + \beta_n x^n + \varepsilon$$

The capacity of the preceding model can be simply measured through n. During the training process, if the network parameters β_{k+1}, $\cdots$, β_n are all 0, then the actual capacity of the network degenerates to the function capacity of the kth polynomial. Therefore, by limiting the sparsity of network parameters, the actual capacity of the network can be constrained.

This constraint is generally achieved by adding additional parameter sparsity penalties to the loss function. The optimization goal before the constraint added is:

$$min \ L\big(f_\theta(x), y\big), (x, y) \in D^{train}$$

After adding additional constraints to the parameters of the model, the goal of optimization becomes:

$$min \ L\big(f_\theta(x), y\big) + \lambda \cdot \Omega(\theta), (x, y) \in D^{train}$$

where $\Omega(\theta)$ represents the sparsity constraint function on the network parameters θ. Generally, the sparsity constraint of the parameter θ is achieved by constraining the L norm of the parameter, that is:

$$\Omega(\theta) = \sum_{\theta_i} \|\theta_i\|_l$$

where $\|\theta_i\|_l$ represents the l norm of the parameter θ_i.

In addition to minimizing the original loss function $L(x, y)$, the new optimization goal also needs to constrain the sparsity $\Omega(\theta)$ of the network parameters. The optimization algorithm will reduce the network parameter sparsity $\Omega(\theta)$ as much as possible while reducing $L(x, y)$. Here λ is the weight parameter to balance the importance of $L(x, y)$ and $\Omega(\theta)$. Larger λ means that the sparsity of the network is more important; smaller λ means that the training error of the network is more important. By selecting the appropriate λ, you can get better training performance, while ensuring the sparsity of the network, which lead to a good generalization ability.

Commonly used regularization methods are L0, L1, and L2 regularization.

9.5.1 L0 Regularization

L0 regularization refers to the regularization calculation method using the L0 norm as the sparsity penalty term $\Omega(\theta)$, namely:

$$\Omega(\theta) = \sum_{\theta_i} \|\theta_i\|_0$$

The L0 norm $\|\theta_i\|_0$ is defined as the number of non-zero elements in θ_i. The constraint of $\sum_{\theta_i} \|\theta_i\|_0$ can force the connection weights in the network to be mostly 0, thereby reducing the actual amount of network parameters and network capacity. However, because the L0 norm is not derivable, gradient descent algorithm cannot be used for optimization. L0 norm is not often used in neural networks.

9.5.2 L1 Regularization

The regularization calculation method using the L1 norm as the sparsity penalty term $\Omega(\theta)$ is called L1 regularization, that is:

$$\Omega(\theta) = \sum_{\theta_i} \|\theta_i\|_1$$

The L1 norm $\|\theta_i\|_1$ is defined as the sum of the absolute values of all elements in the tensor θ_i. L1 regularization is also called Lasso regularization, which is continuously derivable and widely used in neural networks.

L1 regularization can be implemented as follows:

```
# Create weights w1,w2
w1 = tf.random.normal([4,3])
w2 = tf.random.normal([4,2])
# Calculate L1 regularization term
loss_reg = tf.reduce_sum(tf.math.abs(w1))\
    + tf.reduce_sum(tf.math.abs(w2))
```

9.5.3 L2 Regularization

The regularization calculation method using the L2 norm as the sparsity penalty term $\Omega(\theta)$ is called L2 regularization, that is:

$$\Omega(\theta) = \sum_{\theta_i} \|\theta_i\|_2$$

The L2 norm $\|\theta_i\|_2$ is defined as the sum of squares of all elements in the tensor θ_i. L2 regularization is also called ridge regularization, which is continuous and derivable like L1 regularization, and is widely used in neural networks.

The L2 regularization term is implemented as follows:

```
# Create weights w1,w2
w1 = tf.random.normal([4,3])
w2 = tf.random.normal([4,2])
# Calculate L2 regularization term
loss_reg = tf.reduce_sum(tf.square(w1))\
      + tf.reduce_sum(tf.square(w2))
```

9.5.4 Regularization Effect

Continue to take the crescent-shaped two-class data as an example. Under the condition that the other hyperparameters such as the network structure remain unchanged, the L2 regularization term is added to the loss function, and different regularization hyperparameter λ are used to obtain regularization effects of different degrees.

After training 500 Epochs, we obtain the classification decision boundaries of the learning model, as shown in Figure 9-16, Figure 9-17, Figure 9-18, and Figure 9-19. The distribution represents the classification effect when the regularization coefficient $\lambda = 0.00001, 0.001, 0.1, and\ 0.13$ is used. It can be seen that as the regularization coefficient increases, the network penalties for parameter sparsity become larger, thus forcing the optimization algorithm to search for models that make the network capacity smaller. When $\lambda = 0.00001$, the regularization effect is relatively weak, and the network is overfitting. However, when在$\lambda = 0.1$, the network has been optimized to the appropriate capacity, and there is no obvious overfitting or underfitting.

In actual training, it is generally preferred to try smaller regularization coefficients to observe whether the network is overfitting. Then try to increase the parameter λ gradually to increase the sparsity of the network parameters and improve the generalization ability. However, excessively large λ may cause the network to not converge and need to be adjusted according to the actual task.

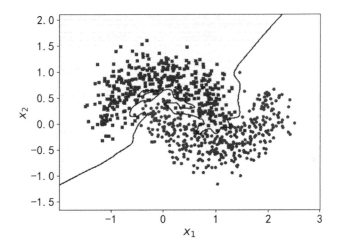

Figure 9-16.   *Regularization parameter:0.00001*

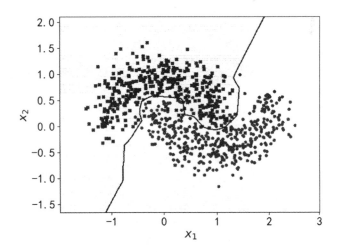

Figure 9-17.   *Regularization parameter:0.001*

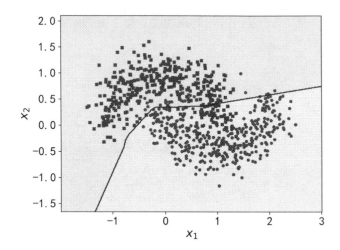

Figure 9-18. *Regularization parameter:0.1*

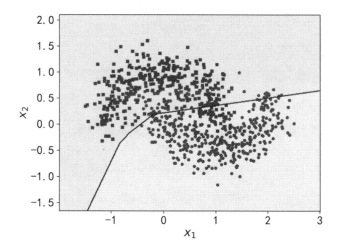

Figure 9-19. *Regularization parameter:0.13*

Under different regularization coefficients, we counted the value range of each connection weight in the network. Consider the weight matrix W of second layer of the network, whose shape is [256,256], that is, to convert a vector with an input length of 256 to an output vector of 256. From the perspective of the weight connection of the fully connected layer, the weight W include $256 \cdot 256$ connection lines. We correspond them to the XY grids in Figure 9-20, Figure 9-21, Figure 9-22, and Figure 9-23, where the X axis range is [0,255] and the range of the Y axis is [0,255]. All integer points of the XY grid respectively represent each position of the weight tensor W of shape [256,256], and each grid point indicates the weight of the current connection. From the figure, we can see the influence of different degrees of regularization constraints on the network weights. When $\lambda = 0.00001$, the effect of regularization is relatively weak, and the weight values in the network are relatively large, and are mainly distributed in interval $[-1.6088, 1.1599]$. After increasing the value to $\lambda = 0.13$, the network weight values are constrained in a smaller range $[-0.1104, 0.0785]$. As shown in Table 9-1, the sparseness of the weights after regularization can also be observed.

Table 9-1. *Weight variation after regularization*

λ	min(W)	max(W)	mean(W)
0.00001	-1.6088	1.1599	0.0026
0.001	-0.1393	0.3168	0.0003
0.1	-0.0969	0.0832	0
0.13	-0.1104	0.0785	0

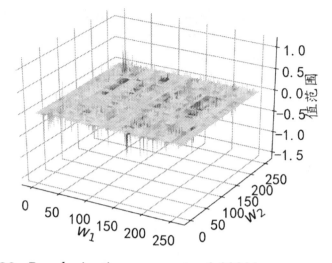

Figure 9-20. *Regularization parameter:0.00001*

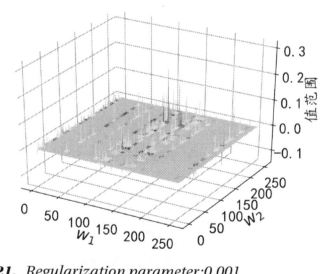

Figure 9-21. *Regularization parameter:0.001*

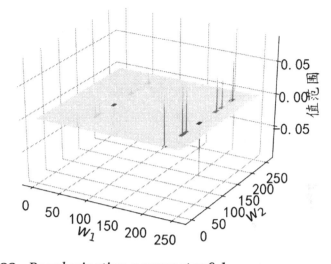

Figure 9-22.   *Regularization parameter:0.1*

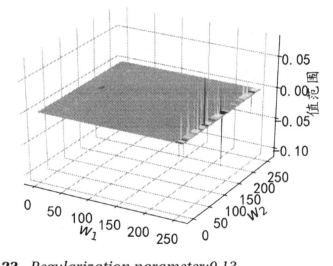

Figure 9-23.   *Regularization parameter:0.13*

9.6 Dropout

In 2012, Hinton et al. used the dropout method in their paper "Improving neural networks by preventing co-adaptation of feature detectors" to improve model performance. Dropout method reduces the number of parameters of the model that actually participates in the calculation during each training by randomly disconnecting the neural network. However, during testing, dropout method will restore all connections to ensure the best performance during model testing.

Figure 9-24 is a schematic diagram of the connection status of a fully connected layer network during a certain forward calculation. Figure 9-24(a) is a standard fully connected neural network. The current node is connected to all input nodes in the previous layer. In the network layer to which the dropout function is added, as shown in Figure 9-24(b), whether each connection is disconnected conforms to a certain preset probability distribution, such as a Bernoulli distribution with a disconnect probability Figure 9-24(b) shows a specific sampling result. The dotted line indicates that the sampling result is a disconnected line, and the solid line indicates the sampling result is not disconnected.

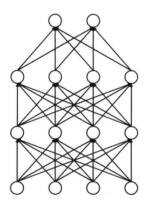 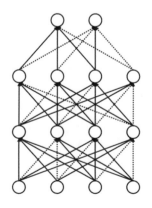

(a) Fully connected network   (b) With dropout

Figure 9-24. *Dropout diagram*

In TensorFlow, you can implement the dropout function through the tf.nn.dropout(x, rate) function, where the rate parameter sets the probability p of disconnection. For example:

```
# Add dropout operation with disconnection rate of 0.5
x = tf.nn.dropout(x, rate=0.5)
```

You can also use dropout as a network layer and insert a Dropout layer in the middle of the network. For example:

```
# Add Dropout layer with disconnection rate of 0.5
model.add(layers.Dropout(rate=0.5))
```

In order to explore the influence of the Dropout layer on network training, we maintained the hyperparameters such as the number of network layers unchanged, and observed the impact of dropout on network training by inserting different numbers of Dropout layers in the five fully connected layers. As shown in Figure 9-25, Figure 9-26, Figure 9-27, and Figure 9-28, the distribution draws the decision boundary effect of the network model without adding Dropout layers, adding one, two, and four Dropout layers. It can be seen that when the Dropout layer is not added, the network model has the same result as the previous observation. With the increase of the Dropout layer, the actual capacity of the network model during training decreases and the generalization ability becomes stronger.

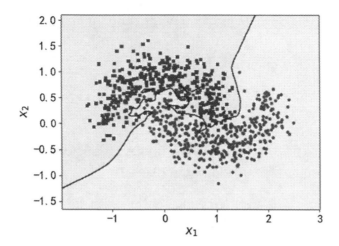

Figure 9-25. *Without Dropout layer*

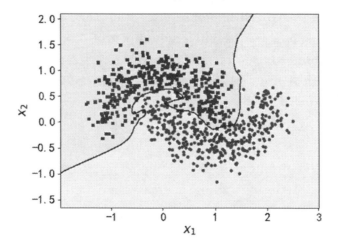

Figure 9-26. *With one Dropout layer*

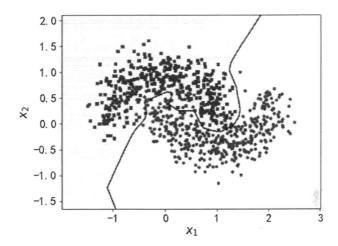

Figure 9-27. *With two Dropout layer*

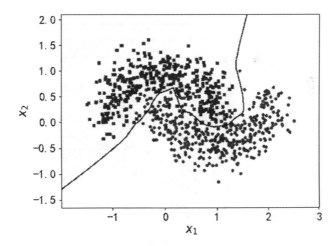

Figure 9-28. *With four Dropout layer*

9.7 Data Augmentation

In addition to the methods described previously, which can effectively detect and suppress overfitting, increasing the size of the dataset is the most important way to solve overfitting problem. However, collecting sample data and labels is often costly. For a limited dataset, the number

345

of training samples can be increased through data augmentation technology to obtain a certain degree of performance improvement. Data augmentation refers to changing the characteristics of the sample based on a priori knowledge while keeping the sample label unchanged, so that the newly generated sample also conforms or approximately conforms to the true distribution of the data.

Taking image data as an example, let's introduce how to do data augmentation. The size of the pictures in the dataset is often inconsistent. In order to facilitate the processing of the neural network, the pictures need to be rescaled to a fixed size, as shown in Figure 9-29, which is a fixed size 224 × 224 picture after rescaling. For the person in the picture, according to a priori knowledge, we know that rotation, scaling, translation, cropping, changing the angle of view, and blocking a certain local area will not change the main category label of the picture, so for the picture data, there are a variety of data augmentation methods.

Figure 9-29. *A picture after rescaling to 224 × 224 pixels*

TensorFlow provides common image processing functions, located in the tf.image submodule. Through the tf.image.resize function, we can zoom the pictures. We generally implement data augmentation in the preprocessing step. After reading the picture from the file system, the image data augmentation operation can be performed. For example:

```
def preprocess(x,y):
    # Preprocess function
    # x: picture path, y:picture label
    x = tf.io.read_file(x)
    x = tf.image.decode_jpeg(x, channels=3) # RGBA
    # rescale pictures to 244x244
    x = tf.image.resize(x, [244, 244])
```

9.7.1 Rotation

Rotating pictures is a very common way of augmenting picture data. By rotating the original picture at a certain angle, new pictures at different angles can be obtained, and the label information of these pictures remains unchanged, as shown in Figure 9-30.

Figure 9-30. *Image rotation*

Through tf.image.rot90(x, k = 1), the picture can be rotated by 90 degrees counterclockwise k times, for example:

```
# Picture rotates 180 degrees counterclockwise
x = tf.image.rot90(x,2)
```

9.7.2 Flip

The flip of the picture is divided into flip along the horizontal axis and along the vertical axis, as shown in Figure 9-31 and Figure 9-32, respectively. In TensorFlow, you can use tf.image.random_flip_left_right and tf.image.random_flip_up_down to randomly flip the image in the horizontal and vertical directions, for example:

```
# Random horizontal flip
x = tf.image.random_flip_left_right(x)
# Random vertical flip
x = tf.image.random_flip_up_down(x)
```

Figure 9-31. *Horizontal flip*

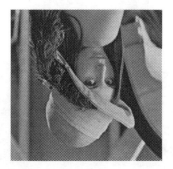

Figure 9-32. *Vertical flip*

9.7.3 Cropping

By removing part of the edge pixels in the left, right, or up and down directions of the original image, the main body of the image can be kept unchanged, and new image samples can be obtained at the same time. When actually cropping, the picture is generally scaled to a size slightly larger than the network input size, and then cropped to a suitable size. For example, if the input size of the network is 224 × 224, then you can use the resize function to rescale the picture to 244 × 244, and then randomly crop to the size 224 × 224. The code is implemented as follows:

```
# Rescale picture to larger size
x = tf.image.resize(x, [244, 244])
# Then randomly crop the picture to the desired size
x = tf.image.random_crop(x, [224,224,3])
```

Figure 9-33 is a picture zoomed to 244 × 244, Figure 9-34 is an example of random cropping to 244 × 244, and Figure 9-35 is also an example of random cropping.

Figure 9-33. *Before cropping*

Figure 9-34. *After cropping and rescaling-1*

Figure 9-35. *After cropping and rescaling-2*

9.7.4 Generate Data

By training the generative model on the original data and learning the distribution of the real data, the generative model can be used to obtain new samples. This method can also improve network performance to a certain extent. For example, conditional generation adversarial network (conditional GAN, CGAN for short) can generate labeled sample data, as shown in Figure 9-36.

Figure 9-36. *CGAN generated numbers*

9.7.5 Other Methods

In addition to the typical picture data augmentation methods described previously, the picture data can be arbitrarily transformed to obtain new pictures based on a priori knowledge without changing the picture tag information. Figure 9-37 demonstrates the picture data after superimposing Gaussian noise on the original picture, Figure 9-38 demonstrates the new picture obtained by changing the viewing angle of the picture, and Figure 9-39 demonstrates the new picture obtained by randomly blocking parts of the original picture.

Figure 9-37. *Adding Gaussian noise*

Figure 9-38. *Changing viewing angle*

Figure 9-39. *Randomly blocking parts*

9.8 Hands-On Overfitting

Earlier, we used a large amount of crescent-shaped two-class datasets to demonstrate the performance of the network model under various measures to prevent overfitting. In this section, we will complete the exercise based on the overfitting and underfitting models of the two classification datasets of crescent shape.

9.8.1 Build the Dataset

The feature vector length of the sample dataset we used is 2, and the label is 0 or 1, which represents two categories. With the help of the make_moons tool provided in the scikit-learn library, we can generate a training set of any number of data. First open the cmd command terminal and install the scikit-learn library. The command is as follows:

```
# Install scikit-learn library
pip install -U scikit-learn
```

To demonstrate the phenomenon of overfitting, we only sampled 1000 samples, and added Gaussian noise with a standard deviation of 0.25 as in the following:

```
# Import libraries
from sklearn.datasets import make_moons
# Randomly choose 1000 samples, and split them into training
and testing sets
X, y = make_moons(n_samples = N_SAMPLES, noise=0.25, random_
state=100)
X_train, X_test, y_train, y_test = train_test_split(X, y,
                                    test_size = TEST_SIZE,
                                    random_state=42)
```

353

The make_plot function can easily draw the distribution map of the data according to the coordinate X of the sample and the label y of the sample:

```
def make_plot(X, y, plot_name, file_name, XX=None, YY=None,
preds=None):
    plt.figure()
    # sns.set_style("whitegrid")
    axes = plt.gca()
    axes.set_xlim([x_min,x_max])
    axes.set_ylim([y_min,y_max])
    axes.set(xlabel="$x_1$", ylabel="$x_2$")
    # Plot prediction surface
    if(XX is not None and YY is not None and preds is
    not None):
        plt.contourf(XX, YY, preds.reshape(XX.shape), 25,
        alpha = 0.08, cmap=cm.Spectral)
        plt.contour(XX, YY, preds.reshape(XX.shape),
        levels=[.5], cmap="Greys", vmin=0, vmax=.6)
    # Plot samples
    markers = ['o' if i == 1 else 's' for i in y.ravel()]
    mscatter(X[:, 0], X[:, 1], c=y.ravel(), s=20,
            cmap=plt.cm.Spectral, edgecolors='none',
            m=markers)
    # Save the figure
    plt.savefig(OUTPUT_DIR+'/'+file_name)
```

Draw the distribution of 1000 samples for sampling, as shown in Figure 9-40, the red square points are one category, and the blue circles are another category.

```
# Plot data points
make_plot(X, y, None, "dataset.svg")
```

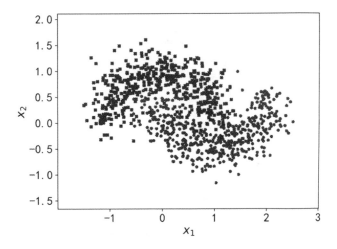

Figure 9-40. *Moon-shape two-class data points*

9.8.2 Influence of the Number of Network Layers

In order to explore the degree of overfitting at different network depths, we conducted a total of five training experiments. When $n \in [0, 4]$, build a fully connected layer network with $n + 2$ layers, and train 500 Epochs through the Adam optimizer to obtain the separation curve of the network on the training set, as shown in Figures 9.12, 9.13, 9.14, and 9.15 .

```
for n in range(5): # Create 5 different network with
different layers
    model = Sequential()
    # Create 1st layer
    model.add(Dense(8, input_dim=2,activation='relu'))
    for _ in range(n): # Add nth layer
        model.add(Dense(32, activation='relu'))
    model.add(Dense(1, activation='sigmoid')) # Add last layer
```

355

```
model.compile(loss='binary_crossentropy', optimizer='adam',
metrics=['accuracy']) # Configure and train
history = model.fit(X_train, y_train, epochs=N_EPOCHS,
verbose=1)
# Plot boundaries for different network
preds = model.predict_classes(np.c_[XX.ravel(),
YY.ravel()])
title = "Network layer ({})".format(n)
file = "NetworkCapacity%f.png"%(2+n*1)
make_plot(X_train, y_train, title, file, XX, YY, preds)
```

9.8.3 Impact of Dropout

In order to explore the impact of the Dropout layer on network training, we conducted a total of five experiments. Each experiment used a seven-layer fully connected layer network for training, but inserted 0~4 Dropout layers in the fully connected layer at intervals and passed Adam The optimizer trains 500 Epochs. The network training results are shown in Figures 9.25, 9.26, 9.27, and 9.28.

```
for n in range(5): # Create 5 different networks with different
number of Dropout layers
model = Sequential()
    # Create 1st layer
    model.add(Dense(8, input_dim=2,activation='relu'))
    counter = 0
    for _ in range(5): # Total number of layers is 5
        model.add(Dense(64, activation='relu'))
        if counter < n: # Add n Dropout layers
            counter += 1
            model.add(layers.Dropout(rate=0.5))
    model.add(Dense(1, activation='sigmoid')) # Output layer
```

```
model.compile(loss='binary_crossentropy', optimizer='adam',
metrics=['accuracy']) # Configure and train
# Train
history = model.fit(X_train, y_train, epochs=N_EPOCHS,
verbose=1)
# Plot decision boundaries for different number of
Dropout layers
preds = model.predict_classes(np.c_[XX.ravel(),
YY.ravel()])
title = "Dropout({})".format(n)
file = "Dropout%f.png"%(n)
make_plot(X_train, y_train, title, file, XX, YY, preds)
```

9.8.4 Impact of Regularization

In order to explore the influence of regularization coefficients on network model training, we adopted the L2 regularization method to construct a five-layer neural network, in which the weight tensor W of the second, third, and fourth neural network layers are added with L2 regularization constraints terms as follows:

```
def build_model_with_regularization(_lambda):
    # Create networks with regularization terms
    model = Sequential()
    model.add(Dense(8, input_dim=2,activation='relu')) #
    without regularization
    model.add(Dense(256, activation='relu', # With L2
    regularization
                    kernel_regularizer=regularizers.l2
                    (_lambda)))
    model.add(Dense(256, activation='relu', # With L2
    regularization
```

```
                     kernel_regularizer=regularizers.l2
                     (_lambda)))
    model.add(Dense(256, activation='relu', # With L2
    regularization
                     kernel_regularizer=regularizers.l2
                     (_lambda)))
    # Output
    model.add(Dense(1, activation='sigmoid'))
    model.compile(loss='binary_crossentropy', optimizer='adam',
    metrics=['accuracy']) # Configure and train
    return model
```

Under the condition of keeping the network structure unchanged, we adjust the regularization coefficient λ = 0.00001, 0.001, 0.1, 0.12, 0.13 to test the training effect of the network and draw the decision boundary curve of the learning model on the training set, as shown in Figure 9-16, Figure 9-17, Figure 9-18, and Figure 9-19.

```
for _lambda in [1e-5,1e-3,1e-1,0.12,0.13]:
    # Create model with regularization term
model = build_model_with_regularization(_lambda)
    # Train model
    history = model.fit(X_train, y_train, epochs=N_EPOCHS,
    verbose=1)
    # Plot weight range
    layer_index = 2
    plot_title = "Regularization-[lambda = {}]".format(str(_
    lambda))
    file_name = " Regularization _" + str(_lambda)
    # Plot weight ranges
    plot_weights_matrix(model, layer_index, plot_title,
    file_name)
```

```
# Plot decision boundaries
preds = model.predict_classes(np.c_[XX.ravel(),
YY.ravel()])
title = " regularization ".format(_lambda)
file = " regularization %f.svg"%_lambda
make_plot(X_train, y_train, title, file, XX, YY, preds)
```

The plot_weights_matrix code of the matrix 3D plot function is as follows:

```
def plot_weights_matrix(model, layer_index, plot_name,
file_name):
    # Plot weight ranges
    # Get weights for certain layers
    weights = model.layers[LAYER_INDEX].get_weights()[0]
    # Get minimum, maximum and mean values
    min_val = round(weights.min(), 4)
    max_val = round(weights.max(), 4)
    mean_val = round(weights.mean(), 4)
    shape = weights.shape
    # Generate grids
    X = np.array(range(shape[1]))
    Y = np.array(range(shape[0]))
    X, Y = np.meshgrid(X, Y)
    print(file_name, min_val, max_val,mean_val)
    # Plot 3D figures
    fig = plt.figure()
    ax = fig.gca(projection='3d')
    ax.xaxis.set_pane_color((1.0, 1.0, 1.0, 0.0))
    ax.yaxis.set_pane_color((1.0, 1.0, 1.0, 0.0))
    ax.zaxis.set_pane_color((1.0, 1.0, 1.0, 0.0))
    # Plot weight ranges
```

```
surf = ax.plot_surface(X, Y, weights, cmap=plt.get_
cmap('rainbow'), linewidth=0)
ax.set_xlabel('x', fontsize=16, rotation = 0)
ax.set_ylabel('y', fontsize=16, rotation = 0)
ax.set_zlabel('weight', fontsize=16, rotation = 90)
# save figure
plt.savefig("./" + OUTPUT_DIR + "/" + file_name + ".svg")
```

9.9 References

[1].   I. Goodfellow, Y. Bengio and A. Courville, Deep
Learning, MIT Press, 2016.

CHAPTER 10

Convolutional Neural Networks

> At present, artificial intelligence has not reached the level of 5 years old human, but the progress in perception is rapid. In the field of machine speech and visual recognition, there is no suspense to surpass humans in five to ten years.
>
> —Xiangyang Shen

We have introduced the basic theory of neural networks, the use of TensorFlow, and the basic fully connected network model and have a more comprehensive and in-depth understanding of neural networks. But for deep learning, we still have a little doubt. The depth of deep learning refers to the deeper layers of the network, generally more than five layers, and most of the neural network layers introduced so far are implemented within five layers. So what is the difference and connection between deep learning and neural networks?

Essentially, deep learning and neural networks refer to the same type of algorithm. In the 1980s, the network model based on the multilayer perceptron (MLP) mathematical model of biological neurons was called a neural network. Due to factors such as limited computing power and small data size at the time, neural networks were generally only able to train to a small number of layers. We call this type of neural network a

L. Long and X. Zeng, *Beginning Deep Learning with TensorFlow*,
https://doi.org/10.1007/978-1-4842-7915-1_10

361

shallow neural network (shallow neural network). It is not easy for shallow neural networks to extract high-level features from data, and the general expression ability is not good. Although it has achieved good results in simple tasks such as digital picture recognition, it is quickly surpassed by the new support vector machine proposed in the 1990s.

Geoffrey Hinton, a professor at the University of Toronto in Canada, has long insisted on the research of neural networks. However, due to the popularity of support vector machines at that time, research related to neural networks encountered many obstacles. In 2006, Geoffrey Hinton proposed a layer-by-layer pre-training algorithm in [1], which can effectively initialize the deep belief networks (DBN) network, thereby making it possible to train large-scale, deep layers (millions of parameters) of networks. In the paper, Geoffrey Hinton called the neural network deep neural network, and the related research is also called deep learning (deep learning). From this point of view, deep learning and neural networks are essentially consistent in their designation, and deep learning focuses more on deep neural networks. The "depth" of deep learning will be most vividly reflected in the relevant network structure in this chapter.

Before learning a deeper network model, let us first consider such a question: The theoretical research of neural networks was basically in place in the 1980s, but why did it fail to fully exploit the great potential of deep networks? Through the discussion of this question, we lead to the core content of this chapter: convolutional neural networks. This is also a type of neural network that can easily reach hundreds of layers.

10.1 Problems with Fully Connected N

First, let's analyze the problems of the fully connected network. Consider a simple four-layer fully connected layer network. The input is a handwritten digital picture vector of 784 nodes after leveling. The number of nodes in the middle three hidden layers is 256, and the number of nodes in the output layer is ten, as shown in Figure 10-1.

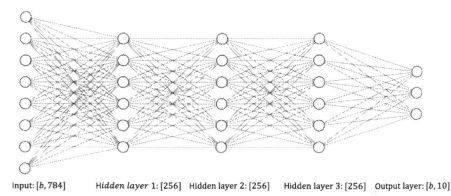

Input: [b, 784]   Hidden layer 1: [256]   Hidden layer 2: [256]   Hidden layer 3: [256]   Output layer: [b, 10]

Figure 10-1. *Simplified diagram of four-layer fully connected network structure*

We can quickly build this network model through TensorFlow: add 4 dense layers, and use the Sequential container to encapsulate it as a network object:

```
import tensorflow as tf
from tensorflow import keras
from tensorflow.keras import layers,Sequential,losses,
optimizers,datasets
# Create 4-layer fully connected network
model = keras.Sequential([
    layers.Dense(256, activation='relu'),
    layers.Dense(256, activation='relu'),
    layers.Dense(256, activation='relu'),
    layers.Dense(10),
])
# build model and print the model info
model.build(input_shape=(4, 784))
model.summary()
```

Use the summary() function to print out the statistical results of the parameters of each layer from the model, as shown in Table 10-1. How are the parameters of the network calculated? The weight scalar of each connecting line is considered as a parameter, so for a fully connected layer with n input nodes and m output nodes, there are a total of $n \cdot m$ parameters contained in the tensor W, and m parameters are contained in the vector b. Therefore, the total number of parameters of the fully connected layer is $n \cdot m + m$. Taking the first layer as an example, the input feature length is 784, the output feature length is 256, and the parameter amount of the current layer is $784 \cdot 256 + 256 = 200960$. The same method can be used to calculate the parameter amounts of the second, third, and fourth layers, which are 65792, 65792, *and* 2570, respectively. The total parameter amount is about 340,000. In a computer, if you save a single weight as a float-type variable, you need to occupy at least 4 bytes of memory (float takes more memory in Python), then 340,000 parameters require at least about 1.34MB of memory. In other words, storing the network parameters alone requires 1.34MB of memory. In fact, the network training process also needs to cache the computation graph, gradient information, input and intermediate calculation results, etc., where gradient-related operations take up a lot of resources.

Table 10-1. *Network parameter statistics*

Layer	Hidden Layer 1	Hidden Layer 2	Hidden Layer 3	Output Layer
Number of parameters	200960	65792	65792	2570

So how much memory does it take to train such a network? We can simply simulate resource consumption on modern GPU devices. In TensorFlow, if you do not set the GPU memory occupation method, all GPU memory will be occupied by default. Here, the TensorFlow memory usage is set to be allocated on demand, and the GPU memory resources occupied by it are observed as follows:

```
# List all GPU devices
gpus = tf.config.experimental.list_physical_devices('GPU')
if gpus:
  try:
    # Set GPU occupation as on demand
    for gpu in gpus:
      tf.config.experimental.set_memory_growth(gpu, True)
  except RuntimeError as e:
    # excepting handling
    print(e)
```

The preceding code is inserted after the TensorFlow library imported and before the model created. TensorFlow is configured to apply for GPU memory resources as needed through tf.config.experimental.set_memory_growth(gpu, True). In this way, the amount of GPU memory occupied by TensorFlow is the amount required for the operation. When the batch size is set to 32, we observed that GPU memory occupied about 708MB and CPU memory occupied about 870MB during training. Because the deep learning frameworks have different design considerations, this number is for reference only. Even so, we can feel that the computational cost of the four-layer fully connected layer is not small.

Back to the 1980s, what is the concept of 1.3MB network parameters? In 1989, Yann LeCun used a 256KB memory computer to implement his algorithm in the paper on handwritten zip code recognition [2]. This computer was also equipped with an AT&T DSP-32C DSP computing card (floating point computing capability is about 25 MFLOPS). For the 1.3MB network parameters, the computer with 256KB memory cannot even load the network parameters, let alone network training. It can be seen that the higher memory usage of the fully connected layer severely limits the development of the neural network towards a larger scale and deeper layers.

10.1.1 Local Correlation

Next, we explore how to avoid the defect of excessively large parameters of the fully connected network. For the convenience of discussion, we take the scene of picture type data as an example. For 2D image data, before entering the fully connected layer, the matrix data needs to be flattened into a 1D vector, and then each pixel is connected to each output node in pairs as shown in Figure 10-2.

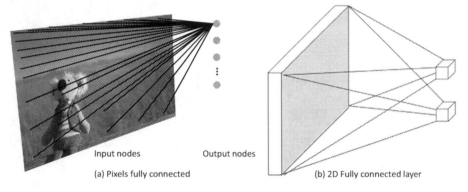

Input nodes   Output nodes

(a) Pixels fully connected   (b) 2D Fully connected layer

Figure 10-2. *2D feature fully connected diagram*

It can be seen that each output node of the network layer is connected to all input nodes for extracting the feature information of all input nodes. This dense connection method is the root cause of the large number of parameters and the high computational cost of the fully connected layer. The fully connected layer is also called dense connection layer (dense layer), and the relationship between output and input is:

$$o_j = \sigma\left(\sum_{i \in nodes(I)} w_{ij} x_i + b_j \right)$$

where *nodes(I)* represents the set of nodes in layer I.

So, is it necessary to connect the output node with all the input nodes? Is there an approximate simplified model? We can analyze the importance distribution of input nodes to output nodes, only consider a more important part of the input node, and discard the less important part of the node, so that the output node only needs to be connected to some input nodes, expressed as:

$$o_j = \sigma\left(\sum_{i \in top(I,j,k)} w_{ij} x_i + b_j \right)$$

where $top(I, j, k)$ represents the top k node set in layer I that has the highest importance for the number node in layer J. In this way, the weighted connections of the fully connected layer can be reduced from $\|I\| \cdot \|J\|$ to $k \cdot \|J\|$, where $\|I\|$ *and* $\|J\|$ represent the number of nodes in the I and J layers respectively.

Then the problem changes to exploring the importance distribution of the input node of layer I to the number output node j. However, it is very difficult to find out the importance distribution of each intermediate node. We can use prior knowledge to further simplify this problem.

In real life, there are a lot of data that use location or distance as a measure of importance distribution. For example, people who live closer to themselves are more likely to have greater influence on themselves (location correlation), and stock trend predictions should pay more attention to the recent trend (time correlation); each pixel of the picture is more related to the surrounding pixels (location correlation). Taking 2D image data as an example, if we simply think that the pixels with Euclidean distance from the current pixel is less than or equal to $\dfrac{k}{\sqrt{2}}$ are more important, and those with the Euclidean distance is greater than $\dfrac{k}{\sqrt{2}}$ are less important, then we can easily simplify the problem of finding the importance distribution of each pixel. As shown in Figure 10-3, the pixels where the solid grid is located are used as reference points and the pixels whose Euclidean distance is less than or equal to $\dfrac{k}{\sqrt{2}}$ are represented by a rectangular grid. The pixels in the grid are more important, and the pixels outside the grid are less important. This window is called the receptive field, which characterizes the importance distribution of each pixel to the central pixel. The pixels within the grid will be considered, and the pixels outside the grid will be ignored for the central pixel.

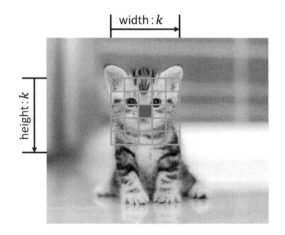

Figure 10-3. *Importance distribution of pixels*

This hypothetical characteristic of distance-based importance distribution is called local correlation. It only focuses on some nodes that are close to itself and ignores nodes that are far away. Under this assumption of importance distribution, the connection mode of the fully connected layer becomes as shown in Figure 10-4. The output node j is only connected to the local area (receptive field) centered by j and has no connection to other pixels.

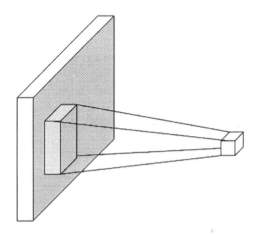

Figure 10-4. *Locally connected network*

Using the idea of local correlation, we record the height and width of the receptive field window as k (the height and width of the receptive field may not be equal; for convenience, we only consider the case where the height and width are equal). The current node is connected with all pixels in the receptive field, regardless of other pixels outside. The input and output relationship of the network layer is expressed as follows:

$$o_j = \sigma\left(\sum_{dist(i,j)\le\frac{k}{\sqrt{2}}} w_{ij}x_i + b_j \right)$$

where $dist(i,j)$ represents the Euclidean distance between i and j nodes.

10.1.2 Weight Sharing

Each output node is only connected to $k \times k$ input nodes in the receptive field, and the number of output layer nodes is $\|J\|$. So the number of the parameters of the current layer is $k \times k \times \|J\|$. Comparing to the fully connected layer, because k is usually small, such as 1, 3, and 5, so $k \times k \ll \|I\|$, which means it successfully reduced the amount of parameters.

Can the amount of parameters be further reduced, for example, can we only need $k \times k$ parameters to complete the calculation of the current layer? The answer is yes. Through the idea of weight sharing, for each output node o_j, the same weight matrix W is used, then no matter how many output nodes $\|J\|$ will be, the number of network layer parameters is always $k \times k$. As shown in Figure 10-5, when calculating the output pixel at the upper left corner, the weight matrix is used:

$$W = \begin{bmatrix} w_{11} & w_{12} & w_{13} & w_{21} & w_{22} & w_{23} & w_{31} & w_{32} & w_{33} \end{bmatrix}$$

Multiply and accumulate with the pixels inside the corresponding receptive field as the output value of the upper left pixel. When calculating the lower right receptive field, share the weight parameters W, that is, use the same weight parameters W to multiply and accumulate to get the output of the lower right pixel value. There are only $3 \times 3 = 9$ parameters in the network layer at this time, and it has nothing to do with the number of input and output nodes.

Figure 10-5. *Weight sharing matrix diagram*

By applying the idea of local correlation and weight sharing, we have successfully reduced the number of network parameters from $\|I\| \times \|J\|$ to $k \times k$ (to be precise, under the conditions of a single input channel and a single convolution kernel). This kind of weighted "local connection layer" network is actually a convolutional neural network. Next, we will introduce convolution operations from a mathematical perspective, and then formally learn the principles and implementation of convolutional neural networks.

10.1.3 Convolution Operation

Under the a priori of local correlation, we propose a simplified "local connection layer." For all pixels in the window $k \times k$, feature information is extracted by multiplying and accumulating weights, and each output node extracts features corresponding to the receptive field area. information. This operation is actually a standard operation in the field of signal processing: discrete convolution operation. Discrete convolution operation has a wide range of applications in computer vision. Here is a mathematical explanation of the convolutional neural network layer.

In the field of signal processing, the convolution operation of 1D continuous signals is defined as the integration of two functions: function $f(\tau)$, function $g(\tau)$, where $g(\tau)$ becomes $g(n - \tau)$ after flipping and translation. The 1D continuous convolution is defined as:

$$(f \otimes g)(n) = \int_{-\infty}^{\infty} f(\tau)g(n-\tau)d\tau$$

Discrete convolution replaces the integral operation with the accumulation operation:

$$(f \otimes g)(n) = \sum_{\tau=-\infty}^{\infty} f(\tau)g(n-\tau)$$

As for why convolution is defined in this way, I will not elaborate on it due to space limitations. We focus on 2D discrete convolution operations. In computer vision, the convolution operation is based on 2D picture function $f(m, n)$ and 2D convolution kernel $g(m, n)$, where $f(i, j)$ and $g(i, j)$ only exists in the effective area of the respective window, and the other areas are regarded as 0, as shown in Figure 10-6. The 2D discrete convolution is defined as:

$$[f \otimes g](m,n) = \sum_{i=-\infty}^{\infty} \sum_{j=-\infty}^{\infty} f(i,j) g(m-i, n-j)$$

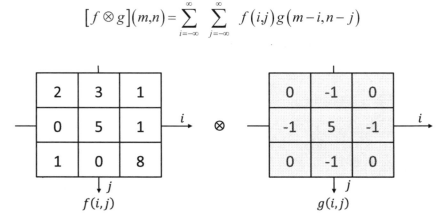

Figure 10-6. *2D image function f(i, j) and convolution kernel function g(i, j)*

Let's introduce the 2D discrete convolution operation in detail. First, invert the convolution kernel function $g(i, j)$ (invert each time along the x and y directions) to become $g(-i, -j)$. When $(m, n) = (-1, -1)$; it means that the convolution kernel function $g(-1 - i, -1 - j)$ is flipped and then shifted one unit to the left and the upward. At this time:

$$[f \otimes g](-1, -1) = \sum_{i=-\infty}^{\infty} \sum_{j=-\infty}^{\infty} f(i, j) g(-1 - i, -1 - j)$$

$$= \sum_{i \in [-1,1]} \sum_{j \in [-1,1]} f(i, j) g(-1 - i, -1 - j)$$

The 2D function only has valid values when $i \in [-1, 1], j \in [-1, 1]$. In other positions, it is 0. According to the calculation formula, we can get $[f \otimes g](0, -1) = 7$, as shown in Figure 10-7.

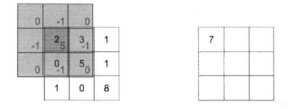

Figure 10-7. *Discrete convolution operation-1*

Similarly, when $(m, n) = (0, -1)$: $[f \otimes g]$
$(0, -1) = \sum_{i \in [-1, 1]} \sum_{j \in [-1, 1]} f(i, j) g(0 - i, -1 - j)$

That is, after the convolution kernel is flipped, the unit is shifted upwards and the corresponding position is multiplied and accumulated, $[f \otimes g](0, -1) = 7$, as shown in Figure 10-8.

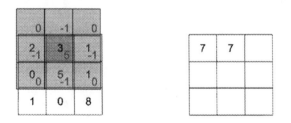

Figure 10-8. *Discrete convolution operation-2*

When $(m, n) = (1, -1)$:

$$[f \otimes g](1, -1) = \sum_{i \in [-1,1]} \sum_{j \in [-1,1]} f(i, j) g(1 - i, -1 - j)$$

That is, after the convolution kernel is flipped, it is translated to the right and upward by one unit, and the corresponding position is multiplied and accumulated, $[f \otimes g](1, -1) = 1$, as shown in Figure 10-9.

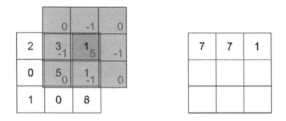

Figure 10-9. *Discrete convolution operation-3*

When $(m, n) = (-1, 0)$:

$$[f \otimes g](-1,0) = \sum_{i\in[-1,1]} \sum_{j\in[-1,1]} f(i,j)g(-1-i,-j)$$

That is, after the convolution kernel is flipped, it is translated to the left by one unit, and the corresponding position is multiplied and accumulated, $[f \otimes g](-1,0) = 1$, as shown in Figure 10-10.

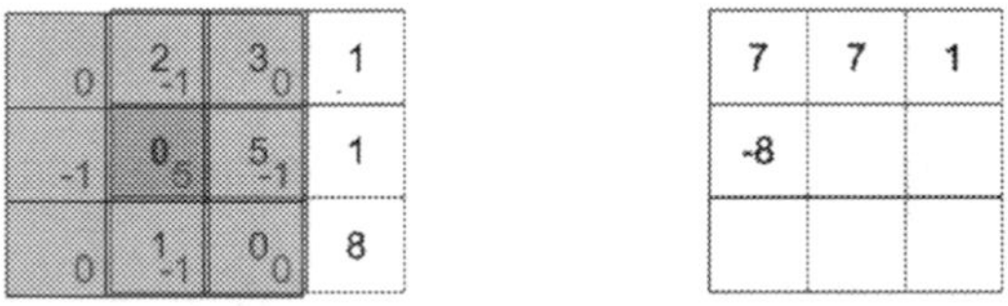

Figure 10-10. *Discrete convolution operation-4*

Cyclic calculation in this way, we can get all the values of the function $[f \otimes g](m, m)$, $m \in [-1, 1]$, $n \in [-1, 1]$, as shown in Figure 10-11.

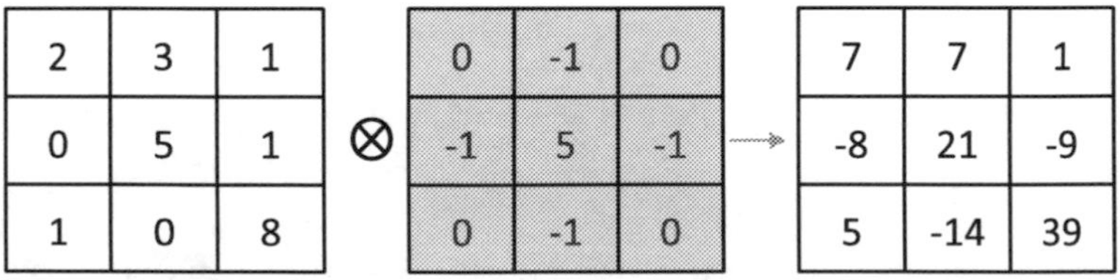

2	3	1		0	-1	0		7	7	1
0	5	1	$\otimes$	-1	5	-1	$\longrightarrow$	-8	21	-9
1	0	8		0	-1	0		5	-14	39

Figure 10-11. *2D discrete convolution operation*

So far, we have successfully completed the convolution operation of the picture function and the convolution kernel function to obtain a new feature map.

Recalling the operation of "weight multiplying and accumulating", we record it as $[f \cdot g](m, n) : [f \cdot g](m, n) = \sum_{i \in [-w/2, w/2]} \sum_{j \in [-h/2, h/2]} f(i, j) g(i - m, j - m)$

Comparing it carefully with the standard 2D convolution operation, it is not difficult to find that the convolution kernel function $g(m, n)$ in "weight multiply-accumulate" has not been flipped. For neural networks, the goal is to learn a function $g(m, n)$ to make L as small as possible. As for whether it is exactly the "convolution kernel" function defined in the convolution operation, it is not very important, because we will not directly use it. In deep learning, the function $g(m, n)$ is collectively called a convolution kernel (Kernel), sometimes called filter, weight, etc. Since the function $g(m, n)$ is always used to complete the convolution operation, the convolution operation has actually realized the idea of weight sharing.

Let's summarize the 2D discrete convolution operation process: each time by moving the convolution kernel and multiplying and accumulating with the receptive field pixels at the corresponding position of the picture, the output value at this position is obtained. The convolution kernel is a weight matrix W with rows and columns as size of k. The window corresponding to the size k on the feature map is the receptive field. The receptive field and the weight matrix are multiplied and accumulated to obtain the output value at this position. Through weight sharing, we gradually move the convolution kernel from the upper left to the right and downward to extract the pixel features at each position until the bottom right, completing the convolution operation. It can be seen that the two ways of understanding are the same. From a mathematical point of view, the convolutional neural network is to complete the discrete convolution operation of the 2D function; from the perspective of local correlation and weight sharing, the same effect can be obtained. Through these two perspectives, we can not only intuitively understand the calculation process of the convolutional neural network, but also rigorously derive from the mathematical point of view. It is based on convolution operations that convolutional neural networks can be so named.

In the field of computer vision, 2D convolution operations can extract useful features of data and perform convolution operations on input images with specific convolution kernels to obtain output images with different characteristics. As shown in Table 10-2, some common convolution kernels and corresponding effects are listed.

Table 10-2. *Common convolution kernels and their effect*

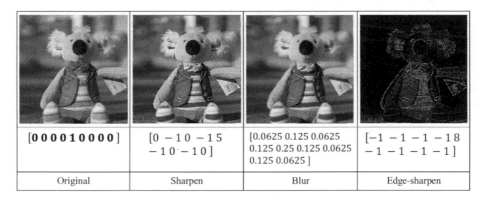

[0 0 0 0 1 0 0 0 0]	[0 − 1 0 − 1 5 − 1 0 − 1 0]	[0.0625 0.125 0.0625 0.125 0.25 0.125 0.0625 0.125 0.0625]	[−1 − 1 − 1 − 1 8 − 1 − 1 − 1 − 1]
Original	Sharpen	Blur	Edge-sharpen

10.2 Convolutional Neural Network

The convolutional neural network makes full use of the idea of local correlation and weight sharing, which greatly reduces the amount of network parameters, thereby improving training efficiency and making it easier to realize ultra-large-scale deep networks. In 2012, Alex Krizhevsky of the University of Toronto in Canada applied the deep convolutional neural network to the large-scale image recognition challenge ILSVRC-2012, and achieved a Top-5 error rate of 15.3% on the ImageNet dataset, ranking first. Comparing to the second place, Alex reduced the Top-5 error rate by 10.9% [3]. This huge breakthrough has attracted strong industry attention. Convolutional neural networks quickly became the new favorite in the field of computer vision. Subsequently, in a series of tasks, convolution-based neural network models have been proposed one after another and have achieved tremendous improvements in the original performance.

Now let's introduce the specific calculation process of the convolutional neural network layer. Taking 2D image data as an example, the convolutional layer accepts input feature maps X with height h and width w, and the number of channels c_{in}. Under the action of c_{out} convolution kernels with height h and width w and the number of channels c_{in}, feature maps with the height h' and width w' and c_{out} channels are generated. It should be noted that the height and width of the convolution kernel can be unequal. In order to simplify the discussion, we only consider the equal height and width cases, and then it can be easily extended to the case of unequal height and width.

We start with the discussion of the single-channel input and single-convolution kernel and then generalize to the multi-channel input and single-convolution kernel and finally discuss the most commonly used and most complex convolutional layer implementation of multi-channel input and multiple convolution kernels.

10.2.1 Single-Channel Input and Single Convolution Kernel

First, we discuss single-channel input $c_{in} = 1$, such as a gray-scale image with only one channel of gray value and a single convolution kernel $c_{out} = 1$. Take the input matrix X with size 5×5 and the convolution kernel matrix with size 3×3 as examples, as shown in Figure 10-12. The receptive field of the same size as the convolution kernel (the green box above the input *X*) is first moved to the top left of the input *X*. Select the receptive field element on the input and multiply it by the corresponding element of the convolution kernel (the middle box in the picture):

$$\left[1-1\,0-1-2\,2\,1\,2-2\right]\odot\left[-1\,1\,2\,1-1\,3\,0-1-2\right]=\left[-1-1\,0-1\,2\,6\,0-2\,4\right]$$

The $\odot$ symbol indicates the Hadamard Product, that is, the corresponding element of the matrix is multiplied. The symbol @ (matrix multiplication) is another common forms of matrix operations. After the operation of the matrix, all 9 values are added:

$$-1-1+0-1+2+6+0-2+4=7$$

We get the scalar 7 and write to the position of the first row and first column of the output matrix, as shown in Figure 10-12.

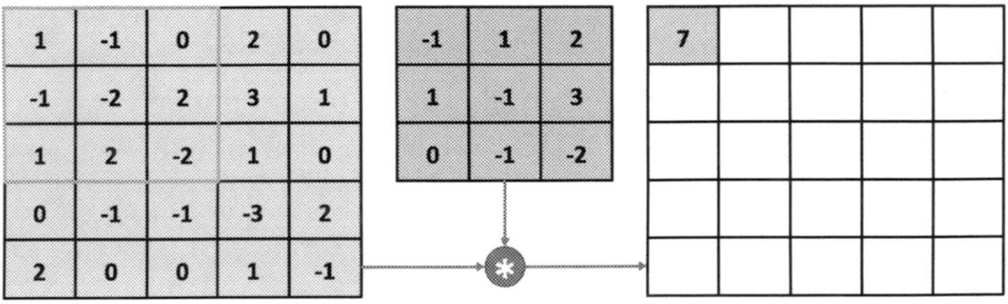

1	-1	0	2	0
-1	-2	2	3	1
1	2	-2	1	0
0	-1	-1	-3	2
2	0	0	1	-1

-1	1	2
1	-1	3
0	-1	-2

7				

Figure 10-12. *3 × 3 convolution operation-1*

After the feature extraction of the first receptive field area is completed, the receptive field window moves one step unit (Strides, denoted as *s*, default is 1) to the right and select the nine receptive field elements in the green box in Figure 10-13. Similarly, multiplying and accumulating the corresponding elements of the convolution kernel, we can get the output 10, which is written to the first row and second column position.

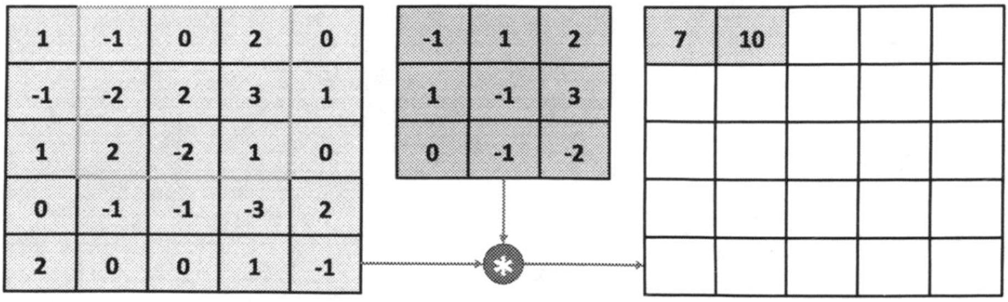

1	-1	0	2	0
-1	-2	2	3	1
1	2	-2	1	0
0	-1	-1	-3	2
2	0	0	1	-1

-1	1	2
1	-1	3
0	-1	-2

$\circledast$

7	10			

Figure 10-13. 3×3 convolution operation-2

Move the receptive field window to the right by one step unit again, select the element in the green box in Figure 10-14, multiply and accumulate with the convolution kernel, get the output 3, and write to the first row and third column of the output, as shown in Figure 10-14.

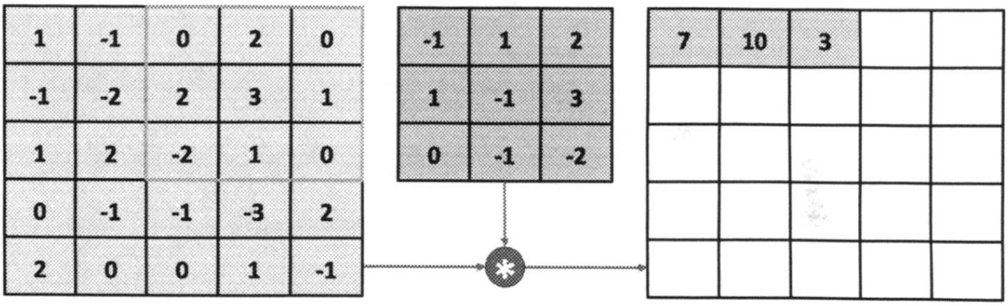

1	-1	0	2	0
-1	-2	2	3	1
1	2	-2	1	0
0	-1	-1	-3	2
2	0	0	1	-1

-1	1	2
1	-1	3
0	-1	-2

$\circledast$

7	10	3		

Figure 10-14. 3×3 convolution operation-3

At this point, the receptive field has moved to the far right of the effective pixel input, and it cannot continue to move to the right (without filling the invalid element), so the receptive field window moves down by one step unit ($s = 1$) and returns to the beginning of the current line, continue to select the new receptive field element area, as shown in Figure 10-15, and the convolution kernel operation results in output -1. Because the receptive field moves down by one step, so the output value -1 is written in the second row and the first column position.

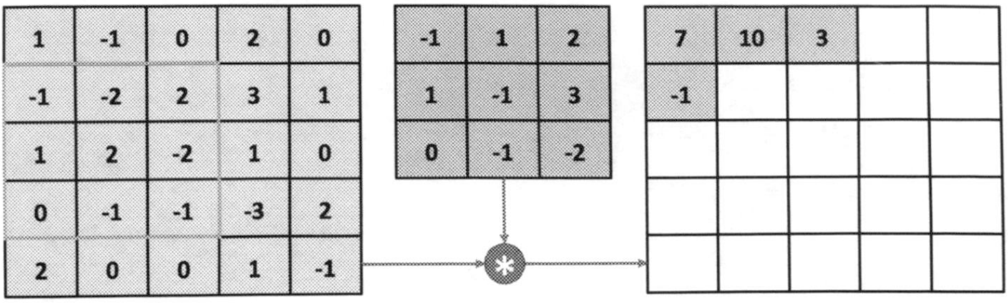

1	-1	0	2	0
-1	-2	2	3	1
1	2	-2	1	0
0	-1	-1	-3	2
2	0	0	1	-1

-1	1	2
1	-1	3
0	-1	-2

7	10	3		
-1				

Figure 10-15. *3 × 3 convolution operation-4*

According to the preceding method, each time the receptive field moves right by one step ($s = 1$), if it exceeds the input boundary, it moves down by one step ($s = 1$) and returns to the beginning of the line until the receptive field moves to the rightmost and bottommost position, as shown in Figure 10-16. Each selected receptive field element is multiplied by the corresponding element of the convolution kernel and written to the corresponding position of the output. In the end, we get a 3 × 3 matrix, which is slightly smaller than the input 5 × 5, this is because the receptive field cannot exceed the element boundary. It can be observed that the size of the output matrix of the convolution operation is determined by the size k of the convolution kernel, the height h and width w of the input X, the moving step s, and whether boundaries are filled.

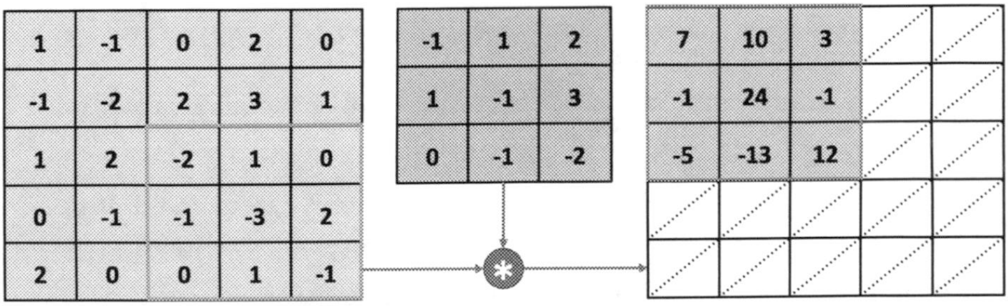

1	-1	0	2	0
-1	-2	2	3	1
1	2	-2	1	0
0	-1	-1	-3	2
2	0	0	1	-1

-1	1	2
1	-1	3
0	-1	-2

7	10	3		
-1	24	-1		
-5	-13	12		

Figure 10-16. *3 × 3 convolution operation-5*

Now we have introduced the calculation process of single-channel input and single convolution kernel. The actual number of input channels of the neural network is often large. Next, we will learn the convolution operation method of multi-channel input and a single convolution kernel.

10.2.2 Multi-channel Input and Single Convolution Kernel

Multi-channel input convolutional layers are more common. For example, a color image contains three channels (R/G/B). The pixel value on each channel indicates the intensity of the R/G/B color. In the following, we take three-channel input and a single convolution kernel as an example to extend the convolution operation of single-channel input to multi-channel. As shown in Figure 10-17, the leftmost 5 × 5 matrix of each row represents the input channels 1~3, the 3 × 3 matrix in the second column represents the channels 1~3 of the convolution kernel, and the matrix in the third column represents the middle matrix of the calculation on the current channel; the rightmost matrix represents the final output of the convolutional layer operation.

In the case of multi-channel input, the number of channels of the convolution kernel needs to match the number of input channels. The *ith* channel of the convolution kernel and the *ith* channel of the input X are calculated to obtain the first intermediate matrix, which can be then regarded as the case of single input and single convolution kernel. The corresponding elements of the intermediate matrix of all channels are added again as the final output.

The specific calculation process is as follows: in the initial state, as shown in Figure 10-17, the receptive field window on each channel synchronously falls on the leftmost and topmost positions on the corresponding channel. The receptive field area elements and the convolution kernel on each channel multiply and accumulate the matrix

above the corresponding channel to obtain the intermediate variables
of the output 7, -11, and -1 on the three channels, and then we can
add these intermediate variables to get the output -5 and write it to the
corresponding position.

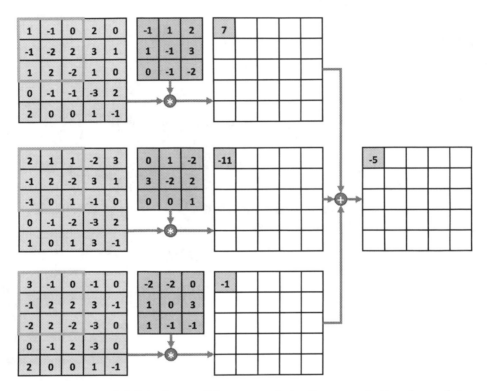

Figure 10-17. *Multi-channel input and single convolution kernel-1*

Then, the receptive field window moves synchronously to the right
by one step ($s = 1$) on each channel. At this time, the receptive field area
elements are shown in Figure 10-18. The receptive field on each channel is
multiplied by the matrix on the corresponding channel of the convolution
kernel and is then accumulated to get the intermediate variables 10, 20,
and 20. We then add them up to get the output 50 and write the element
position of the first row and second column.

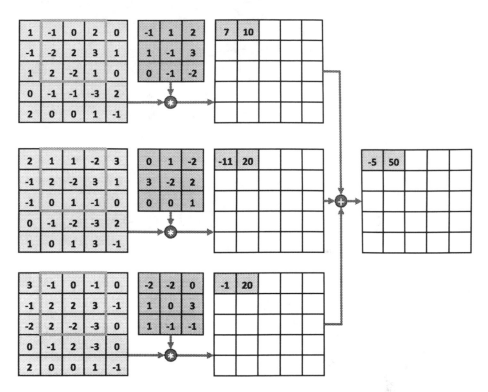

Figure 10-18. *Multi-channel input and single convolution kernel-2*

In this way, the receptive field window is moved synchronously to the rightmost and bottommost positions. All the convolution operations of the input and the convolution kernel are completed, and the resulting 3 × 3 output matrix is shown in Figure 10-19.

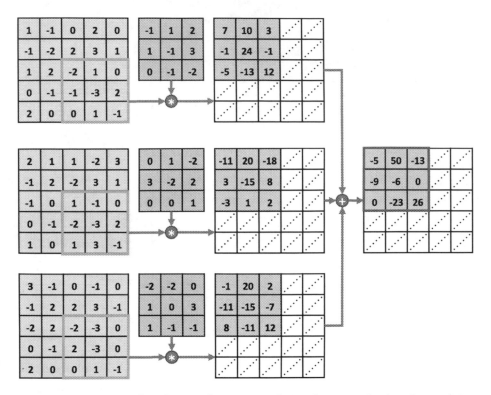

Figure 10-19. *Multi-channel input and single convolution kernel-3*

The entire calculation diagram is shown in Figure 10-20. The receptive field at each input channel is multiplied by the corresponding channel of the convolution kernel to obtain intermediate variables equal to the number of channels. All of these intermediate variables are added to obtain the output value in the current position. The number of input channels determines the number of convolution kernel channels. A convolution kernel can only get one output matrix, regardless of the number of input channels.

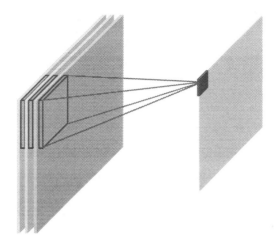

Figure 10-20. *Multi-channel input and single convolution kernel diagram*

Generally speaking, a convolution kernel can only complete the extraction of a certain logical feature. When multiple logical features need to be extracted at the same time, it can be achieved by adding multiple convolution kernels to improve the expression ability of the neural network. This is the case of multi-channel input and multi-convolution kernels.

10.2.3 Multi-channel Input and Multi-convolution Kernel

Multi-channel input and multi-convolution kernels are the most common forms of convolutional neural networks. We have already introduced the operation process of single convolution kernels. Each convolution kernel and input are convolved to obtain an output matrix. When there are multiple convolution kernels, the ith ($i \in [1, n]$, n is the number of convolution kernels) convolution kernel and input X get the ith output matrix (also called the channel i of output tensor O), and finally all the output matrix in the channel dimension stitch together (stack operation to create a new dimension – the number of output channels) to generate an output tensor O that contains n channels.

Take a convolutional layer with three channels of input and two convolution kernels as an example. The first convolution kernel and input X get the first output channel, and the second convolution kernel and input X get the second output channel, as shown in Figure 10-21. The two output channels are stitched together to form the final output O. The size k, stride size s, and padding settings of each convolution kernel are uniformly set, so as to ensure that each output channel has the same size to meet the conditions of stitching.

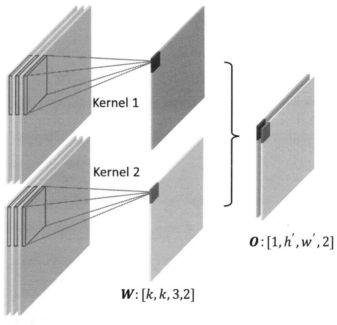

Figure 10-21. *Diagram of multi-convolution kernels*

10.2.4 Stride Size

In convolution operation, how to control the density of receptive field layout? For inputs with high information density, such as pictures with a large number of objects, in order to maximize the useful information, it is desirable to arrange the receptive field windows more densely during network design. For inputs with lower information density, such as a picture of the ocean, we can reduce the number of receptive fields in an appropriate amount. The control method of receptive field density is generally realized by moving strides.

The stride size refers to the unit of length for each movement of the receptive field window. For 2D input, it is divided into movement lengths in the x (right) direction and y (downward) direction. In order to simplify the discussion, we only consider the case of same stride size for both directions, which is also the most common setting in neural networks. As shown in Figure 10-22, the position of the receptive field window represented by the solid green line is the current position, and the dashed green line represents the position of the last receptive field. The movement length from the last position to the current position is the definition of the stride size. In Figure 10-22, the stride length of the receptive field in the x direction is 2, which is expressed as $s = 2$.

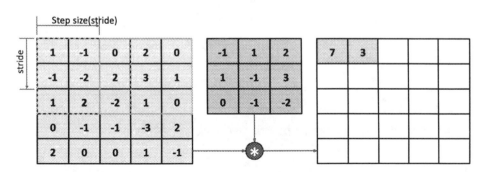

Figure 10-22. *Diagram of step size(namely stride)*

When the receptive field reaches to the right boundary of the input X, it moves down one stride ($s = 2$) and returns to the beginning of the line as shown in Figure 10-23.

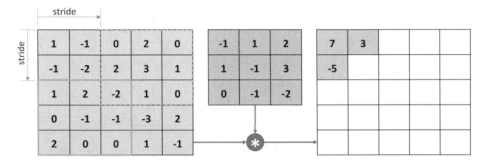

Figure 10-23. *Convolution operation stride size demnostration-1*

Circulate back and forth until the bottom and right edges are reached as shown in Figure 10-24. The final output height and width of the convolutional layer are only 2 × 2. Compared with the previous situation ($s = 1$), the output height and width are reduced from 3 × 3 to 2 × 2 and the number of receptive fields is reduced to only 4.

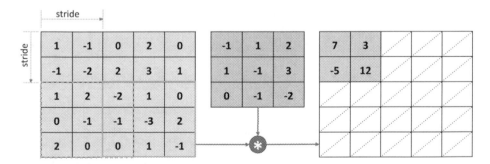

Figure 10-24. *Convolution operation stride size demnostration-2*

It can be seen that by setting the stride size, the extraction of information density can be effectively controlled. When the stride size is small, the receptive field moving window is small, which is helpful to extract more feature information and the size of the output tensor is larger; when the stride size is larger, the receptive field moving window is larger which is helpful to reduce the calculation cost and filter redundant information, and of course, the size of the output tensor is also smaller.

10.2.5 Padding

After the convolution operation, the height and width of the output will generally be smaller than the height and width of the input. Even when the stride size is 1, the height and width of the output will be slightly smaller than the input height and width. When designing a network model, it is sometimes desired that the height and width of the output can be the same as the height and width of the input, thereby facilitating the design of network parameters and residual connection. In order to make the height and width of the output equal to that of the input, it is common to increase the input by padding several invalid elements on the height and width of the original input. By carefully designing the number of filling units, the height and width of the output after the convolution operation can be equal to the original input, or even larger.

As shown in Figure 10-25, we can fill an indefinite number at the top, bottom, left, or right boundaries. The default filled number is 0, and it can also be filled with customized data. In Figure 10-25, one row is filled in the upper and lower directions, and two columns are filled in the left and right directions.

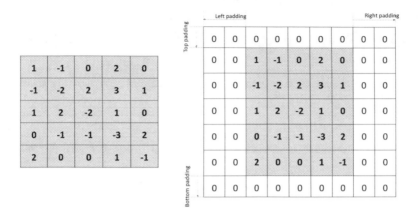

Figure 10-25. *Matrix padding diagram*

So how to calculate the convolutional layer after filling? We can simply replace the input *X* with the new tensor *X′* obtained after filling. As shown in Figure 10-26, the initial position of the receptive field is at the upper left of *X′*. Similar as before, the output 1 is obtained and written to the corresponding position of the output tensor.

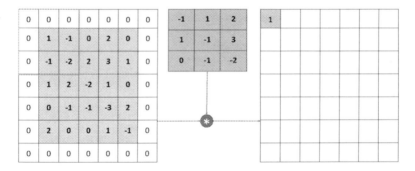

Figure 10-26. *Convolution operation after padding-1*

Move the stride by one unit and repeat the operation to get the output 0, as shown in Figure 10-27.

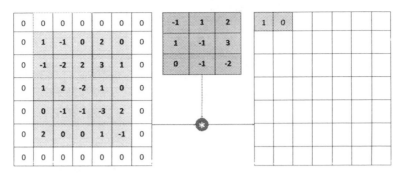

Figure 10-27. *Convolution operation after padding-2*

Looping back and forth, the resulting output tensor is shown in Figure 10-28.

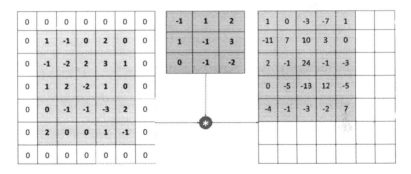

Figure 10-28. *Convolution operation after padding-3*

Through the carefully designed padding scheme, that is, filling one unit $(p = 1)$ up, down, left, and right, we can get the result O that has the same height and width of the input. Without padding, as shown in Figure 10-29, we can only get the output slightly smaller than the input.

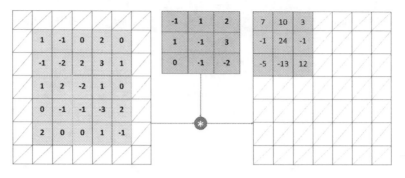

Figure 10-29. *Convolution output without padding*

The output size $[b, h', w', c_{out}]$ of the convolutional neural layer is determined by the number of convolution kernels c_{out}, the size of the convolution kernel k, the stride size s, the number of padding p (only considering the same number of top and bottom paddings p_h, and the same number of left and right paddings p_w), and the height h and width w of the input X. The mathematical relationship between can be expressed as:

$$h' = \left\lfloor \frac{h + 2 \cdot p_h - k}{s} \right\rfloor + 1$$

$$w' = \left\lfloor \frac{w + 2 \cdot p_w - k}{s} \right\rfloor + 1$$

where p_h and p_w indicate the padding quantity in the height and width directions, respectively, and $\lfloor \cdot \rfloor$ indicates rounding down. Taking the preceding example as an example, $h = w = 5$, $k = 3$, $p_h = p_w = 1$, $s = 1$, the output are:

$$h' = \left\lfloor \frac{5 + 2 * 1 - 3}{1} \right\rfloor + 1 = \lfloor 4 \rfloor + 1 = 5$$

$$w' = \left\lfloor \frac{5 + 2 * 1 - 3}{1} \right\rfloor + 1 = \lfloor 4 \rfloor + 1 = 5$$

In TensorFlow, when $\hat{t:s} = 1$, if you want the output O and input X to be equal in height and width, you only need to simply set the parameter padding="SAME" to make TensorFlow automatically calculate the number of padding, which is very convenient.

10.3 Convolutional Layer Implementation

In TensorFlow, you can either build a neural network through a low-level implementation of custom weights, or you can directly call a high-level API of convolutional layers to quickly build a complex network. We mainly take 2D convolution as an example to introduce how to implement a convolutional neural network layer.

10.3.1 Custom Weights

In TensorFlow, the 2D convolution operation can be easily realized through the tf.nn.conv2d function. tf.nn.conv2d performs a convolution operation based on input $X:[b, h, w, c_{in}]$ and convolution kernel $W:[k, k, c_{in}, c_{out}]$ to get the output $O : [b, h', w', c_{out}]$, where c_{in} represents the number of input channels, c_{out} indicates the number of convolution kernels which is also the number of output channels.

```
In [1]:
x = tf.random.normal([2,5,5,3]) # input with 3 channels with
height and width 5
# Create w using [k,k,cin,cout] format, 4 3x3 kernels
w = tf.random.normal([3,3,3,4])
# Stride is 1, padding is 0,
out = tf.nn.conv2d(x,w,strides=1,paddi
ng=[[0,0],[0,0],[0,0],[0,0]])
Out[1]: #  shape of output tensor
TensorShape([2, 3, 3, 4])
```

The format of the padding parameter is:

```
padding=[[0,0],[top,bottom],[left,right],[0,0]]
```

For example, if one unit is filled up in all directions (top, bottom, left, and right), the padding parameter is as follows:

```
In [2]:
x = tf.random.normal([2,5,5,3]) # input with 3 channels with
height and width 5
# Create w using [k,k,cin,cout] format, 4 3x3 kernels
w = tf.random.normal([3,3,3,4])
# Stride is 1, padding is 0,
out = tf.nn.conv2d(x,w,strides=1,paddi
ng=[[0,0],[1,1],[1,1],[0,0]])
Out[2]: # shape of output tensor
TensorShape([2, 5, 5, 4])
```

In particular, by setting the parameters padding='SAME' and strides=1, we can get the same size for the input and output of the convolutional layer, wherein the specific number of padding is automatically calculated by TensorFlow. For example:

```
In [3]:
x = tf.random.normal([2,5,5,3]) # input
w = tf.random.normal([3,3,3,4]) # 4 3x3 kernels
# Stride is 1,padding is "SAME"
# padding="SAME" gives use same size only when stride=1
out = tf.nn.conv2d(x,w,strides=1,padding='SAME')
Out[3]: TensorShape([2, 5, 5, 4])
```

When $s > 1$, setting padding='SAME' would cause the output height and width to decrease $\dfrac{1}{s}$ of original size. For example:

```
In [4]:
x = tf.random.normal([2,5,5,3])
w = tf.random.normal([3,3,3,4])
out = tf.nn.conv2d(x,w,strides=3,padding='SAME')
Out [4]:TensorShape([2, 2, 2, 4])
```

The convolutional neural network layer is the same as the fully connected layer, and the network can be set with a bias vector. The tf.nn. conv2d function does not implement the calculation of the bias vector. We can add the bias manually. For example:

```
# Create bias tensor
b = tf.zeros([4])
# Add bias to convolution output. It'll broadcast to size of
[b,h',w',cout]
out = out + b
```

10.3.2 Convolutional Layer Classes

Through the convolution layer classes layers.Conv2D, you can directly define the convolution kernel W and bias tensor b and directly call the class instance to complete the forward calculation of the convolution layer. In TensorFlow, the naming of APIs has certain rules. Objects with uppercase letters generally represent classes, and all lowercases generally represent functions, such as layers.Conv2D represents convolutional layer classes, and nn.conv2d represents convolution functions. Using the class method will automatically create the required weight tensor and bias vector. The user does not need to memorize the definition format of the convolution kernel tensor, so it is easier and more convenient to use, but we also lose some flexibility. The function interface needs to define weights and bias by itself, which is more flexible.

When creating a new convolutional layer class, you only need to specify the number of convolution kernel parameters filters, the size of the convolution kernel kernel_size, the stride, padding, etc. A convolutional layer with 4 3 × 3 convolution kernels is created as follows (the step stride is 1, and the padding scheme is'SAME'):

```
layer = layers.Conv2D(4,kernel_size=3,strides=1,padding='SAME')
```

If the height and width of the convolution kernel are not equal, and the stride along different directions is not equal neither, it is necessary to design the kernel_size parameter in the tuple format (k_h, k_w) and the strides parameter (s_h, s_w). Create 4 3 × 4 convolution kernels as follows ($s_h = 2$ in the vertical direction, and $s_w = 1$ in the horizontal direction):

```
layer = layers.Conv2D(4,kernel_size=(3,4),strides=(2,1),paddi
ng='SAME')
```

After the creation is complete, the forward calculation can be completed by calling the instance (__call__ method), for example:

```
In [5]:
layer = layers.Conv2D(4,kernel_size=3,strides=1,padding='SAME')
out = layer(x) # forward calculation
out.shape # shape of output
Out[5]:TensorShape([2, 5, 5, 4])
```

In class Conv2D, the convolution kernel tensor W and bias b are saved, and the list of W and b can be returned directly through the class member trainable_variables. For example:

```
In [6]:
# Return all trainable variables
layer.trainable_variables
Out[6]:
```

```
[<tf.Variable 'conv2d/kernel:0' shape=(3, 3, 3, 4)
dtype=float32, numpy=
 array([[[[ 0.13485974, -0.22861657,  0.01000655,  0.11988598],
         [ 0.12811887,  0.20501086, -0.29820845, -0.19579397],
         [ 0.00858489, -0.24469738, -0.08591779,
           -0.27885547]], ...
 <tf.Variable 'conv2d/bias:0' shape=(4,) dtype=float32,
 numpy=array([0., 0., 0., 0.], dtype=float32)>]
```

This layer.trainable_variables class member is very useful in obtaining the variables to be optimized in the network layer. You can also directly call class instance layer.kernel, layer.bias to access W and b.

10.4 Hands-On LeNet-5

In the 1990s, Yann LeCun et al. proposed a neural network for recognition of handwritten digits and machine-printed character pictures, which was named LeNet-5 [4]. The proposal of LeNet-5 enabled the convolutional neural network to be successfully commercialized at that time and was widely used in tasks such as postcode and check number recognition. Figure 10-30 is the network structure diagram of LeNet-5. It accepts digital and character pictures of size 32 × 32 as input and then passes through the first convolution layer to obtain the tensor with shape $[b, 28, 28, 6]$. After a downsampling layer, the tensor size is reduced to $[b, 14, 14, 6]$. After the second convolutional layer, the tensor shape becomes $[b, 10, 10, 16]$. After similar downsampling layer, the tensor size is reduced to $[b, 5, 5, 16]$. Before entering the fully connected layer, the tensor is converted to shape $[b, 400]$ and feed into two fully connected layers with the number of input nodes 120 and 84, respectively. A tensor with shape $[b, 84]$ is obtained and finally goes through the Gaussian connections layer.

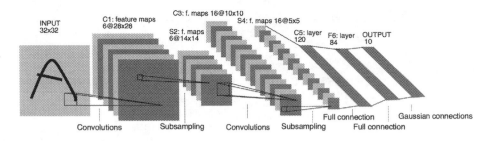

Figure 10-30. *LeNet-5 structure [4]*

It now appears that the LeNet-5 network has fewer layers (two convolutional layers and two fully connected layers), fewer parameters, and lower computational cost, especially with the support of modern GPUs, which can be trained in minutes.

We have made a few adjustments based on LeNet-5 to make it easier to implement using modern deep learning frameworks. First, we adjust the input shape from 32×32 to 28×28, and then implement the two downsampling layers as the maximum pooling layer (reducing the height and width of the feature map, which will be introduced later), and finally replacing the Gaussian connections layer with a fully connected layer. The modified network is also referred to as the LeNet-5 network hereinafter. The network structure diagram is shown in Figure 10-31.

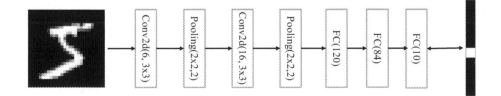

Figure 10-31. *Modified LeNet-5 structure*

We train the LeNet-5 network based on the MNIST handwritten digital picture dataset and test its final accuracy. We have already introduced how to load the MNIST dataset in TensorFlow, so I won't go into details here.

First create LeNet-5 through the Sequential container as follows:

```python
from tensorflow.keras import Sequential

network = Sequential([
    layers.Conv2D(6,kernel_size=3,strides=1), # Convolutional
    layer with 6 3x3 kernels
    layers.MaxPooling2D(pool_size=2,strides=2), # Pooling layer
    with size 2
    layers.ReLU(), # Activation function
    layers.Conv2D(16,kernel_size=3,strides=1), # Convolutional
    layer with 16 3x3 kernels

    layers.MaxPooling2D(pool_size=2,strides=2), # Pooling layer
    with size 2
    layers.ReLU(), # Activation function
    layers.Flatten(), # Flatten layer

    layers.Dense(120, activation='relu'), # Fully-
    connected layer
    layers.Dense(84, activation='relu'), # Fully-
    connected layer
    layers.Dense(10) # Fully-connected layer
                    ])
# build the network
network.build(input_shape=(4, 28, 28, 1))
# network summary
network.summary()
```

The summary () function counts the parameters of each layer and prints out the network structure information and details of the parameters of each layer, as shown in Table 10-3, we can compare with the parameter scale of the fully connected network 10.1.

Table 10-3. *Network parameter statistics*

Layer	Convolutional layer 1	Convolutional layer 2	Fully connected layer 1	Fully connected layer 2	Fully connected layer 3
Parameter amount	60	880	48120	10164	850

It can be seen that the parameter amount of the convolutional layer is very small, and the main parameter amount is concentrated in the fully connected layer. Because the convolutional layer reduces the input feature dimension a lot, the parameter amount of the fully connected layer is not too large. The parameter amount of the entire model is about 60K, and the number of fully connected network parameters in Table 10.1 reaches 340,000, so convolutional neural networks can significantly reduce the amount of network parameters while increasing the depth of the network.

In the training phase, first add a dimension ($[b, 28,28,1]$) to the original input of shape $[b, 28, 28]$ in the dataset and send it to the model for forward calculation to obtain the output tensor with shape $[b, 10]$. We create a new cross-entropy loss function class for processing classification tasks. By setting the from_logits=True flag, the softmax activation function is implemented in the loss function, and there is no need to manually add the loss function, which improves numerical stability. The code is as follows:

```
from tensorflow.keras import losses, optimizers
# Create loss function
criteon = losses.CategoricalCrossentropy(from_logits=True)
```

The training implementation is as follows:

```
# Create Gradient tape environment
with tf.GradientTape() as tape:
    # Expand input dimension =>[b,28,28,1]
    x = tf.expand_dims(x,axis=3)
```

```
# Forward calculation, [b, 784] => [b, 10]
out = network(x)
# One-hot encoding, [b] => [b, 10]
y_onehot = tf.one_hot(y, depth=10)
# Calculate cross-entropy
loss = criteon(y_onehot, out)
```

After obtaining the loss value, the gradient between the loss and the network parameter network.trainable_variables is calculated by TensorFlow's gradient recorder tf.GradientTape(), and the network weight parameter is automatically updated by the optimizer object as in the following:

```
# Calcualte gradient
grads = tape.gradient(loss, network.trainable_variables)
# Update paramaters
optimizer.apply_gradients(zip(grads, network.trainable_
variables))
```

The training can be completed after repeating the preceding steps several times.

In the testing phase, since there is no need to record gradient information, the code generally does not need to be written in the environment "with tf.GradientTape() as tape". After the output obtained by the forward calculation passes the Softmax function, we get the probability P that the network predicts that the current picture x belongs to the category i ($i \in [0, 9]$). Use the argmax function to select the index of the element with the highest probability as the current prediction category, compare it with the real label, and calculate the number of True samples in the comparison result. The number of samples with correct predictions divided by the total sample number gives us the test accuracy of the network.

```
        # Use correct to record the number of correct
        predictions
        # Use total to record the total number
        correct, total = 0,0
        for x,y in db_test: # Loop through all samples
            # Expand dimension =>[b,28,28,1]
            x = tf.expand_dims(x,axis=3)
            # Forward calculation to get probability, [b, 784]
            => [b, 10]
            out = network(x)
            # Technically, we should pass out to softmax()
            function firs.
    # But because softmax() doesn't change the order the numbers,
    we omit the softmax() part.
            pred = tf.argmax(out, axis=-1)
            y = tf.cast(y, tf.int64)
            # Calculate the correct prediction number
            correct += float(tf.reduce_sum(tf.cast
            (tf.equal(pred, y),tf.float32)))
            # Total sample number
            total += x.shape[0]
        # Calculate accuracy
        print('test acc:', correct/total)
```

After cyclically training 30 Epochs on the dataset, the training accuracy of the network reached 98.1%, and the test accuracy also reached 97.7%. For the simple handwritten digital picture recognition tasks, the old LeNet-5 network can already achieve good results, but for slightly more complex tasks, such as color animal picture recognition, LeNet-5 performance will drop sharply.

10.5 Representation Learning

We have introduced the working principle and implementation method of the convolutional neural network layer. The complex convolutional neural network model is also based on the stacking of convolutional layers. In the past, researchers have discovered that the deeper the network layer, the stronger the model's expressive ability, and the more likely it is to achieve better performance. So what are the characteristics of the stacked convolutional network, so that the deeper the layer, the stronger the network's expressive ability?

In 2014, Matthew D. Zeiler et al. [5] tried to use visual methods to understand exactly what convolutional neural networks learned. By mapping the feature map of each layer back to the input picture using the "Deconvolutional Network," we can view the learned feature distribution, as shown in Figure 10-32. It can be observed that the features of the second layer correspond to the extraction of the underlying images such as edges, corners, and colors; the third layer starts to capture the middle features of texture; the fourth and fifth layers present some features of the object, such as puppy faces, bird's feet, and other high-level features. Through these visualizations, we can experience the feature learning process of the convolutional neural network to a certain extent.

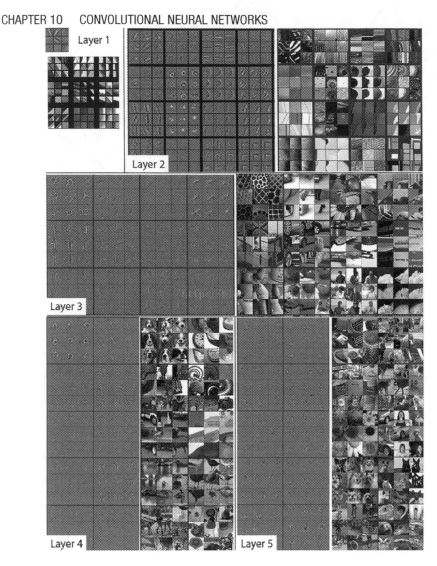

Figure 10-32. *Visualization of convolutional neural network features [5]*

The image recognition process is generally considered to be a representation learning process. Starting from the original pixel features received, it gradually extracts low-level features such as edges and corners, then mid-level features such as textures, and then high-level features such as object parts. The last network layer learns classification logic based on these learned abstract feature representations. The higher the layer and the more accurate the learned features, the more favorable the classification of the classifier is, thereby obtaining better performance. From the perspective of representation learning, convolutional neural networks extract features layer by layer, and the process of network training can be considered as a feature learning process. Based on the learned high-level abstract features, classification tasks can be conveniently performed.

Applying the idea of representation learning, a well-trained convolutional neural network can often learn better features. This feature extraction method is generally universal. For example, learning the representation of head, foot, body, and other characteristics on cat and dog tasks can also be used to some extent on other animals. Based on this idea, the first few feature extraction layers of the deep neural network trained on task A can be migrated to task B, and only the classification logic of task B (represented as the last layer of the network) needs to be trained. This method is a type of transfer learning, also known as fine-tuning.

10.6 Gradient Propagation

After completing the handwritten digital image recognition exercise, we have a preliminary understanding of the use of convolutional neural networks. Now let's solve a key problem. The convolutional layer implements discrete convolution operations by moving the receptive field. So how does its gradient propagation work?

Consider a simple case where the input is a 3×3 single-channel matrix, and a 2×2 convolution kernel is used to perform the convolution operation. We then calculate the error between the flattened output and the corresponding label, as shown in Figure 10-33. Let's discuss the gradient update method for this case.

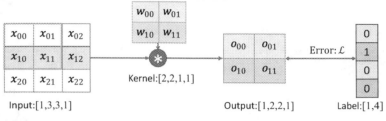

Input:[1,3,3,1]   Output:[1,2,2,1]   Label:[1,4]

Figure 10-33. *Gradient propagation example for the convolutional layer*

First derive the expression of the output tensor O:

$$o_{00} = x_{00}w_{00} + x_{01}w_{01} + x_{10}w_{10} + x_{11}w_{11} + b$$

$$o_{01} = x_{01}w_{00} + x_{02}w_{01} + x_{11}w_{10} + x_{12}w_{11} + b$$

$$o_{10} = x_{10}w_{00} + x_{11}w_{01} + x_{20}w_{10} + x_{21}w_{11} + b$$

$$o_{11} = x_{11}w_{00} + x_{12}w_{01} + x_{21}w_{10} + x_{22}w_{11} + b$$

Taking w_{00} gradient calculation as an example, decompose by chain rule:

$$\frac{\partial L}{\partial w_{00}} = \sum_{i \in \{00,01,10,11\}} \frac{\partial L}{\partial o_i} \frac{\partial o_i}{\partial w_{00}}$$

where $\dfrac{\partial L}{\partial O_i}$ can be directly derived from the error function. Let's

consider $\dfrac{\partial O_i}{\partial w_i}$:

$$\frac{\partial o_{00}}{\partial w_{00}} = \frac{\partial \left(x_{00}w_{00} + x_{01}w_{01} + x_{10}w_{10} + x_{11}w_{11} + b \right)}{w_{00}} = x_{00}$$

Similarly, one can derive:

$$\frac{\partial o_{01}}{\partial w_{00}} = \frac{\partial \left(x_{01}w_{00} + x_{02}w_{01} + x_{11}w_{10} + x_{12}w_{11} + b \right)}{w_{00}} = x_{01}$$

406

$$\frac{\partial o_{10}}{\partial w_{00}} = \frac{\partial \left(x_{10}w_{00} + x_{11}w_{01} + x_{20}w_{10} + x_{21}w_{11} + b \right)}{w_{00}} = x_{10}$$

$$\frac{\partial o_{11}}{\partial w_{00}} = \frac{\partial \left(x_{11}w_{00} + x_{12}w_{01} + x_{21}w_{10} + x_{22}w_{11} + b \right)}{w_{00}} = x_{11}$$

It can be observed that the method of cyclically moving the receptive field does not change the derivatization of the network layer, and the derivation of the gradient is not complicated. But when the number of network layers increases, the artificial gradient derivation will become very cumbersome. But don't worry, the deep learning framework can help us automatically complete the gradient calculation and update of all parameters, we only need to design the network structure.

10.7 Pooling Layer

In the convolutional layer, the height and width of the feature map can be reduced by adjusting the stride size parameter s, thereby reducing the amount of network parameters. In fact, in addition to setting the stride size, there is a special network layer that can reduce the parameter amount as well, which is known as the pooling layer.

The pooling layer is also based on the idea of local correlation. By sampling or aggregating information from a group of locally related elements, we can obtain new element values. In particular, the max pooling layer selects the largest element value from the local related element set, and the average pooling layer calculates the average value from the local related element set. Taking a 5×5 max pooling layer as an example, suppose the receptive field window size $k = 2$ and stride $s = 1$, as shown in Figure 10-34. The green dotted box represents the position of the first receptive field, and the set of receptive field elements is:

$$\{1,-1,-1,-2\}$$

According to max pooling, we have:

$$x' = max\big(\{1,-1,-1,-2\}\big) = 1$$

If the average pooling operation is used, the output value would be:

$$x' = avg\big(\{1,-1,-1,-2\}\big) = -0.75$$

After calculating the receptive field of the current position, similar to the calculation step of the convolutional layer, the receptive field is moved to the right by several units according to the stride size. The output becomes:

$$x' = max\big(-1,0,-2,2\big) = 2$$

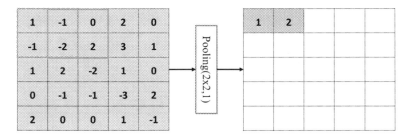

Figure 10-34. *Max pooling example-1*

In the same way, gradually move the receptive field window to the far right and calculate the output $x' = max(2,0,3,1) = 1$. At this time, the window has reached the input edge. The receptive field window moves down by one stride and returns to the beginning of the line, as shown in Figure 10-35.

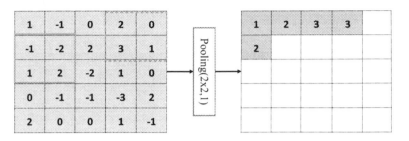

Figure 10-35. *Max pooling example-2*

Loop back and forth until we reach the bottom and right, we get the output of the max pooling layer as shown in Figure 10-36. The length and width are slightly smaller than the input height and width.

1	-1	0	2	0
-1	-2	2	3	1
1	2	-2	1	0
0	-1	-1	-3	2
2	0	0	1	-1

Pooling(2x2,1)

1	2	3	3	
2	2	3	3	
2	2	1	2	
2	0	1	2	

Figure 10-36. *Max pooling example-3*

Because the pooling layer has no parameters to learn, the calculation is simple, and the size of the feature map can be effectively reduced; it is widely used in computer vision-related tasks.

By carefully designing the height, width k, and stride parameter s of the receptive field of the pooling layer, various dimensionality reduction operations can be realized. For example, a common pooling layer setting is $k = 2$, $s = 2$, which can achieve the purpose of outputting only half of the input height and width. As shown in Figure 10-37 and Figure 10-38, the receptive field $k = 3$, stride size $s = 2$, input X has height and width 5×5, but the output only has height and width 2×2.

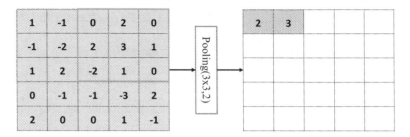

Figure 10-37. *Pooling layer example (half size output)-1*

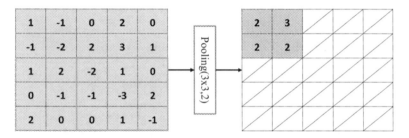

Figure 10-38. *Pooling layer example (half size output)-2*

10.8 BatchNorm Layer

With the advent of convolutional neural networks, the amount of network parameters has been greatly reduced, making it possible for deep networks with dozens of layers. However, before the emergence of the residual network, the increasing number of neural network layers makes the training very unstable, and sometimes the network does not update or even does not converge for a long time. At the same time, the network is more sensitive to hyperparameters, and the slight change of hyperparameters will change training trajectory of the network completely.

In 2015, Google researcher Sergey Ioffe et al. proposed a method of parameter normalization and designed the Batch Normalization (abbreviated as BatchNorm, or BN) layer [6]. The proposal of the BN layer makes the setting of network hyperparameters more free, such as a larger learning rate, and more random network initialization. In the meantime, the network has a faster convergence speed and better performance. After the BN layer was proposed, it was widely used in various deep network models. The convolutional layer, BN layer, ReLU layer, and pooling layer once became the standard unit blocks of network models. The stacking Conv-BN-ReLU-Pooling method often generates good model performance.

Why do we need to normalize the data in the network? It is difficult to explain this problem thoroughly from a theoretical level, even the explanation given by the author of the BN layer may not convince everyone. Rather than entangle the reasons, it is better to experience the benefits of data normalization through specific questions.

Consider the Sigmoid activation function and its gradient distribution. As shown in Figure 10-39, the derivative value of the Sigmoid function in the interval $x \in [-2, 2]$ is distributed in the interval $[0.1, 0.25]$. When $x > 2$ or $x < -2$, the derivative of the Sigmoid function becomes very small, approaching 0, which is prone to gradient dispersion. In order to avoid the gradient dispersion phenomenon of the Sigmoid function due to too large or too small input, it is very important to normalize the function input to a small interval near 0. It can be seen from Figure 10-39 that after normalization, the value is mapped near 0, and the derivative value here is not too small, so that gradient dispersion is not easy to appear. This is an example of the benefit of normalization.

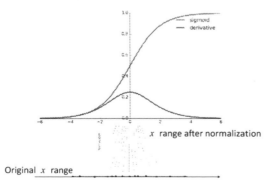

***Figure 10-39.** Sigmoid function and its derivative*

Let's look at another example. Consider a linear model with two input nodes, as shown in Figure 10-40(a):

$$L = a = x_1 w_1 + x_2 w_2 + b$$

Discuss the optimization problems under the following two input distributions:

- $x_1 \in [1, 10], \ x_2 \in [1, 10]$

- $x_1 \in [1, 10], \ x_2 \in [100, 1000]$

Because the model is relatively simple, two types of contour maps of the loss function can be drawn. Figure 10-40(b) is a schematic diagram of an optimized trajectory when $x_1 \in [1, 10]$ *and* $x_2 \in [100, 1000]$, and Figure 10-40(c) is a schematic diagram of an optimized trajectory when $x_1 \in [1, 10]$ *and* $x_2 \in [1, 10]$. The center of the ring in the figure is the global extreme point.

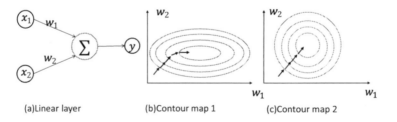

(a)Linear layer   (b)Contour map 1   (c)Contour map 2

Figure 10-40. *An example of data normalization*

Consider:

$$\frac{\partial L}{\partial w_1} = x_1$$

$$\frac{\partial L}{\partial w_2} = x_2$$

When the input distributions are similar, and the partial derivative values are the same, the optimized trajectory of the function is shown in Figure 10-40(c); when the input distributions differ greatly, for example $x_1 \ll x_2$,

$$\frac{\partial L}{\partial w_1} \ll \frac{\partial L}{\partial w_2}$$

The equipotential line of the loss function is steeper on the axis, and a possible optimization trajectory is shown in Figure 10-40(b). Comparing the two optimized trajectories, it can be observed that when the distributions of x_1 and x_2 are similar, the convergence in Figure 10-40(c) is faster and the optimized trajectory is more ideal.

Through the preceding two examples, we can empirically conclude: when the network layer input distribution is similar, and the distribution is in a small range (such as near 0), it favors the function optimization more. So how to ensure that the input distribution is similar? Data normalization can achieve this purpose, and data can be mapped to:

$$\hat{x} = \frac{x - \mu_r}{\sqrt{\sigma_r^2 + \dot{o}}}$$

where μ_r is the mean and σ_r^2 the variance of all data, ϵ is a small number, such as $1e-8$.

In the batch-based training phase, how to obtain all the input statistics μ_r and σ_r^2 of each network layer? Consider the mean μ_B and variance σ_B^2 within the batch:

$$\mu_B = \frac{1}{m} \sum_{i=1}^{m} x_i$$

$$\sigma_B^2 = \frac{1}{m} \sum_{i=1}^{m} (x_i - \mu_B)^2$$

It can be regarded as approximate of μ_r *and* σ_r^2, where m is the number of batch samples. Therefore, in the training phase, through normalization:

$$\hat{x}_{train} = \frac{x_{train} - \mu_B}{\sqrt{\sigma_B^2 + \epsilon}}$$

and approximate the overall mean μ_r and variance σ_r^2 using each batch's mean μ_B and variance σ_B^2.

In the test phase, we can normalize the test data using:

$$\hat{x}_{test} = \frac{x_{test} - \mu_r}{\sqrt{\sigma_r^2 + \epsilon}}$$

The preceding operation does not introduce additional variables to be optimized, and the mean and variance are obtained through existing data, and do not need to participate in gradient update. In fact, in order to improve the expressive ability of the BN layer, the author of the BN layer introduced the "scale and shift" technique to map and transform the variables again:

$$\tilde{x} = \hat{x} \cdot \gamma + \beta$$

where the parameter γ scales the normalized variable again, and the parameter β realizes the translation operation. The difference is that the parameters γ *and* β are automatically optimized by the backpropagation algorithm to achieve the purpose of scaling and panning data distribution "on demand" at the network layer.

Let's learn how to implement the BN layer in TensorFlow.

10.8.1 Forward Propagation

We denote the input of the BN layer as x and the output as $\tilde{x}$. The forward propagation process is discussed in training phase and testing phase.

Training phase: first calculate the current batch's mean μ_B and variance σ_B^2, and then normalize the data according to:

$$\tilde{x}_{train} = \frac{x_{train} - \mu_B}{\sqrt{\sigma_B^2 + \epsilon}} \cdot \gamma + \beta$$

We then use:

$$\mu_r \leftarrow momentum \cdot \mu_r + (1 - momentum) \cdot \mu_B$$

$$\sigma_r^2 \leftarrow momentum \cdot \sigma_r^2 + (1 - momentum) \cdot \sigma_B^2$$

to iteratively update the statistical values μ_r and σ_r^2 of the global training data, where momentum is a hyperparameter that needs to be set to balance the update amplitude: when *momentum* = 0, μ_r and σ_r^2 are directly set as μ_B and σ_B^2 of the latest batch; when *momentum* = 1, μ_r and σ_r^2 remain unchanged. In TensorFlow, momentum is set to 0.99 by default.

Test phase: the BN layer uses

$$\tilde{x}_{test} = \frac{x_{test} - \mu_r}{\sqrt{\sigma_r^2 + \epsilon}} * \gamma + \beta$$

to calculate $\tilde{x}_{test}$, where μ_r, σ_r^2, γ, β come from the statistics or optimization results of the training phase, and are used directly in the test phase, and these parameters are not updated.

10.8.2 Backward Propagation

In the backward update phase, the back propagation algorithm solves the gradients $\frac{\partial L}{\partial \gamma}$ and $\frac{\partial L}{\partial \beta}$ of the loss function and automatically optimizes the parameters *γ and β*according to the gradient update rule.

It should be noted that for 2D feature map input X: $[b, h, w, c]$, the BN layer does not calculate μ_B *and* σ_B^2 of every point; instead, it calculates μ_B *and* σ_B^2 on each channel on the channel axis c, so μ_B *and* σ_B^2 are the mean and variance of all other dimensions on each channel. Taking the input of shape [100,32,32,3] as an example, the mean value on the channel axis c is calculated as follows:

```
In [7]:
x=tf.random.normal([100,32,32,3])
# Combine other dimensions except the channel dimension
x=tf.reshape(x,[-1,3])
# Calculate mean
ub=tf.reduce_mean(x,axis=0)
ub
Out[7]:
<tf.Tensor: id=62, shape=(3,), dtype=float32,
numpy=array([-0.00222636, -0.00049868, -0.00180082],
dtype=float32)>
```

The has c channels, so c averaged values are generated.

In addition to the method of statistical data on the axis c, we can also easily extend the method to other dimensions, as shown in Figure 10-41:

- Layer Norm:Calculate the mean and variance of all features of each sample.

- Instance Norm:Calculate the mean and variance of features on each channel of each sample.

- Group Norm:Divide c channel into several groups, and count the feature mean and variance in the channel group of each sample.

The normalization method mentioned previously is proposed by several independent papers, and it has been verified that it is equivalent or superior to the BatchNorm algorithm in some applications. It can be seen that the research of deep learning algorithms is not difficult. As long as you think more and practice your engineering ability, everyone will have the opportunity to publish innovative results.

416

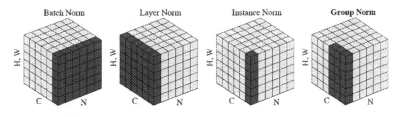

Figure 10-41. *Different normalization illustration [7]*

10.8.3 Implementation of BatchNormalization layer

In TensorFlow, the BN layer can be easily implemented through the layers. BatchNormalization() class:

```
# Create BN layer
layer=layers.BatchNormalization()
```

Different from the fully connected layer and the convolutional layer, the behavior of the BN layer in the training phase and the test phase is different. It is necessary to distinguish the training mode from the test mode by setting the training flag.

Take the network model of LeNet-5 as an example, add the BN layer after the convolutional layer; the code is as follows:

```
network = Sequential([
    layers.Conv2D(6,kernel_size=3,strides=1),
    # Insert BN layer
    layers.BatchNormalization(),
    layers.MaxPooling2D(pool_size=2,strides=2),
    layers.ReLU(),
    layers.Conv2D(16,kernel_size=3,strides=1),
    # Insert BN layer
    layers.BatchNormalization(),
    layers.MaxPooling2D(pool_size=2,strides=2),
```

```
    layers.ReLU(),
    layers.Flatten(),
    layers.Dense(120, activation='relu'),
    layers.Dense(84, activation='relu'),
    layers.Dense(10)
                ])
```

In the training phase, you need to set the network parameter training=True to distinguish whether the BN layer is a training or testing model. The code is as follows:

```
with tf.GradientTape() as tape:
    # Insert channel dimension
    x = tf.expand_dims(x,axis=3)
    # Forward calculation, [b, 784] => [b, 10]
    out = network(x, training=True)
```

In the testing phase, you need to set training=False to avoid wrong behavior in the BN layer. The code is as follows:

```
for x,y in db_test:
    # Insert channel dimension
    x = tf.expand_dims(x,axis=3)
    # Forward calculation
    out = network(x, training=False)
```

10.9 Classical Convolutional Network

Since the introduction of AlexNet [3] in 2012, a variety of deep convolutional neural network models have been proposed, among which the more representative ones are the VGG series [8], the GoogLeNet series [9], the ResNet series [10], and the DenseNet series [11]. The overall trend of their network layers is gradually increasing. Take the classification performance of the network model on the ImageNet dataset of the ILSVRC challenge as an example. As shown in Figure 10-42, the network models before the emergence of AlexNet were all shallow neural networks, and the Top-5 error rate was above 25%. The AlexNet 8-layer deep neural network reduced the Top-5 error rate to 16.4%, and the performance was greatly improved. The subsequent VGG and GoogleNet models continued to reduce the error rate to 6.7%; the emergence of ResNet increased the number of network layers to 152 layers for the first time. The error rate is also reduced to 3.57%.

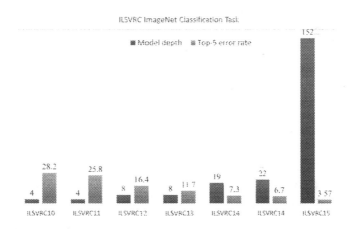

Figure 10-42. *Model performance on classification tasks of ImageNet dataset*

This section will focus on the characteristics of these network models.

10.9.1 AlexNet

In 2012, Alex Krizhevsky, the champion of the ImageNet dataset classification task of the ILSVRC12 challenge, proposed an eight-layer deep neural network model AlexNet, which receives the input size of 224 × 224 color image data and gets the probability distribution of 1000 categories after five convolutional layers and three fully connected layers. In order

to reduce the dimensionality of the feature map, AlexNet added the Max
Pooling layer after the first, second, and fifth convolutional layers. As
shown in Figure 10-43, the number of parameters of the network reached
60 million. In order to train the model on NVIDIA GTX 580 GPU (3GB GPU
memory) at the time, Alex Krizhevsky disassembled the convolutional
layer and the first two fully connected layers on two GPUs for training
separately, and merged the last layer into one GPU to do backward update.
AlexNet achieved a Top-5 error rate of 15.3% in ImageNet, which is 10.9%
lower than the second place.

The innovations of AlexNet are:

- The number of layers has reached eight.

- Uses the ReLU activation function. Most of previous
  neural networks use the Sigmoid activation function,
  which is relatively complicated to calculate and is
  prone to gradient dispersion.

- Introduces the Dropout layer. Dropout improves
  the generalization ability of the model and prevents
  overfitting.

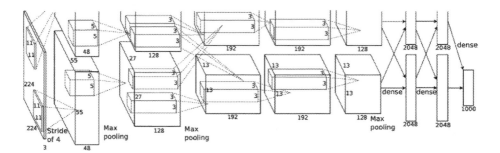

Figure 10-43. *AlexNet architecture [3]*

10.9.2 VGG Series

The superior performance of the AlexNet model has inspired the industry to move in the direction of deeper network models. In 2014, the runner-up of the ImageNet classification task of the ILSVRC14 challenge, the VGG Lab of the University of Oxford, proposed a series of network models such as VGG11, VGG13, VGG16, and VGG19 (Figure 10-45), and increased the network depth to up to 19 layers [8]. Take VGG16 as an example, it accepts color picture data with size of 224 × 224, and then passes through 2 Conv-Conv-Pooling units and 3 Conv-Conv-Conv-Pooling units, and finally outputs the probability of current picture belonging to 1000 categories through a 3 fully connected layers as shown in Figure 10-44. VGG16 achieved a Top-5 error rate of 7.4% on ImageNet, which is 7.9% lower than AlexNet's error rate.

The innovations of the VGG series network are:

- The number of layers is increased to 19.

- Uses a smaller 3x3 convolution kernel, which has fewer parameters and lower computational cost compared to the 7x7 convolution kernel in AlexNet.

- Uses a smaller pooling layer window 2 × 2 and stride size $s = 2$, while $s = 2$and pooling window is 3x3 in AlexNet.

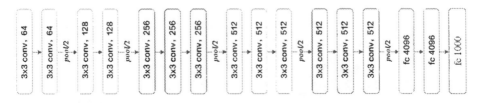

Figure 10-44. *VGG16 architecture*

ConvNet Configuration					
A	A-LRN	B	C	D	E
11 weight layers	11 weight layers	13 weight layers	16 weight layers	16 weight layers	19 weight layers
input (224 × 224 RGB image)					
conv3-64	conv3-64 **LRN**	conv3-64 **conv3-64**	conv3-64 conv3-64	conv3-64 conv3-64	conv3-64 conv3-64
maxpool					
conv3-128	conv3-128	conv3-128 **conv3-128**	conv3-128 conv3-128	conv3-128 conv3-128	conv3-128 conv3-128
maxpool					
conv3-256 conv3-256	conv3-256 conv3-256	conv3-256 conv3-256	conv3-256 conv3-256 **conv1-256**	conv3-256 conv3-256 **conv3-256**	conv3-256 conv3-256 conv3-256 **conv3-256**
maxpool					
conv3-512 conv3-512	conv3-512 conv3-512	conv3-512 conv3-512	conv3-512 conv3-512 **conv1-512**	conv3-512 conv3-512 **conv3-512**	conv3-512 conv3-512 conv3-512 **conv3-512**
maxpool					
conv3-512 conv3-512	conv3-512 conv3-512	conv3-512 conv3-512	conv3-512 conv3-512 **conv1-512**	conv3-512 conv3-512 **conv3-512**	conv3-512 conv3-512 conv3-512 **conv3-512**
maxpool					
FC-4096					
FC-4096					
FC-1000					
soft-max					

Figure 10-45. *VGG series network architecture [8]*

10.9.3 GoogLeNet

The number of 3x3 convolution kernel has less parameters, the computational cost is lower, and the performance is even better. Therefore, the industry began to explore the smallest convolution kernel: the 1x1 convolution kernel. As shown in Figure 10-46, the input is a three-channel 5x5 picture, and the convolution operation is performed with a single 1x1 convolution kernel. The data of each channel is calculated with the convolution kernel of the corresponding channel to obtain the intermediate matrix of the three channels, and the corresponding positions are added to get the final output tensor. For the input shape of

$[b, h, w, c_{in}]$, the output of the 1x1 convolutional layer is $[b, h, w, c_{out}]$, where c_{in} is the number of channels of input data, c_{out} is the number of channels of output data, and is also the number of 1x1 convolution kernels. A special feature of the 1x1 convolution kernel is that it can only transform the number of channels without changing the width and height of the feature map.

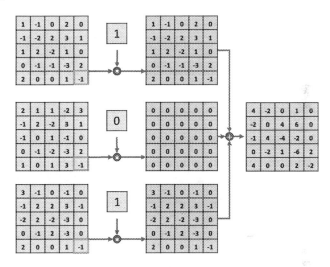

Figure 10-46. *1 × 1 convolutional kernel example*

In 2014, Google, the champion of the ILSVRC14 challenge, proposed a large number of network models using 3x3 and 1x1 convolution kernels: GoogLeNet, with a network layer number of 22 [9]. Although the number of layers of GoogLeNet is much larger than that of AlexNet, its parameter amount is only half of AlexNet, and its performance is much better than AlexNet. On the ImageNet dataset classification task, GoogLeNet achieved a Top-5 error rate of 6.7%, which is 0.7% lower than VGG16 in error rate.

The GoogLeNet network adopts the idea of modular design and forms a complex network structure by stacking a large number of Inception modules. As shown in Figure 10-47, the input of the Inception module is

X, and then passes through four sub-networks, and finally are spliced and merged on the channel axis to form the output of the Inception module. The four sub-networks are:

- 1 × 1 convolutional layer.

- 1 × 1 convolutional layer, and then through a 3x3 convolutional layer.

- 1 × 1 convolutional layer, and then through a 5x5 convolutional layer.

- 3 × 3 maximum pooling layer, and then through the 1x1 convolutional layer.

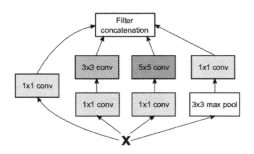

Figure 10-47. *Inception module*

The network structure of GoogLeNet is shown in Figure 10-48. The network structure in the red box is the network structure in Figure 10-47.

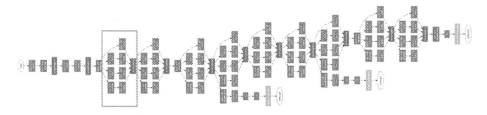

Figure 10-48. *GoogLeNet architecture [9]*

10.10 Hands-On CIFAR10 and VGG13

MNIST is one of the most commonly used datasets for machine learning, but because handwritten digital pictures are very simple, and the MNIST dataset only saves image gray information, it is not suitable for inputting a network model designed as RGB three-channel. This section will introduce another classic image classification dataset: CIFAR10.

The CIFAR10 dataset was released by Canadian Institute for Advanced Research. It contains color pictures of ten categories of objects such as airplanes, cars, birds, and cats. Each category has collected 6,000 large and small pictures, totaling 60,000 pictures. Among them, 50,000 sheets are used as training datasets, and 10,000 sheets are used as test datasets. Each type of sample is shown in Figure 10-49.

Figure 10-49. *CIFAR10 Data Set[1]*

---

[1] Image source: `www.cs.toronto.edu/~kriz/cifar.html`

In TensorFlow, similarly, there is no need to manually download, parse, and load the CIFAR10 dataset. The training set and test set can be directly loaded through the datasets.cifar10.load_data() function. For example,

```
# Load CIFAR10 data set
(x,y), (x_test, y_test) = datasets.cifar10.load_data()
# Delete one dimension of y, [b,1] => [b]
y = tf.squeeze(y, axis=1)
y_test = tf.squeeze(y_test, axis=1)
# Print the shape of training and testing sets
print(x.shape, y.shape, x_test.shape, y_test.shape)
# Create training set and preprocess
train_db = tf.data.Dataset.from_tensor_slices((x,y))
train_db = train_db.shuffle(1000).map(preprocess).batch(128)
# Create testing set and preprocess
test_db = tf.data.Dataset.from_tensor_slices((x_test,y_test))
test_db = test_db.map(preprocess).batch(128)
# Select a Batch
sample = next(iter(train_db))
print('sample:', sample[0].shape, sample[1].shape,
      tf.reduce_min(sample[0]), tf.reduce_max(sample[0]))
```

TensorFlow will automatically download the dataset to the path C:\Users\username\.keras\datasets, and users can view it, or manually delete the unnecessary dataset cache. After the preceding code runs, the shape of X and y in the training set is (50000, 32, 32, 3) and (50000), and the shape of X and y in the test set is (10000, 32, 32, 3) and (10000), which indicates the size of the picture is 32 × 32, those are color pictures, the number of samples in the training set is 50,000, and the number of samples in the test set is 10,000.

CIFAR10 image recognition task is not simple. This is mainly due to the fact that the image content of CIFAR10 requires a lot of details to be presented, and the resolution of the saved images is only 32 × 32, which makes the subject information blurry and even difficult for human eyes to distinguish. The expression ability of shallow neural networks is limited and is difficult to reach better performance. In this section, we will modify the VGG13 network structure according to the characteristics of our data set to complete CIFAR10 image recognition as follows:

- Adjust the network input to 32 × 32. The original network input is 224 × 224, resulting in too large input feature dimensions and too large network parameters.

- The dimensions of the three fully connected layers are [256,64,10] for the setting of ten classification tasks.

Figure 10-50 is the adjusted VGG13 network structure, which we collectively call the VGG13 network model.

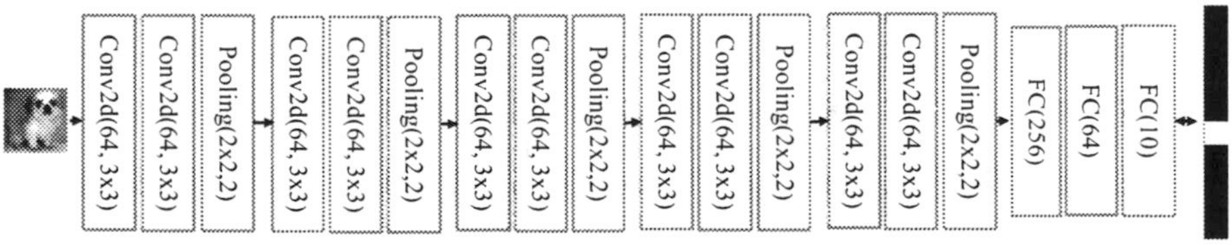

Figure 10-50. *Adjusted VGG13 model structure*

We implement the network as two sub-networks: convolutional sub-network and fully connected sub-network. The convolution sub-network is composed of five sub-modules, each of which contains the Conv-Conv-MaxPooling unit structure. The code is as follows:

```
conv_layers = [
    # Conv-Conv-Pooling unit 1
    # 64 3x3 convolutional kernels with same input and
    output size
    layers.Conv2D(64, kernel_size=[3, 3], padding="same",
    activation=tf.nn.relu),
```

```
layers.Conv2D(64, kernel_size=[3, 3], padding="same",
activation=tf.nn.relu),
# Reduce the width and height size to half of its original
layers.MaxPool2D(pool_size=[2, 2], strides=2,
padding='same'),

# Conv-Conv-Pooling unit 2, output channel increases to
128, half width and height
layers.Conv2D(128, kernel_size=[3, 3], padding="same",
activation=tf.nn.relu),
layers.Conv2D(128, kernel_size=[3, 3], padding="same",
activation=tf.nn.relu),
layers.MaxPool2D(pool_size=[2, 2], strides=2,
padding='same'),

# Conv-Conv-Pooling unit 3, output channel increases to
256, half width and height

layers.Conv2D(256, kernel_size=[3, 3], padding="same",
activation=tf.nn.relu),
layers.Conv2D(256, kernel_size=[3, 3], padding="same",
activation=tf.nn.relu),
layers.MaxPool2D(pool_size=[2, 2], strides=2,
padding='same'),

# Conv-Conv-Pooling unit 4, output channel increases to
512, half width and height
layers.Conv2D(512, kernel_size=[3, 3], padding="same",
activation=tf.nn.relu),
layers.Conv2D(512, kernel_size=[3, 3], padding="same",
activation=tf.nn.relu),
layers.MaxPool2D(pool_size=[2, 2], strides=2,
padding='same'),
```

```
    # Conv-Conv-Pooling unit 5, output channel increases to
    512, half width and height
    layers.Conv2D(512, kernel_size=[3, 3], padding="same",
    activation=tf.nn.relu),
    layers.Conv2D(512, kernel_size=[3, 3], padding="same",
    activation=tf.nn.relu),
    layers.MaxPool2D(pool_size=[2, 2], strides=2,
    padding='same')
]
conv_net = Sequential(conv_layers)
```

The fully connected sub-network contains three fully connected layers, each layer adds a ReLU nonlinear activation function, except for the last layer. The code is shown as follows:

```
# Create 3 fully connected layer sub-network
fc_net = Sequential([
    layers.Dense(256, activation=tf.nn.relu),
    layers.Dense(128, activation=tf.nn.relu),
    layers.Dense(10, activation=None),
])
```

After the subnet is created, use the following code to view the parameters of the network:

```
# build network and print parameter info
conv_net.build(input_shape=[4, 32, 32, 3])
fc_net.build(input_shape=[4, 512])
conv_net.summary()
fc_net.summary()
```

The total number of parameters of the convolutional network is about 940,000, the total number of parameters of the fully connected network is about 177,000, and the total number of parameters of the network is about 950, 000, which is much less than the original version of VGG13.

Since we implemented the network as two sub-networks, when performing gradient update, it is necessary to merge the parameter of the two sub-networks as in the following:

```
# merge parameters of two sub-networks
variables = conv_net.trainable_variables + fc_net.trainable_
variables
# calculate gradient for all parameters
grads = tape.gradient(loss, variables)
# update gradients
optimizer.apply_gradients(zip(grads, variables))
```

Run the cifar10_train.py file to start training the model. After training 50 Epochs, the test accuracy of the network reached 77.5%.

10.11 Convolutional Layer Variants

The research of convolutional neural networks has produced a variety of excellent network models, and various variants of convolutional layers have been proposed. This section will focus on several typical convolutional layer variants.

10.11.1 Dilated/Atrous Convolution

In order to reduce the number of parameters of the network, the design of the convolution kernel usually chooses a smaller 1×1 and 3×3 receptive field size. The small convolution kernel makes the network's receptive field area limited when extracting features, but increasing the receptive field area will increase the amount of network parameters and computational costs, so it is necessary to weigh the design.

Dilated/Atrous Convolution is a better solution to this problem. Dilated/Atrous Convolution adds a dilation rate parameter to the receptive field of ordinary convolution to control the sampling step size of the receptive field area, as shown in Figure 10-51. When the sampling step dilation rate of the receptive field is 1, the distance between the sampling points of each receptive field is 1, and the dilated convolution at this time degenerates to ordinary convolution; when the dilation rate is 2, one point is sampled every two units in the receptive field. As shown in the green grid in the green box in the middle of Figure 10-51, the distance between each sampling grid is 2. Similarly, the dilation rate on the right side of Figure 10-51 is 3, and the sampling step is 3. Although the increase in dilation rate will increase the area of the receptive field, the actual number of points involved in the calculation remains unchanged.

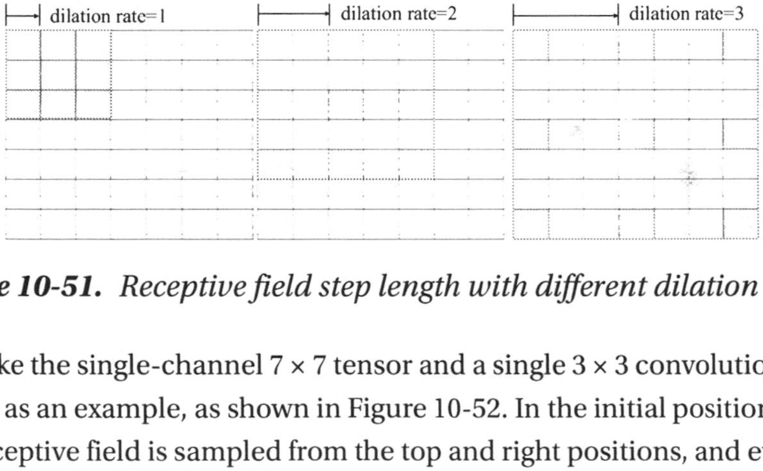

Figure 10-51. *Receptive field step length with different dilation rate*

Take the single-channel 7 × 7 tensor and a single 3 × 3 convolution kernel as an example, as shown in Figure 10-52. In the initial position, the receptive field is sampled from the top and right positions, and every other point is sampled. A total of 9 data points are collected, as shown in the green box in Figure 10-52. These 9 data points are multiplied by the convolution kernel and written into the corresponding position of the output tensor.

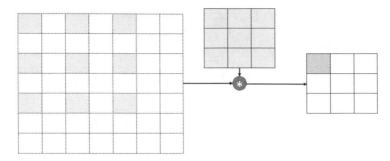

Figure 10-52. *Dilated convolution sample-1*

The convolution kernel window moves one unit to the right according to the step size $s = 1$, as shown in Figure 10-53. The same interval sampling is carried out. A total of 9 data points are sampled. The multiplication and accumulation operation is completed with the convolution kernel, and the output tensor is written to corresponding position until the convolution kernel moves to the bottom and rightmost position. It should be noted that the moving step size s of the convolution kernel window and the sampling step size dilation rate of the receptive field region are different concepts.

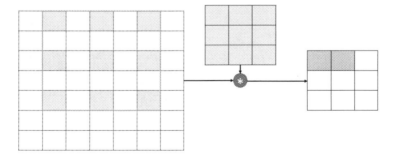

Figure 10-53. *Dilated convolution sample-2*

Dilated convolution provides a larger receptive field window without increasing network parameters. However, when setting up a network model using hollow convolution, the dilation rate parameter needs to be carefully designed to avoid grid effects. At the same time, a larger dilation rate parameter is not conducive to tasks such as small object detection and semantic segmentation.

In TensorFlow, you can choose to use normal convolution or dilated convolution by setting the dilation_rate parameter of the layers.Conv2D() class. For example

```
In [8]:
x = tf.random.normal([1,7,7,1]) # Input
# Dilated convolution, 1 3x3 kernel
layer = layers.Conv2D(1,kernel_
size=3,strides=1,dilation_rate=2)
out = layer(x) # forward calculation
out.shape
Out[8]: TensorShape([1, 3, 3, 1])
```

When the dilation_rate parameter is set to the default value 1, the normal convolution method is used for calculation; when the dilation_rate parameter is greater than 1, the dilated convolution method is sampled for calculation.

10.11.2 Transposed Convolution

Transposed convolution (or fractionally strided convolution, sometimes it is also called deconvolution). In fact, deconvolution is mathematically defined as the inverse process of convolution, but transposed convolution cannot recover the input of the original convolution, so it is not appropriate to call it deconvolution) by filling a large amount of padding between the inputs to achieve the effect that the output height and width are greater than the input height and width, so as to achieve the purpose of upsampling, as shown in Figure 10-54. We first introduce the calculation process of transposed convolution, and then introduce the relationship between transposed convolution and ordinary convolution.

To simplify the discussion, we only discuss the input with $h = w$, that is, the case where the input height and width are equal.

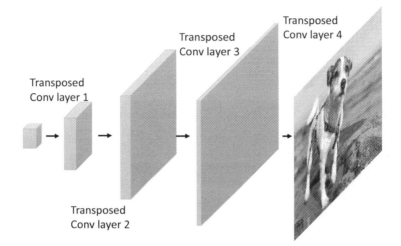

Figure 10-54. *Transposed convolution for upsampling*

$o + 2p - k = n * s$

Consider the following example: the single-channel feature map has 2×2 input, and the transposed convolution kernel is 3×3, $s = 2$, and padding $p = 0$. First, evenly insert $s - 1$ blank data points between the input data points, the resulting matrix is 3×3, as shown in the second matrix in Figure 10-55. Filling the corresponding rows/columns around the 3×3 matrix according to the filling amount $k - p - 1 = 3 - 0 - 1 = 2$. At this time, the height and width of the input tensor are 7×7, as shown in the third matrix in Figure 10-55.

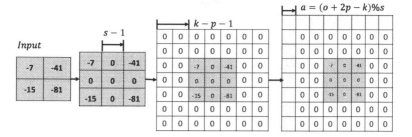

Figure 10-55. *Input and padding example*

On the 7 × 7 input tensor, apply the 3 × 3 convolution kernel operations with stride size $s' = 1$ and padding $p = 0$ (note that the step size s' of the ordinary convolution at this stage is always 1, which is different from the step size s of the transposed convolution). According to the ordinary convolution calculation formula, the output size is:

$$o = \frac{i + 2 * p - k}{s'} + 1 = \frac{7 + 2 * 0 - 3}{1} + 1 = 5$$

It means 5 × 5 output size. We directly follow this calculation process to give the final transposed convolution output and input relationship. When $o + 2p - k$ is a multiple of s, the relationship is satisfied $o = (i - 1)s + k - 2p$

Transposed convolution is not the inverse process of ordinary convolution, but there is a certain connection between the two, and transposed convolution is also implemented based on ordinary convolution. Under the same setting, the input x is obtained after the ordinary convolution operation $o = Conv(x)$, and sending o to the transposed convolution operation gives $x' = ConvTranspose(o)$, where $x' \neq x$, but with same shape. We can use ordinary convolution operations with input as 5 × 5, stride size $s = 2$, padding $p = 0$, and 3 × 3 convolution kernel to verify the demonstration, as shown in Figure 10-56.

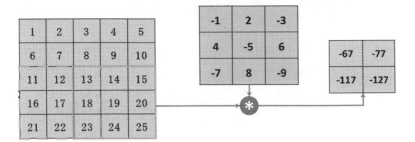

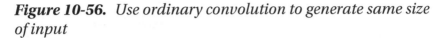

Figure 10-56. *Use ordinary convolution to generate same size of input*

It can be seen that the output with size 5 × 5 of the transposed convolution is sent to the ordinary convolution under the same set conditions, and the output of size 2 × 2 can be obtained. This size is exactly the input size of the transposed convolution. At the same time, we also observe that the output matrix is not exactly the input matrix fed into the transposed convolution. Transposed convolution and ordinary convolution are not mutually inverse processes and cannot recover the input content of the other party, but can only recover tensors of equal size. Therefore, it is not appropriate to call it deconvolution.

Based on TensorFlow to implement the transposed convolution operation of the preceding example, the code is as follows:

```
In [8]:
# Create matrix X with size 5x5
x = tf.range(25)+1
# Reshape X to certain shape
x = tf.reshape(x,[1,5,5,1])
x = tf.cast(x, tf.float32)
# Create constant matrix
w = tf.constant([[-1,2,-3.],[4,-5,6],[-7,8,-9]])
# Reshape dimension
w = tf.expand_dims(w,axis=2)
```

```
w = tf.expand_dims(w,axis=3)
# Regular convolution calculation
out = tf.nn.conv2d(x,w,strides=2,padding='VALID')
out
Out[9]: # Output size is 2x2
<tf.Tensor: id=14, shape=(1, 2, 2, 1), dtype=float32, numpy=
array([[[[ -67.],
         [ -77.]],

        [[-117.],
         [-127.]]]], dtype=float32)>
```

Now we use the output of ordinary convolution as the input of transposed convolution to verify whether the output of transposed convolution is 5 × 5; the code is as follows:

```
In [10]:
# Transposed convolution calculation
xx = tf.nn.conv2d_transpose(out, w, strides=2,
    padding='VALID',
    output_shape=[1,5,5,1])
Out[10]: # Output size is 5x5
<tf.Tensor: id=117, shape=(5, 5), dtype=float32, numpy=
array([[   67.,  -134.,    278.,  -154.,    231.],
       [ -268.,   335.,   -710.,   385.,   -462.],
       [  586.,  -770.,   1620.,  -870.,   1074.],
       [ -468.,   585.,  -1210.,   635.,   -762.],
       [  819.,  -936.,   1942., -1016.,   1143.]],
       dtype=float32)>
```

It can be seen that transposed convolution can recover the input of ordinary convolution of the same size, but the output of transposed convolution is not equivalent to the input of ordinary convolution.

$o + 2p - k \neq n * s$

Let us analyze a detail of the relationship between input and output in the convolution operation in more depth. Consider the output expression of the convolution operation:

$$o = \left\lfloor \frac{i + 2 * p - k}{s} \right\rfloor + 1$$

When the stride size $s > 1$, the round-down operation of $\left\lfloor \dfrac{i + 2 * p - k}{s} \right\rfloor$ makes multiple input sizes i correspond to the same output size o. For example, consider the convolution operation with input size 6×6, convolution kernel size 3×3, and stride size 1. The code is as follows:

```
In [11]:
x = tf.random.normal([1,6,6,1])
# 6x6 input
out = tf.nn.conv2d(x,w,strides=2,padding='VALID')
out.shape
x = tf.random.normal([1,6,6,1])...
Out[12]: # Output size 2x2, same as when the input size is 5x5
<tf.Tensor: id=21, shape=(1, 2, 2, 1), dtype=float32, numpy=
array([[[[ 20.438847 ],
        [ 19.160788 ]],

       [[  0.8098897],
        [-28.30303  ]]]], dtype=float32)>
```

In this case, the convolutional output of the same size 2×2 can be obtained as shown in Figure 10-56. Therefore, convolution operations with different input sizes may obtain the same output. Considering that the input and output relationship between convolution and transposed convolution is interchangeable, from the perspective of transposed convolution, after the input size i is subjected to the transposed convolution operation, different output size o may be obtained. Therefore, by filling the a rows and a columns in Figure 10-55 to achieve different sizes of output o, so as to restore the normal convolution with different sizes of input, the relationship of a is:

$$a = (o + 2p - k)\%s$$

The output of the transposed convolution becomes:

$$o = (i - 1)s + k - 2p + a$$

In TensorFlow, there is no need to manually specify a. We just specify the output size. TensorFlow will automatically derive the number of rows and columns that need to be filled, provided that the output size is legal. For example:

```
In [13]:
# Get output of size 6x6
xx = tf.nn.conv2d_transpose(out, w, strides=2,
    padding='VALID',
    output_shape=[1,6,6,1])
xx
Out[13]:
<tf.Tensor: id=23, shape=(1, 6, 6, 1), dtype=float32, numpy=
array([[[[ -20.438847 ],
        [  40.877693 ],
        [ -80.477325 ],
        [  38.321575 ],
        [ -57.48236  ],
        [   0.        ]],...
```

The tensor with height and width 5×5 can also be obtained by changing the parameter output_shape=[1,5,5,1].

Matrix Transposition

The transposition W'^T of transposed convolution means that the sparse matrix W' generated by the convolution kernel matrix W needs to be transposed first, and then the matrix multiplication operation is performed, while the ordinary convolution does not have the step of transposition. This is why it is called transposed convolution.

Consider the ordinary Conv2d operation: X and W, the convolution kernel needs to be cyclically moved in the row and column directions according to the strides to obtain the data of the receptive field involved in the operation, and the "multiply and accumulate" value at each window is calculated serially, which is extremely inefficient. In order to speed up the operation, mathematically, the convolution kernel W can be rearranged into a sparse matrix W' according to strides, and then the operation $W' @ X'$ is completed once (in fact, the matrix W' is too sparse, resulting in many useless 0-multiplication operations, and many deep learning frameworks do not use this implementation).

Take the following convolution kernel as an example: the input X of 4 rows and 4 columns, the height and width as 3×3, stride of 1, and no padding. First, X will be flattened to X', as shown in Figure 10-57.

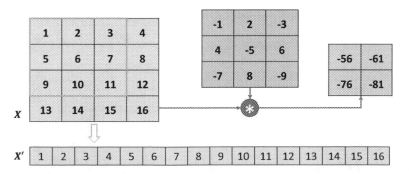

Figure 10-57. *Transposed convolution X*

Then convert the convolution kernel W into a sparse matrix W', as shown in Figure 10-58.

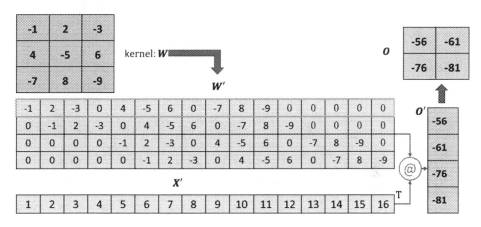

Figure 10-58. *Transposed convolution W*

At this time, ordinary convolution operation can be realized by matrix multiplication once:

$$O' = W' @ X'$$

If given O, how to generate a tensor of the same shape and size as X it? Multiply the transposed matrix W' and the rearranged matrix O' as shown in Figure 10-57:

$$X' = W'^T @ O'$$

Reshape X' to the same as the original input size X. For example, the shape of O' is [4, 1], the shape of W'^T is [16, 4], the shape of X' obtained by matrix multiplication is [16, 1], and the tensor with shape [4, 4] can be generated after reshaping. Since transposed convolution needs to be transposed before it can be multiplied with the input matrix of transposed convolution during matrix operation, it is called transposed convolution.

Transposed convolution has the function of "magnifying feature maps" and has been widely used in generating confrontation networks and semantic segmentation. For example, the generator in DCGAN [12] achieves layer-by-layer "magnification" by stacking transposed convolution layers and finally get a very realistic generated picture.

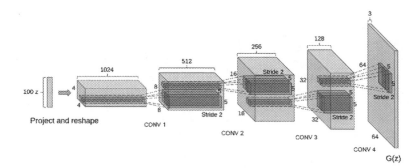

Figure 10-59. *DCGAN architecture [12]*

Transposed Convolution Implementation

In TensorFlow, the transposed convolution operation can be realized through nn.conv2d_transpose() function. We first complete the ordinary convolution operation through nn.conv2d. Note that the definition format of the convolution kernel of transposed convolution is $[k, k, c_{out}, c_{in}]$. For example

```
In [14]:
# Input 4x4
x = tf.range(16)+1
x = tf.reshape(x,[1,4,4,1])
x = tf.cast(x, tf.float32)
# 3x3 kernel
w = tf.constant([[-1,2,-3.],[4,-5,6],[-7,8,-9]])
w = tf.expand_dims(w,axis=2)
```

```
w = tf.expand_dims(w,axis=3)
# Regular convolutional operation
out = tf.nn.conv2d(x,w,strides=1,padding='VALID')
Out[14]:
<tf.Tensor: id=42, shape=(2, 2), dtype=float32, numpy=
array([[-56., -61.],
       [-76., -81.]], dtype=float32)>
```

With strides=1, padding='VALID', and the convolution kernel unchanged, we try to restore the height and width tensor of the same size as the input x through the transposed convolution operation of the convolution kernel w and the output. The code is as follows:

```
In [15]: # Restore 4x4 input
xx = tf.nn.conv2d_transpose(out, w, strides=1, padding='VALID',
output_shape=[1,4,4,1])
tf.squeeze(xx)
Out[15]:
<tf.Tensor: id=44, shape=(4, 4), dtype=float32, numpy=
array([[  56.,  -51.,   46.,  183.],
       [-148.,  -35.,   35., -123.],
       [  88.,   35.,  -35.,   63.],
       [ 532.,  -41.,   36.,  729.]], dtype=float32)>
```

It can be seen that the 4 × 4 feature map is generated by the transposed convolution, but the data of the feature map is not the same as the input x.

When using tf.nn.conv2d_transpose for transposed convolution operation, you need to manually set the output height and width. tf.nn. conv2d_transpose does not support customized padding settings, it can only be set to VALID or SAME.

When padding='VALID' is set, the output size is:

$$o = (i-1)s + k$$

When padding='SAME' is set, the output size is:

$$o = i \cdot s$$

If the reader is temporarily unable to understand the principle details of transposed convolution, he/she can keep the preceding two expressions in mind. For example, when calculating the 2 × 2 transposed convolution input and the 3 × 3 convolution kernel, strides=1, padding='VALID', the output size is:

$$h' = w' = (2-1) \cdot 1 + 3 = 4$$

When calculating 2 × 2 transposed convolution input and the 3 × 3 convolution kernel, strides=3, padding='SAME', the output size is:

$$h' = w' = 2 \cdot 3 = 6$$

Transposed convolution can also be the same as other layers. Create a transposed convolution layer through the layers.Conv2DTranspose class, and then call the instance to complete the forward calculation:

```
In [16]:
layer = layers.Conv2DTranspose(1,kernel_size=3,strides=1,paddin
g='VALID')
xx2 = layer(out)
xx2
Out[16]:
<tf.Tensor: id=130, shape=(1, 4, 4, 1), dtype=float32, numpy=
array([[[[  9.7032385 ],
        [  5.485071  ],
        [ -1.6490463 ],
        [  1.6279562 ]],...
```

10.11.3 Separate Convolution

Here we take depth-wise separable convolution as an example. When the ordinary convolution is operating on multi-channel input, each channel of the convolution kernel and each channel of the input are respectively convolved to obtain a multi-channel feature map, and then the corresponding elements are added to produce the final result of a single convolution kernel output as shown in Figure 10-60.

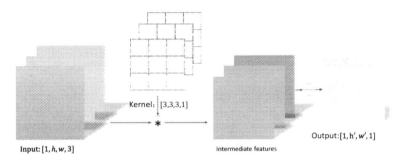

Kernel: [3,3,3,1]

Output:[1, h', w', 1]

Input: [1, h, w, 3]   Intermediate features

Figure 10-60. *Schematic diagram of ordinary convolution calculation*

The calculation process of separate convolution is different. Each channel of the convolution kernel is convolved with each input channel to obtain the intermediate features of multiple channels, as shown in Figure 10-61. This multi-channel intermediate feature tensor is then subjected to the ordinary convolution operation of multiple 1×1 convolution kernels to obtain multiple outputs with constant height and width. These outputs are spliced on the channel axis to produce the final separated convolutional layer output. It can be seen that the separated convolution layer includes a two-step convolution operation. The first convolution operation is a single convolution kernel, and the second convolution operation includes multiple convolution kernels.

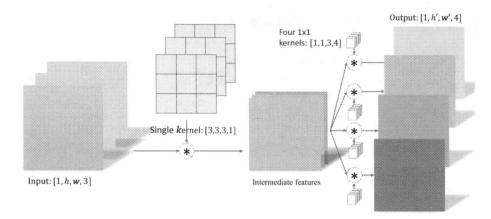

Figure 10-61. *Schematic diagram of depth separable convolution calculation*

So what are the advantages of using separate convolution? An obvious advantage is that for the same input and output, the parameters of the separable convolution are about 1/3 of the ordinary convolution. Consider the example of ordinary convolution and separate convolution in the preceding figure. The parameter quantity of ordinary convolution is:

$$3 \cdot 3 \cdot 3 \cdot 4 = 108$$

The first part of the parameter of the separated convolution is:

$$3 \cdot 3 \cdot 3 \cdot 1 = 27$$

The second part of the parameter is:

$$1 \cdot 1 \cdot 3 \cdot 4 = 14$$

The total parameter amount of the separated convolution is only 39, but it can realize the same input and output size transformation of the ordinary convolution. Separate convolution has been widely used in areas sensitive to computational cost, such as Xception and MobileNets.

10.12 Deep Residual Network

The emergence of network models such as AlexNet, VGG, and GoogLeNet has brought the development of neural networks to a stage of dozens of layers. Researchers have found that the deeper the network, the more likely it is to obtain better generalization capabilities. But as the model deepens, the network becomes more and more difficult to train, which is mainly caused by gradient dispersion and gradient explosion. In a neural network with a deeper number of layers, when the gradient information is transmitted from the last layer of the network to the first layer of the network layer by layer, there will be a phenomenon that the gradient is close to 0 or the gradient value is very large during the transfer process. The deeper the network layer, the more serious this phenomenon may be.

So how to solve the gradient dispersion and gradient explosion phenomenon of deep neural networks? A very natural idea is that since shallow neural networks are not prone to these gradients, you can try to add a fallback mechanism to the deep neural networks. When the deep neural network can easily fall back to the shallow neural network, the deep neural network can obtain model performance equivalent to that of the shallow neural network, but not worse.

By adding a direct connection between the input and output – Skip Connection – the neural network has the ability to fall back. Taking the VGG13 deep neural network as an example, assuming that the gradient dispersion phenomenon is observed in the VGG13 model, and the ten-layer network model does not observe the gradient dispersion phenomenon, then you can consider adding Skip Connection to the last two convolutional layers, as shown in Figure 10-62. In this way, the network model can automatically choose whether to complete the feature transformation through these two convolutional layers, or skip these two convolutional layers and choose Skip Connection, or combine the output of the two convolutional layers and Skip Connection .

Figure 10-62. *Architecture of VGG13 with Skip Connection*

In 2015, He Kaiming and others from Microsoft Research Asia published a Skip Connection-based deep residual network (residual neural network, referred to as ResNet) algorithm [10], and proposed 18 layers, 34 layers, 50 layers, 101 layers, and 152 layers network, that is, ResNet-18, ResNet-34, ResNet-50, ResNet-101 and ResNet-152 models, and even successfully trained a very deep neural network with 1202 layers. ResNet has achieved the best performance on tasks such as classification and detection on the ImageNet dataset of the ILSVRC 2015 Challenge. The ResNet papers have so far received more than 25,000 citations, which shows the influence of ResNet in the artificial intelligence community.

10.12.1 ResNet Principle

ResNet implements the fallback mechanism by adding Skip Connection between the input and output of the convolutional layers, as shown in Figure 10-63. The input x passes through two convolutional layers to obtain the output $F(x)$ after feature transformation, and the corresponding element of $F(x)$ is added to x to get the final output:

$$H(x) = x + F(x)$$

$H(x)$ is called residual block (ResBlock for short). Since the convolutional neural network surrounded by Skip Connection needs to learn the mapping $F(x) = H(x) - x$, it is called the residual network.

In order to satisfy the addition of the input x and the output $F(x)$ of the convolutional layer, the input shape needs to be exactly the same as the shape of the output $F(x)$. When the shapes are inconsistent, the input x is generally transformed to the same shape of $F(x)$ by adding additional convolution operations on Skip Connection, as shown in the function $identity(x)$ in Figure 10-63, where $identity(x)$ mainly takes the 1×1 convolutional operation to adjust the input number of channels.

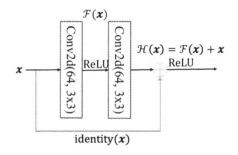

Figure 10-63. *Residual module*

Figure 10-64 compares the 34-layer deep residual network, the 34-layer ordinary deep network, and the 19-layer VGG network structure. It can be seen that the deep residual network reaches a deeper network layer by stacking residual modules, thereby obtaining a deep network model with stable training and superior performance.

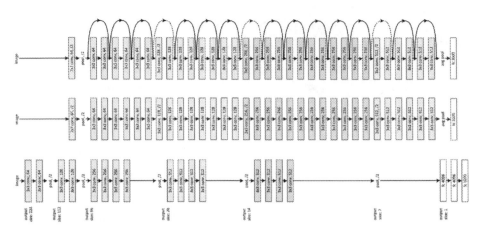

Figure 10-64. *Network architecture comparison [10]*

10.12.2 ResBlock Implementation

The deep residual network does not add a new network layer type but only adds a Skip Connection between the input and the output, so there is no underlying implementation for ResNet. The residual module can be implemented in TensorFlow by calling the ordinary convolutional layer.

First, create a new class. Initialize the convolutional layer and activation function layer needed in the residual block, and then create a new convolutional layer; the code is as follows:

```
class BasicBlock(layers.Layer):
    # Residual block
    def __init__(self, filter_num, stride=1):
        super(BasicBlock, self).__init__()
        # Create Convolutional Layer 1
        self.conv1 = layers.Conv2D(filter_num, (3, 3),
        strides=stride, padding='same')
        self.bn1 = layers.BatchNormalization()
        self.relu = layers.Activation('relu')
        # Create Convolutional Layer 2
```

```
self.conv2 = layers.Conv2D(filter_num, (3, 3),
strides=1, padding='same')
self.bn2 = layers.BatchNormalization()
```

When the shape of $F(x)$ and x is different, it cannot be added directly. We need to create a new convolutional layer *identity*(x) to complete the shape conversion of x. Following the preceding code, the implementation is as follows:

```
if stride != 1: # Insert identity layer
    self.downsample = Sequential()
    self.downsample.add(layers.Conv2D(filter_num,
    (1, 1), strides=stride))
else: # connect directly
    self.downsample = lambda x:x
```

During forward propagation, you only need to add $F(x)$ and *identity*(x) and add the ReLU activation function. The forward calculation function code is as follows:

```
def call(self, inputs, training=None):
    # Forward calculation
    out = self.conv1(inputs) # 1st Conv layer
    out = self.bn1(out)
    out = self.relu(out)
    out = self.conv2(out) # 2nd Conv layer
    out = self.bn2(out)
    #  identity() conversion
    identity = self.downsample(inputs)
    # f(x)+x
    output = layers.add([out, identity])
    # activation function
    output = tf.nn.relu(output)
    return output
```

10.13 DenseNet

The idea of Skip Connection has achieved great success on ResNet. Researchers have begun to try different Skip Connection schemes, among which DenseNet [11] is more popular. DenseNet aggregates the feature map information of all the previous layers with the output of the current layer through Skip Connection. Unlike ResNet's corresponding position addition method, DenseNet uses splicing operations in the channel axis dimension to aggregate feature information.

As shown in Figure 10-65, the input X_0 is passed through the convolutional layer H_1 and the output X_1 is spliced with the channel axis to obtain the aggregated feature tensor, which is sent to the convolutional layer H_2 to obtain the output X_2. Similarly, X_2 is spliced with X_1 and X_0 and sent to the next layer. Repeat this way until the output of the last layer X_4 and the feature information of all previous layers: $\{X_i\}_{i=0,1,2,3}$ are aggregated to the final output of the module. Such a densely connected module based on Skip Connection is called dense block.

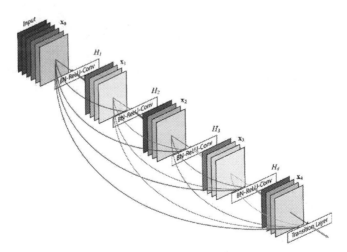

Figure 10-65. *Dense block architecture[2]*

---

[2] Image source: https://github.com/liuzhuang13/DenseNet

DenseNet constructs a complex deep neural network by stacking multiple dense blocks, as shown in Figure 10-66.

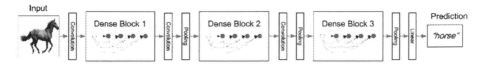

Figure 10-66. *A typical DenseNet architecture*[3]

Figure 10-67 compares the performance of different versions of DenseNet, the performance comparison of DenseNet and ResNet, and the training curves of DenseNet and ResNet.

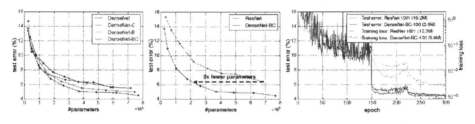

Figure 10-67. *Comparison of DenseNet and ResNet performance [11]*

10.14 Hands-On CIFAR10 and ResNet18

In this section, we will implement the 18-layer deep residual network ResNet18, train, and test it on the CIFAR10 image dataset. We will compare its performance with the 13-layer ordinary neural network VGG13.

The standard ResNet18 accepts image data of size 224 × 224. We adjust ResNet18 appropriately so that its input size is 32 × 32 and its output dimension is 10. The adjusted ResNet18 network structure is shown in Figure 10-68.

---

[3] Image source: `https://github.com/liuzhuang13/DenseNet`

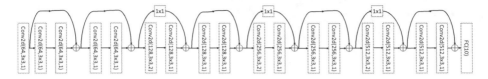

Figure 10-68. *Adjusted ResNet18 architecture*

First implement the residual module of the two convolutional layers in the middle, and residual block of Skip Connection 1x1 convolutional layer as in the following:

```
class BasicBlock(layers.Layer):
    # Residual block
    def __init__(self, filter_num, stride=1):
        super(BasicBlock, self).__init__()
        # 1st conv layer
        self.conv1 = layers.Conv2D(filter_num, (3, 3),
        strides=stride, padding='same')
        self.bn1 = layers.BatchNormalization()
        self.relu = layers.Activation('relu')
        # 2nd conv layer
        self.conv2 = layers.Conv2D(filter_num, (3, 3),
        strides=1, padding='same')
        self.bn2 = layers.BatchNormalization()

        if stride != 1:
            self.downsample = Sequential()
            self.downsample.add(layers.Conv2D(filter_num,
            (1, 1), strides=stride))
        else:
            self.downsample = lambda x:x

    def call(self, inputs, training=None):
        # Forward calculation
```

```
# [b, h, w, c], 1st conv layer
out = self.conv1(inputs)
out = self.bn1(out)
out = self.relu(out)
# 2nd conv layer
out = self.conv2(out)
out = self.bn2(out)
# identity()
identity = self.downsample(inputs)
# Add two layers
output = layers.add([out, identity])
output = tf.nn.relu(output) # activation function

return output
```

When designing a deep convolutional neural network, generally follow the rule of thumb that the height and width of the feature map gradually decrease and the number of channels gradually increases. The extraction of high-level features can be achieved by stacking Res Blocks with gradually increasing channel numbers, and multiple residual modules can be built at once through build_resblock as in the following:

```
def build_resblock(self, filter_num, blocks, stride=1):
    # stack filter_num BasicBlocks
    res_blocks = Sequential()
    # Only 1st BasicBlock's stride may not be 1
    res_blocks.add(BasicBlock(filter_num, stride))

    for _ in range(1, blocks):# Stride of Other BasicBlocks
    are all 1
        res_blocks.add(BasicBlock(filter_num, stride=1))

    return res_blocks
```

Let's implement a general ResNet network model as in the following:

```
class ResNet(keras.Model):
    # General ResNet class
    def __init__(self, layer_dims, num_classes=10):
    # [2, 2, 2, 2]
        super(ResNet, self).__init__()
        self.stem = Sequential([layers.Conv2D(64, (3, 3),
        strides=(1, 1)),
                                layers.BatchNormalization(),
                                layers.Activation('relu'),
                                layers.MaxPool2D(pool_
                                size=(2, 2), strides=(1, 1),
                                padding='same')
                                ])
        # Stack 4 Blocks
        self.layer1 = self.build_resblock(64,  layer_dims[0])
        self.layer2 = self.build_resblock(128, layer_dims[1],
        stride=2)
        self.layer3 = self.build_resblock(256, layer_dims[2],
        stride=2)
        self.layer4 = self.build_resblock(512, layer_dims[3],
        stride=2)

        # Pooling layer => 1x1
        self.avgpool = layers.GlobalAveragePooling2D()
        # Fully connected layer
        self.fc = layers.Dense(num_classes)

    def call(self, inputs, training=None):
        # Forward calculation
        x = self.stem(inputs)
        # 4 blocks
```

```
    x = self.layer1(x)
    x = self.layer2(x)
    x = self.layer3(x)
    x = self.layer4(x)

    # Pooling layer
    x = self.avgpool(x)
    # Fully connected layer
    x = self.fc(x)

    return x
```

Different ResNets can be generated by adjusting the number of stacks and channels of each Res Block, such as with 64-64-128-128-256-256-512-512 channel configuration, a total of eight Res Blocks, you can get ResNet18 network model. Each ResBlock contains two main convolutional layers, so the number of convolutional layers is 8 · 2 = 16, plus the fully connected layer at the end of the network, a total of 18 layers. Creating ResNet18 and ResNet34 can be simply implemented as follows:

```
def resnet18():
    return ResNet([2, 2, 2, 2])

def resnet34():
    return ResNet([3, 4, 6, 3])
```

Next, complete the loading of the CIFAR10 data set as follows:

```
(x,y), (x_test, y_test) = datasets.cifar10.load_data() #
load data
y = tf.squeeze(y, axis=1) # sequeeze data
y_test = tf.squeeze(y_test, axis=1)
print(x.shape, y.shape, x_test.shape, y_test.shape)
```

```
train_db = tf.data.Dataset.from_tensor_slices((x,y)) # create
training set
train_db = train_db.shuffle(1000).map(preprocess).batch(512)

test_db = tf.data.Dataset.from_tensor_slices((x_test,y_test))
#creat testing set
test_db = test_db.map(preprocess).batch(512)
# sample an example
sample = next(iter(train_db))
print('sample:', sample[0].shape, sample[1].shape,
      tf.reduce_min(sample[0]), tf.reduce_max(sample[0]))
```

The data preprocessing logic is relatively simple. We just need to directly map the data range to the interval $[-1, 1]$. Here you can also perform standardization based on the mean and standard deviation of the ImageNet data pictures as in the following:

```
def preprocess(x, y):
    x = 2*tf.cast(x, dtype=tf.float32) / 255. - 1
    y = tf.cast(y, dtype=tf.int32)
    return x,y
```

The network training logic is the same as the normal classification network training part, and 50 Epochs are trained as in the following:

```
for epoch in range(50): # Train epoch
    for step, (x,y) in enumerate(train_db):
        with tf.GradientTape() as tape:
            # [b, 32, 32, 3] => [b, 10], forward
            calculation
            logits = model(x)
            # [b] => [b, 10],one-hot encoding
            y_onehot = tf.one_hot(y, depth=10)
            # Calculate loss
```

```
        loss = tf.losses.categorical_crossentropy(y_
        onehot, logits, from_logits=True)
        loss = tf.reduce_mean(loss)
    # Calculate gradient
    grads = tape.gradient(loss, model.trainable_
    variables)
    # Update parameters
    optimizer.apply_gradients(zip(grads, model.
    trainable_variables))
```

ResNet18 has a total of 11 million network parameters. After 50 Epochs, the accuracy of the network reached 79.3%. Our code here is relatively streamlined. With the support of careful hyperparameters and data enhancement, the accuracy rate can be higher.

10.15 References

[1]. G. E. Hinton, S. Osindero and Y.-W. Teh, "A Fast Learning Algorithm for Deep Belief Nets," *Neural Comput.*, 18, pp. 1527-1554, 7 2006.

[2]. Y. LeCun, B. Boser, J. S. Denker, D. Henderson, R. E. Howard, W. Hubbard and L. D. Jackel, "Backpropagation Applied to Handwritten Zip Code Recognition," *Neural Comput.*, 1, pp. 541-551, 12 1989.

[3]. A. Krizhevsky, I. Sutskever and G. E. Hinton, "ImageNet Classification with Deep Convolutional Neural Networks," *Advances in Neural Information Processing Systems 25*, F. Pereira, C. J. C. Burges, L. Bottou and K. Q. Weinberger, Curran Associates, Inc., 2012, pp. 1097-1105.

[4].   Y. Lecun, L. Bottou, Y. Bengio and P. Haffner, "Gradient-based learning applied to document recognition," *Proceedings of the IEEE*, 1998.

[5].   M. D. Zeiler and R. Fergus, "Visualizing and Understanding Convolutional Networks, *Computer Vision -- ECCV 2014*, Cham, 2014.

[6].   S. Ioffe and C. Szegedy, "Batch Normalization: Accelerating Deep Network Training by Reducing Internal Covariate Shift," *CoRR*, abs/1502.03167, 2015.

[7].   Y. Wu and K. He, "Group Normalization," *CoRR*, abs/1803.08494, 2018.

[8].   K. Simonyan and A. Zisserman, "Very Deep Convolutional Networks for Large-Scale Image Recognition," *CoRR*, abs/1409.1556, 2014.

[9].   C. Szegedy, W. Liu, Y. Jia, P. Sermanet, S. Reed, D. Anguelov, D. Erhan, V. Vanhoucke and A. Rabinovich, "Going Deeper with Convolutions," *Computer Vision and Pattern Recognition (CVPR)*, 2015.

[10].   K. He, X. Zhang, S. Ren and J. Sun, "Deep Residual Learning for Image Recognition," *CoRR*, abs/1512.03385, 2015.

[11].   G. Huang, Z. Liu and K. Q. Weinberger, "Densely Connected Convolutional Networks," *CoRR*, abs/1608.06993, 2016.

[12].   A. Radford, L. Metz and S. Chintala, Unsupervised Representation Learning with Deep Convolutional Generative Adversarial Networks, 2015.

CHAPTER 11

Recurrent Neural Network

> The powerful rise of artificial intelligence may be the best thing in human history, or it may be the worst thing.
>
> —Steven Hawking

Convolutional neural network uses the local correlation of data and the idea of weight sharing to greatly reduce the amount of network parameters. It is very suitable for pictures with spatial and local correlation. It has been successfully applied to a series of tasks in the field of computer vision. In addition to the spatial dimension, natural signals also have a temporal dimension. Signals with a time dimension are very common, such as the text we are reading, the speech signal emitted when we speak, and the stock market that changes over time. This type of data does not necessarily have local relevance, and the length of the data in the time dimension is also variable. Convolutional neural networks are not good at processing such data.

So analyzing and recognizing this type of signals is a task that must be solved in order to push artificial intelligence to general artificial intelligence. The recurrent neural network that will be introduced in this

© Liangqu Long and Xiangming Zeng 2022
L. Long and X. Zeng, *Beginning Deep Learning with TensorFlow*,
https://doi.org/10.1007/978-1-4842-7915-1_11

chapter can better solve such problems. Before introducing the recurrent neural network, let's first introduce the method of representing data in chronological order.

11.1 Sequence Representation Method

Data with order is generally called a sequence, for example, commodity price data that changes over time is a very typical sequence. Considering the price change trend of a commodity A between January and June, we can record it as a one-dimensional vector: $[x_1, x_2, x_3, x_4, x_5, x_6]$, and its shape is [6]. If you want to represent the price change trend of b goods from January to June, you can record it as a 2-dimensional tensor:

$$\left[\left[x_1^{(1)}, x_2^{(1)}, \cdots, x_6^{(1)}\right], \left[x_1^{(2)}, x_2^{(2)}, \cdots, x_6^{(2)}\right], \cdots, \left[x_1^{(b)}, x_2^{(b)}, \cdots, x_6^{(b)}\right]\right]$$

where b represents the number of commodities, and the tensor shape is $[b, 6]$.

In this way, the sequence signal is not difficult to represent, only a tensor with shape [b, s] is needed, where b is the number of sequences and s is the length of the sequence. However, many signals cannot be directly represented by a scalar value. For example, to represent feature vectors of length n generated by each timestamp, a tensor of shape [b, s, n] is required. Consider more complex text data: sentences. The word generated on each timestamp is a character, not a numerical value, and therefore cannot be directly represented by a scalar. We already know that neural networks are essentially a series of math operations such as matrix multiplication and addition. They cannot directly process string data. If you want neural networks to be used for natural language processing tasks, then how to convert words or characters into numerical values becomes particularly critical. Next, we mainly discuss the representation method of text sequence. For other non-numerical signals, please refer to the representation method of text sequence.

For a sentence containing n words, a simple way to represent the words is the one-hot encoding method we introduced earlier. Take English sentences as an example; suppose we only consider the most commonly used 10,000 words, then each word can be expressed as a sparse one-hot vector with one position as 1, and other positions of 0 and a length of 10,000. As shown in Figure 11-1, if only n location names are considered, each location name can be coded as a one-hot vector of length n.

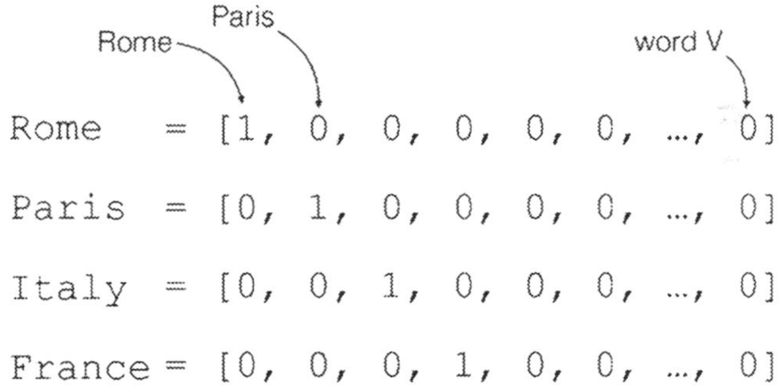

```
                  Paris
       Rome                                      word V
Rome   =  [1,   0,   0,   0,   0,   0,   ...,   0]

Paris  =  [0,   1,   0,   0,   0,   0,   ...,   0]

Italy  =  [0,   0,   1,   0,   0,   0,   ...,   0]

France =  [0,   0,   0,   1,   0,   0,   ...,   0]
```

Figure 11-1. *One-hot encoding of location names*

We call the process of encoding text into numbers as Word Embedding. One-hot encoding is simple and intuitive to implement Word Embedding, and the encoding process does not require learning and training. However, the one-hot encoding vector is high-dimensional and extremely sparse, with a large number of positions as 0s. Therefore, it is computationally expensive and also not conducive to the neural network training. From a semantic point of view, one-hot encoding has a serious problem. It ignores the semantic relevance inherent in words. For example, for the words "like," "dislike," "Rome," "Paris," "like," and "dislike" are strongly related from a semantic point of view. They both indicate the degree of like. "Rome" and " "Paris" is also strongly related. They both indicate two locations in Europe. For a group of such words, if one-hot encoding is used, there is no correlation between the obtained vectors, and the semantic relevance of the original text cannot be well reflected. Therefore, the one-hot encoding has obvious disadvantages.

In the field of natural language processing, there is a special research area about word vector so that the semantic level of relevance can be well reflected through the word vector. One way to measure the correlation between word vectors is the cosine similarity:

$$similarity(a,b) \triangleq \cos\cos(\theta) = \frac{a \cdot b}{|a| \cdot |b|}$$

where a and b represent two word vectors. Figure 11-2 shows the similarity between the words "France" and "Italy," and the similarity between the words "ball" and "crocodile," and θ is the angle between the two word vectors. It can be seen that $\cos\cos(\theta)$ better reflects semantic relevance.

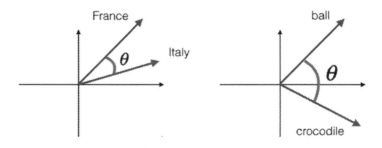

Figure 11-2. *Cosine similarity diagram*

11.1.1 Embedding Layer

In a neural network, the representation vector of a word can be obtained directly through training. We call the representation layer of the word Embedding layer. The Embedding layer is responsible for encoding the word into a word vector v. It accepts the word number i using digital encoding, such as 2 for "I" and 3 for "me". The total number of words in the system is recorded as N_{vocab}, and the output is vector v with length n:

$$v = f_\theta(i|N_{vocab}, n)$$

The Embedding layer is very simple to implement. Build a lookup table with shape $[N_{vocab}, n]$. For any word number i, you only need to query the vector at the corresponding position and return:

$$v = table[i]$$

The Embedding layer is trainable. It can be placed in front of the neural network to complete the conversion of words to vectors. The resulting representation vector can continue to pass through the neural network to complete subsequent tasks, and calculate the error L. The gradient descent algorithm is used to achieve end-to-end training.

In TensorFlow, a Word Embedding layer can be defined by layers. Embedding(N_{vocab}, n), where the N_{vocab} parameter specifies the number of words, and n specifies the length of the word vector. For example:

```
x = tf.range(10) # Generate a digital code of 10 words
x = tf.random.shuffle(x) # Shuffle
# Create a layer with a total of 10 words, each word is
represented by a vector of length 4
net = layers.Embedding(10, 4)
out = net(x) # Get word vector
```

The preceding code creates an Embedding layer of ten words. Each word is represented by a vector of length 4. You can pass in an input with a number code of 0–9 to get the word vectors of these four words. These word vectors are initialized randomly and has not been trained, for example:

```
<tf.Tensor: id=96, shape=(10, 4), dtype=float32, numpy=
array([[-0.00998075, -0.04006485,  0.03493755,  0.03328368],
       [-0.04139598, -0.02630153, -0.01353856,  0.02804044],…
```

We can directly view the query table inside the Embedding layer:

```
In [1]: net.embeddings
Out[1]:
<tf.Variable 'embedding_4/embeddings:0' shape=(10, 4)
dtype=float32, numpy=
array([[ 0.04112223,  0.01824595, -0.01841902,  0.00482471],
       [-0.00428962, -0.03172196, -0.04929272,  0.04603403],…
```

The optimizable property of the net.embeddings tensor is True, which means it can be optimized by the gradient descent algorithm.

```
In [2]: net.embeddings.trainable
Out[2]:True
```

11.1.2 Pre-trained Word Vectors

The lookup table of the Embedding layer is initialized randomly and needs to be trained from scratch. In fact, we can use pre-trained Word Embedding models to get the word representation. The word vector based on pre-trained models is equivalent to transferring the knowledge of the entire semantic space, which can often get better performance.

Currently, the widely used pre-trained models include Word2Vec and GloVe. They have been trained on a massive corpus to obtain a better word vector representation and can directly export the learned word vector table to facilitate migration to other tasks. For example, the GloVe model GloVe.6B.50d has a vocabulary of 400,000, and each word is represented by a vector of length 50. Users only need to download the corresponding model file in order to use it. The "glove6b50dtxt.zip" model file is about 69MB.

So how to use these pre-trained word vector models to help improve the performance of NLP tasks? Very simple. For the Embedding layer, random initialization is no longer used. Instead, we use the pre-trained model parameters to initialize the query table of the Embedding layer. For example:

```
# Load the word vector table from the pre-trained model
embed_glove = load_embed('glove.6B.50d.txt')
# Initialize the Embedding layer directly using the pre-trained
word vector table
net.set_weights([embed_glove])
```

The Embedding layer initialized by the pre-trained word vector model can be set to not participate in training: net.trainable = False, then the pre-trained word vector is directly applied to this specific task. If you also want to learn different representations from the pre-trained word vector model, then the Embedding layer can be included in the backpropagation algorithm by setting net.trainable = True, and gradient descent then can be used to fine-tune the word representation.

11.2 Recurrent Neural Network

Now let's consider how to deal with sequence signals. Taking a text sequence as an example, consider a sentence:

"I hate this boring movie"

Through the Embedding layer, it can be converted into a tensor with shape $[b, s, n]$, where b is the number of sentences, s is the sentence length, and n is the length of the word vector. The preceding sentence can be expressed as a tensor with shape $[1,5,10]$, where 5 represents the length of the sentence word, and 10 represents the length of the word vector.

Next, we will gradually explore a network model that can process sequence signals. We take the sentiment classification task as an example, as shown in Figure 11-3. The sentiment classification task extracts the overall semantic features expressed by the text data and thereby predict the sentiment type of the input text: positive or negative. From the perspective of classification, sentiment classification is a simple two-classification problem. Unlike image classification, because the input is a text sequence, traditional convolutional neural networks cannot achieve good results. So what type of network is good at processing sequence data?

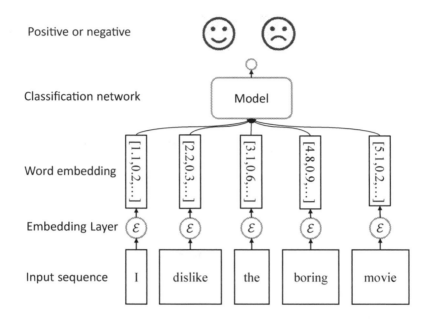

Figure 11-3. *Sentiment classification task*

11.2.1 Is a Fully Connected Layer Feasible?

The first thing we think of is that for each word vector, a fully connected layer network can be used.

$$o = \sigma \left(W_t x_t + b_t \right)$$

Extract semantic features, as shown in Figure 11-4. The word vector of each word is extracted through s fully connected layer classification networks 1. The features of all words are finally merged, and the category probability distribution of the sequence is output through the classification network 2. For a sentence of length s, at least s fully-connected network layers are required.

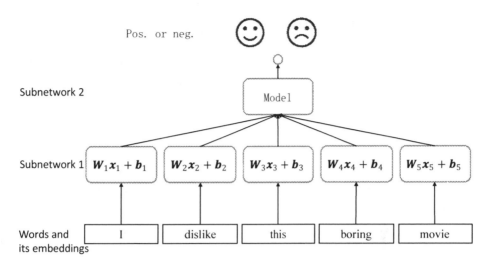

Figure 11-4. *Network architecture 1*

The disadvantages of this scheme are:

- The amount of network parameters is considerable, and the memory usage and calculation cost are high. At the same time, since the length s of each sequence is not the same, the network structure changes dynamically.

- Each fully connected layer sub-network W_i and b_i can only sense the input of the current word vector and cannot perceive the context information before and after, resulting in the lack of overall sentence semantics. Each sub-network can only extract high-level features based on its own input.

We will solve these two disadvantages one by one.

11.2.2 Shared Weight

When introducing convolutional neural networks, we have learned that the reason why convolutional neural networks is better than fully connected networks in processing locally related data is because it makes full use of the idea of weight sharing and greatly reduces the amount of network parameters, which makes the network training more efficient. So, can we learn from the idea of weight sharing when dealing with sequence signals?

In the scheme in Figure 11-4, the network of s fully connected layers does not realize weight sharing. We try to share these s network layer parameters, which is actually equivalent to using a fully connected network to extract the feature information of all words, as shown in Figure 11-5.

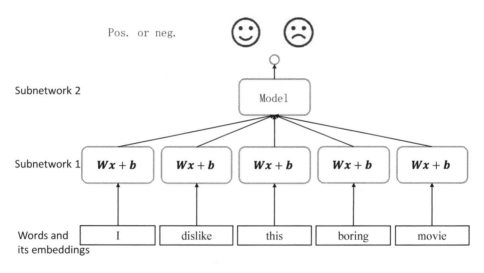

Figure 11-5. *Network architecture 2*

After weight sharing, the amount of parameters is greatly reduced, and network training becomes more stable and efficient. However, this network structure does not consider the order of sequences, and the same output can still be obtained by shuffling the order of the word vectors. Therefore, it cannot obtain effective global semantic information.

11.2.3 Global Semantics

How to give the network the ability to extract overall semantic features? In other words, how can the network extract the semantic information of word vectors in order and accumulate it into the global semantic information of the entire sentence? We thought of the memory mechanism. If the network can provide a separate memory variable, each time the feature of the word vector is extracted and the memory variable is refreshed, until the last input is completed, the memory variable at this time stores the semantic features of all sequences, and because of the order of input sequences, the contents of memory variables are closely related to the sequence order.

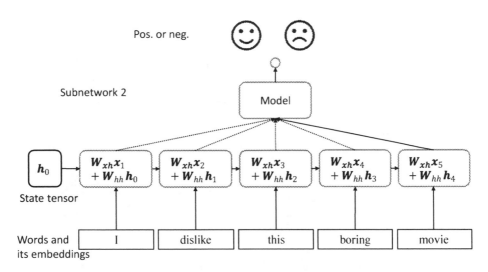

Figure 11-6. *Recurrent neural network (no bias added)*

We implement the preceding memory mechanism as a state tensor h, as shown in Figure 11-6. In addition to the original W_{xh} parameter sharing, an additional W_{hh} parameter is added here. The state tensor h refresh mechanism for each timestamp t is:

$$h_t = \sigma\left(W_{xh}x_t + W_{hh}h_{t-1} + b\right)$$

471

where the state tensor h_0 is the initial memory state, which can be initialized to all 0s. After the input of s word vectors, the final state tensor h_s of the network is obtained. h_s better represents the global semantic information of the sentence. Passing h_s through a fully connected layer classifier can complete the sentiment classification task.

11.2.4 Recurrent Neural Network

Through step-by-step exploration, we finally proposed a "new" network structure, as shown in Figure 11-7. At each time stamp t, the network layer accepts the input x_t of the current time stamp and the network state vector of the previous time stamp h_{t-1}, after:

$$h_t = f_\theta\left(h_{t-1}, x_t\right)$$

After transformation, the new state vector h_t of the current time stamp is obtained and written into the memory state, where f_θ represents the operation logic of the network, and θ is the network parameter set. At each time stamp, the network layer has an output to produce o_t, $o_t = g_\phi(h_t)$, which is to output the state vector of the network after transformation.

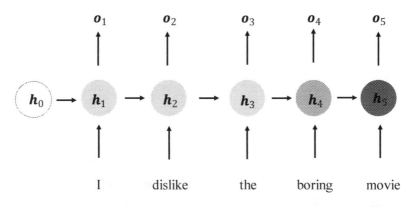

Figure 11-7. Expanded RNN model

472

The preceding network structure is folded on the time stamp, as shown in Figure 11-8. The network cyclically accepts each feature vector x_t of the sequence, refreshes the internal state vector h_t, and forms the output o_t at the same time. For this kind of network structure, we call it the recurrent neural network (RNN).

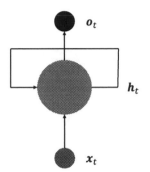

Figure 11-8. *Folded RNN model*

More specifically, if we use the tensors W_{xh}, W_{hh} and bias b to parameterize the f_θ network, and use the following ways to update the memory state, we call this kind of network a basic recurrent neural network, unless otherwise specified; generally speaking, the recurrent neural network refers to this realization.

$$h_t = \sigma\left(W_{xh}x_t + W_{hh}h_{t-1} + b\right)$$

In the recurrent neural network, the activation function uses the Tanh function more, and we can choose not to use the bias b to further reduce the amount of parameters. The state vector h_t can be directly used as output, that is, $o_t = h_t$, or a simple linear transformation of h_t can be done to $o_t = W_{ho}h_t$ to get the network output o_t on each time stamp.

11.3 Gradient Propagation

Through the update expression of the recurrent neural network, it can be seen that the output is derivable to the tensors W_{xh}, W_{hh} and bias b, and the automatic gradient descent algorithm can be used to solve the gradient of the network. Here we simply derive the gradient propagation formula of RNN and explore its characteristics.

Consider the gradient $\dfrac{\partial L}{\partial W_{hh}}$, where L is the error of the network, and only consider the difference between the last output o_t at t and the true value. Since W_{hh} is shared by the weight of each timestamp i, when calculating $\dfrac{\partial L}{\partial W_{hh}}$, it is necessary to sum the gradients on each intermediate timestamp i, using the chain rule to expand as:

$$\frac{\partial L}{\partial W_{hh}} = \sum_{i=1}^{t} \frac{\partial L}{\partial o_t} \frac{\partial o_t}{\partial h_t} \frac{\partial h_t}{\partial h_i} \frac{\partial^+ h_i}{\partial W_{hh}}$$

where $\dfrac{\partial L}{\partial o_t}$ can be obtained directly based on the loss function, in the case of $o_t = h_i$:

$$\frac{\partial o_t}{\partial h_t} = I$$

And the gradient of $\dfrac{\partial^+ h_i}{\partial W_{hh}}$ can also be obtained after expanding h_i:

$$\frac{\partial^+ h_i}{\partial W_{hh}} = \frac{\partial \sigma \left(W_{xh} x_t + W_{hh} h_{t-1} + b \right)}{\partial W_{hh}}$$

Among them $\dfrac{\partial^+ h_i}{\partial W_{hh}}$ only considers the gradient propagation of one time stamp, that is, the "direct" partial derivative, which is different from $\dfrac{\partial L}{\partial W_{hh}}$ that considers the gradient propagation of all timestamps $i = 1, \cdots, t$.

Therefore, we only need to derive the expression of $\dfrac{\partial h_t}{\partial h_i}$ to complete the gradient derivation of the recurrent neural network. Using the chain rule, we divide $\dfrac{\partial h_t}{\partial h_i}$ into the gradient expression of successive timestamps:

$$\frac{\partial h_t}{\partial h_i} = \frac{\partial h_t}{\partial h_{t-1}} \frac{\partial h_{t-1}}{\partial h_{t-2}} \cdots \frac{\partial h_{i+1}}{\partial h_i} = \prod_{k=i}^{t-1} \frac{\partial h_{k+1}}{\partial h_k}$$

Consider:

$$h_{k+1} = \sigma\left(W_{xh} x_{k+1} + W_{hh} h_k + b\right)$$

then:

$$\frac{\partial h_{k+1}}{\partial h_k} = W_{hh}^T diag\left(\sigma'\left(W_{xh} x_{k+1} + W_{hh} h_k + b\right)\right)$$

$$= W_{hh}^T diag\left(\sigma'\left(h_{k+1}\right)\right)$$

where $diag(x)$ takes each element of the vector x as the diagonal element of the matrix and obtains a diagonal matrix with all other elements being 0, for example:

$$diag\left([3,2,1]\right) = [3\,0\,0\,0\,2\,0\,0\,0\,1\,]$$

Therefore,

$$\frac{\partial h_t}{\partial h_i} = \prod_{j=i}^{t-1} diag\left(\sigma'\left(W_{xh} x_{j+1} + W_{hh} h_j + b\right)\right) W_{hh}$$

So far, the gradient derivation of $\dfrac{\partial L}{\partial W_{hh}}$ is completed.

Since deep learning frameworks can help us automatically derive gradients, we only need to understand the gradient propagation mechanism of the recurrent neural network. In the process of deriving $\dfrac{\partial L}{\partial W_{hh}}$, we found that the gradient of $\dfrac{\partial h_t}{\partial h_i}$ includes the continuous multiplication operation of W_{hh}, which is the root cause of the difficulty in training the recurrent neural network the reason. We will discuss it later.

11.4 How to Use RNN Layers

After introducing the principle of the recurrent neural network, let's learn how to implement the RNN layer in TensorFlow. In TensorFlow, the $\sigma(W_{xh}x_t + W_{hh}h_{t-1} + b)$ calculation can be completed by layers. SimpleRNNCell() function. It should be noted that in TensorFlow, RNN stands for recurrent neural network in a general sense. For the basic recurrent neural network we are currently introducing, it is generally called SimpleRNN. The difference between SimpleRNN and SimpleRNNCell is that the layer with cell only completes the forward operation of one timestamp, while the layer without cell is generally implemented based on the cell layer, which has already completed multiple timestamp cycles internally. Therefore, it is more convenient and faster to use.

We first introduce the use of SimpleRNNCell, and then introduce the use of SimpleRNN layer.

11.4.1 SimpleRNNCell

Take a certain input feature length n=4 and cell state vector feature length h=3 as an example. First, we create a SimpleRNNCell without specifying the sequence length s. The code is as follows:

```
In [3]:
cell = layers.SimpleRNNCell(3) # Create RNN Cell, memory vector
length is 3
cell.build(input_shape=(None,4)) # Output feature length n=4
cell.trainable_variables # Print wxh, whh, b tensor
Out[3]:
[<tf.Variable 'kernel:0' shape=(4, 3) dtype=float32,
numpy=...>,
 <tf.Variable 'recurrent_kernel:0' shape=(3, 3) dtype=float32,
 numpy=...>,
 <tf.Variable 'bias:0' shape=(3,) dtype=float32,
 numpy=array([0., 0., 0.], dtype=float32)>]
```

It can be seen that SimpleRNNCell maintains three tensors internally, the kernel variable is the tensor W_{xh}, the recurrent_kernel variable is the tensor W_{hh}, and the bias variable is the bias vector b. However, the memory vector h of RNN is not maintained by SimpleRNNCell, and the user needs to initialize the vector h_0 and record the h_t on each time stamp.

The forward operation can be completed by calling the cell instance:

$$o_t, [h_t] = Cell(x_t, [h_{t-1}])$$

For SimpleRNNCell, $o_t = h_t$, is the same object. There's no additional linear layer conversion. $[h_t]$ is wrapped in a list. This setting is for uniformity with RNN variants such as LSTM and GRU. In the initialization phase of the recurrent neural network, the state vector h_0 is generally initialized to an all-zero vector, for example:

```
In [4]:
# Initialize state vector. Wrap with list, unified format
h0 = [tf.zeros([4, 64])]
x = tf.random.normal([4, 80, 100]) # Generate input tensor, 4
sentences of 80 words
```

```
xt = x[:,0,:] # The first word of all sentences
# Construct a Cell with input feature n=100, sequence length
s=80, state length=64
cell = layers.SimpleRNNCell(64)
out, h1 = cell(xt, h0) # Forward calculation
print(out.shape, h1[0].shape)
Out[4]: (4, 64) (4, 64)
```

It can be seen that after one timestamp calculation, the shape of the output and the state tensor are both [b, h], and the ids of the two are printed as follows:

```
In [5]:print(id(out), id(h1[0]))
Out[5]:2154936585256 2154936585256
```

The two ids are the same, that is, the state vector is directly used as the output vector. For the training of length s, it is necessary to loop through the cell class s times to complete one forward operation of the network layer. For example:

```
h = h0 # Save a list of state vectors on each time stamp
# Unpack the input in the dimension of the sequence length to
get xt:[b,n]
for xt in tf.unstack(x, axis=1):
    out, h = cell(xt, h) # Forward calculation, both out and h
    are covered
# The final output can aggregate the output on each time stamp,
or just take the output of the last time stamp
out = out
```

The output variable out of the last time stamp will be the final output of the network. In fact, you can also save the output on each timestamp, and then sum or average it as the final output of the network.

11.4.2 Multilayer SimpleRNNCell Network

Like the convolutional neural network, although the recurrent neural network has been expanded many times on the time axis, it can only be counted as one network layer. By stacking multiple cell classes in the depth direction, the network can achieve the same effect as a deep convolutional neural network, which greatly improves the expressive ability of the network. However, compared with the number of deep layers of tens or hundreds of convolutional neural networks, recurrent neural networks are prone to gradient diffusion and gradient explosion. Deep recurrent neural networks are very difficult to train. The current common recurrent neural network models generally have number of layers less than 10.

Here we take a two-layer recurrent neural network as an example to introduce the use of cell class to build a multilayer RNN network. First create two SimpleRNNCell cells as follows:

```
x = tf.random.normal([4,80,100])
xt = x[:,0,:] # Take first timestamp of the input x0
# Construct 2 Cells, first cell0, then cell1, the memory state
vector length is 64
cell0 = layers.SimpleRNNCell(64)
cell1 = layers.SimpleRNNCell(64)
h0 = [tf.zeros([4,64])] # initial state vector of cell0
h1 = [tf.zeros([4,64])] # initial state vector of cell1
```

Calculate multiple times on the time axis to realize the forward operation of the entire network. The input xt on each time stamp first passes through the first layer to get the output out0, and then passes through the second layer to get the output out1. The code is as follows:

```
for xt in tf.unstack(x, axis=1):
    # xt is input and output is out0
    out0, h0 = cell0(xt, h0)
```

```
# The output out0 of the previous cell is used as the input
of this cell
out1, h1 = cell1(out0, h1)
```

The preceding method first completes the propagation of the input on one time stamp on all layers and then calculates the input on all time stamps in a loop.

In fact, it is also possible to first complete the calculation of all time stamps input on the first layer, and save the output list of the first layer on all time stamps, and then calculate the propagation of the second layer, the third layer, etc. as in the following:

```
# Save the output above all timestamps of the previous layer
middle_sequences = []
# Calculate the output on all timestamps of the first layer
and save
for xt in tf.unstack(x, axis=1):
    out0, h0 = cell0(xt, h0)
    middle_sequences.append(out0)
# Calculate the output on all timestamps of the second layer
# If it is not the last layer, you need to save the output
above all timestamps
for xt in middle_sequences:
    out1, h1 = cell1(xt, h1)
```

In this way, we need an additional list to save the information of all timestamps in the previous layer: middle_sequences.append(out0). These two methods have the same effect, and you can choose the coding style you like.

It should be noted that each layer of the recurrent neural network at each time stamp has a state output. For subsequent tasks, which state output should we collect and is the most effective? Generally speaking, the state of the last-level cell may preserve the global semantic features of the high-level, so the output of the last-level is generally used as the input of the subsequent task network. More specifically, the state output on the last timestamp of each layer contains the global information of the entire sequence. If you only want to use one state variable to complete subsequent tasks, such as sentiment classification problems, generally the output of the last layer at the last timestamp is most suitable.

11.4.3 SimpleRNN Layer

Through the use of the SimpleRNNCell layer, we can understand every detail of the forward operation of the recurrent neural network. In actual use, for simplicity, we do not want to manually implement the internal calculation process of the recurrent neural network, such as the initialization of the state vector at each layer and the operation of each layer on the time axis. Using the SimpleRNN high-level interface can help us achieve this goal very conveniently.

For example, if we want to complete the forward operation of a single-layer recurrent neural network, it can be easily implemented as follows:

```
In [6]:
layer = layers.SimpleRNN(64) # Create a SimpleRNN layer with a
state vector length of 64
x = tf.random.normal([4, 80, 100])
out = layer(x) # Like regular convolutional networks, one line
of code can get the output
out.shape
Out[6]: TensorShape([4, 64])
```

As you can see, SimpleRNN can complete the entire forward operation process with only one line of code, and it returns the output on the last time stamp by default. If you want to return the output list on all timestamps, you can set return_sequences=True as follows:

```
In [7]:
# When creating the RNN layer, set the output to return all
timestamps
layer = layers.SimpleRNN(64,return_sequences=True)
out = layer(x) # Forward calculation
out # Output, automatic concat operation
Out[7]:
<tf.Tensor: id=12654, shape=(4, 80, 64), dtype=float32, numpy=
array([[[ 0.31804922,  0.7904409 ,  0.13204293,
...,  0.02601025,
          -0.7833339 ,  0.65577114],...>
```

As you can see, the returned output tensor shape is [4,80,64], and the middle dimension 80 is the timestamp dimension. Similarly, we can achieve multilayer recurrent neural networks by stacking multiple SimpleRNNs, such as a two-layer network, and its usage is similar to that of a normal network. For example:

```
net = keras.Sequential([ # Build a 2-layer RNN network
# Except for the last layer, the output of all timestamps needs
to be returned to be used as the input of the next layer
layers.SimpleRNN(64, return_sequences=True),
layers.SimpleRNN(64),
])
out = net(x) # Forward calculation
```

Each layer needs the state output of the previous layer at each time stamp, so except for the last layer, all RNN layers need to return the state output at each time stamp, which is achieved by setting return_ sequences=True. As you can see, using the SimpleRNN layer is similar to the usage of convolutional neural networks, which is very concise and efficient.

11.5 Hands-On RNN Sentiment Classification

Now let's use the basic RNN network to solve the sentiment classification problem. The network structure is shown in Figure 11-9. The RNN network has two layers. The semantic features of the sequence signal are extracted cyclically. The state vector $h_s^{(2)}$ of the last time stamp of the second RNN layer is used as the global semantic feature representation of the sentence. It is sent to the classification network 3 formed by a fully connected layer, and the probability that the sample x is a positive emotion P (x is positive emotion | x) $\in$[0, 1] is obtained.

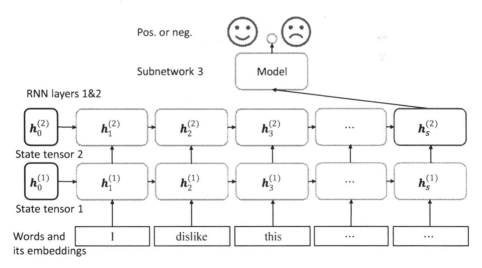

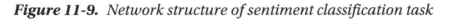

Figure 11-9. *Network structure of sentiment classification task*

11.5.1 Dataset

The classic IMDB movie review dataset is used here to complete the sentiment classification task. The IMDB movie review dataset contains 50,000 user reviews. The evaluation tags are divided into negative and positive. User reviews with IMDB rating <5 are marked as 0, which means negative; user reviews with IMDB rating ≥7 are marked as 1, which means positive. Twenty-five thousand film reviews were used for the training set and 25,000 were used for the test set.

The IMDB dataset can be loaded by datasets tool provided by Keras as follows:

```
In [8]:
batchsz = 128 # Batch size
total_words = 10000 # Vocabulary size N_vocab
max_review_len = 80 # The maximum length of the sentence s, the
sentence part greater than will be truncated, and the sentence
less than will be filled
embedding_len = 100 # Word vector feature length n
# Load the IMDB data set, the data here is coded with numbers,
and a number represents a word
(x_train, y_train), (x_test, y_test) = keras.datasets.imdb.
load_data(num_words=total_words)
# Print the input shape, the shape of the label
print(x_train.shape, len(x_train[0]), y_train.shape)
print(x_test.shape, len(x_test[0]), y_test.shape)
Out[8]:
(25000,) 218 (25000,)
(25000,) 68 (25000,)
```

As you can see, x_train and x_test are one-dimensional arrays with a length of 25,000. Each element of the array is a list of indefinite length, which stores each sentence encoded by numbers. For example, the first sentence of the training set has a total of 218 words, and the first sentence of the test set has 68 words, and each sentence contains the sentence start marker ID.

So how is each word encoded as a number? We can get the coding scheme by looking at its coding table, for example:

```
In [9]:
# Digital code table
word_index = keras.datasets.imdb.get_word_index()
# Print out the words and corresponding numbers in the
coding table
for k,v in word_index.items():
    print(k,v)
Out[10]:
    ...diamiter 88301
    moveis 88302
    mardi 14352
    wells' 11583
    850pm 88303...
```

Since the key of the coding table is a word and the value is an ID, the coding table is flipped and the coding ID of the flag bit is added. The code is as follows:

```
# The first 4 IDs are special bits
word_index = {k:(v+3) for k,v in word_index.items()}
word_index["<PAD>"] = 0  # Fill flag
word_index["<START>"] = 1 # Start flag
word_index["<UNK>"] = 2  # Unknown word sign
word_index["<UNUSED>"] = 3
# Flip code table
reverse_word_index = dict([(value, key) for (key, value) in
word_index.items()])
```

For a digitally encoded sentence, it is converted into string data by the following function:

```
def decode_review(text):
    return ' '.join([reverse_word_index.get(i, '?') for i
    in text])
```

For example, to convert a sentence, the code is as follows:

```
In [11]:decode_review(x_train[0])
Out[11]:
"<START> this film was just brilliant casting location scenery
story direction everyone's...<UNK> father came from...
```

For sentences with uneven lengths, a threshold is artificially set. For sentences larger than this length, select some words to be truncated, you can choose to cut off the beginning of the sentence or the end of the sentence. For sentences less than this length, you can choose to fill at the beginning or end of a sentence. The sentence truncation function can be conveniently realized by the keras.preprocessing.sequence.pad_sequences() function, for example:

```
# Truncate and fill sentences so that they are of equal length,
here long sentences retain the part behind the sentence, and
short sentences are filled in front
x_train = keras.preprocessing.sequence.pad_sequences(x_train,
maxlen=max_review_len)
x_test = keras.preprocessing.sequence.pad_sequences(x_test,
maxlen=max_review_len)
```

After truncating or filling to the same length, wrap it into a dataset object through the Dataset class, and add the commonly used dataset processing flow, the code is as follows:

```
In [12]:
# Build a data set, break up, batch, and discard the last batch
that is not enough batchsz
db_train = tf.data.Dataset.from_tensor_slices((x_train,
y_train))
db_train = db_train.shuffle(1000).batch(batchsz, drop_
remainder=True)
db_test = tf.data.Dataset.from_tensor_slices((x_test, y_test))
db_test = db_test.batch(batchsz, drop_remainder=True)
# Statistical data set attributes
print('x_train shape:', x_train.shape, tf.reduce_max(y_train),
tf.reduce_min(y_train))
print('x_test shape:', x_test.shape)
Out[12]:
x_train shape: (25000, 80) tf.Tensor(1, shape=(), dtype=int64)
tf.Tensor(0, shape=(), dtype=int64)
x_test shape: (25000, 80)
```

It can be seen that the sentence length after truncation and filling
is unified to 80, which is the set sentence length threshold. The drop_
remainder=True parameter discards the last batch, because its real batch
size may be smaller than the preset batch size.

11.5.2 Network Model

We create a custom model class MyRNN, inherited from the model base
class, we need to create a new Embedding layer, two RNN layers, and one
classification layer as follows:

```
class MyRNN(keras.Model):
    # Use Cell method to build a multi-layer network
    def __init__(self, units):
        super(MyRNN, self).__init__()
```

```
# [b, 64], construct Cell initialization state
vector, reuse
self.state0 = [tf.zeros([batchsz, units])]
self.state1 = [tf.zeros([batchsz, units])]
# Word vector encoding [b, 80] => [b, 80, 100]
self.embedding = layers.Embedding(total_words,
embedding_len,
                                  input_length=max_
                                  review_len)
# Construct 2 Cells and use dropout technology to
prevent overfitting
self.rnn_cell0 = layers.SimpleRNNCell(units,
dropout=0.5)
self.rnn_cell1 = layers.SimpleRNNCell(units,
dropout=0.5)
# Construct a classification network to classify the
output features of CELL, 2 classification
# [b, 80, 100] => [b, 64] => [b, 1]
self.outlayer = layers.Dense(1)
```

The word vector is encoded as length n=100, and the state vector length of RNN is h=units. The classification network completes a binary classification task, so the output node is set to 1.

The forward propagation logic is as follows: the input sequence completes the word vector encoding through the Embedding layer, loops through the two RNN layers to extract semantic features, takes the state vector output of the last time stamp of the last layer, and sends it to the classification network. The output probability is obtained after the Sigmoid activation function as in the following:

```
def call(self, inputs, training=None):
    x = inputs # [b, 80]
    # Word vector embedding: [b, 80] => [b, 80, 100]
```

```
x = self.embedding(x)
# Pass 2 RNN CELLs,[b, 80, 100] => [b, 64]
state0 = self.state0
state1 = self.state1
for word in tf.unstack(x, axis=1): # word: [b, 100]
    out0, state0 = self.rnn_cell0(word, state0,
    training)
    out1, state1 = self.rnn_cell1(out0, state1,
    training)
# Last layer's last time stamp as the network output:
[b, 64] => [b, 1]
x = self.outlayer(out1, training)
# Pass through activation function, p(y is pos|x)
prob = tf.sigmoid(x)

return prob
```

11.5.3 Training and Testing

For simplicity, here we use Keras' Compile&Fit method to train the network. Set the optimizer to Adam optimizer, the learning rate is 0.001, the error function uses the two-class cross-entropy loss function BinaryCrossentropy, and the test metric uses the accuracy rate. The code is as follows:

```
def main():
    units = 64 # RNN state vector length n
    epochs = 20 # Training epochs

    model = MyRNN(units) # Create the model
    # Compile
    model.compile(optimizer = optimizers.Adam(0.001),
```

```
                loss = losses.BinaryCrossentropy(),
                metrics=['accuracy'])
    # Fit and validate
    model.fit(db_train, epochs=epochs, validation_data=db_test)
    # Test
    model.evaluate(db_test)
```

After 20 Epoch trainings, the network achieves 80.1% accuracy rate at testing dataset.

11.6 Gradient Vanishing and Gradient Exploding

The training of recurrent neural networks is not stable, and the depth of the network cannot be arbitrarily deepened. Why do recurrent neural networks have difficulty in training? Let's briefly review the key expressions in the gradient derivation:

$$\frac{\partial h_t}{\partial h_i} = \prod_{j=i}^{t-1} diag\left(\sigma'\left(W_{xh}x_{j+1} + W_{hh}h_j + b\right)\right)W_{hh}$$

In other words, the gradient $\frac{\partial h_t}{\partial h_i}$ from time stamp i to time stamp t includes the continuous multiplication operation of W_{hh}. When the largest eigenvalue of W_{hh} is less than 1, multiple consecutive multiplication operations will make the element value of $\frac{\partial h_t}{\partial h_i}$ close to zero; when the value of $\frac{\partial h_t}{\partial h_i}$ is greater than 1, multiple consecutive multiplication operations will make the value of $\frac{\partial h_t}{\partial h_i}$ explosively increase.

We can intuitively feel the generation of gradient vanishing and gradient exploding from the following two examples:

```
In [13]:
W = tf.ones([2,2]) # Create a matrix
eigenvalues = tf.linalg.eigh(W)[0] # Calculate eigenvalue
eigenvalues
Out[13]:
<tf.Tensor: id=923, shape=(2,), dtype=float32, numpy=array(
[0., 2.], dtype=float32)>
```

It can be seen that the maximum eigenvalue of the all-one matrix is 2. Calculate the $W^1 \sim W^{10}$ of the W matrix and draw it as a graph of the power and the L2-norm of the matrix, as shown in Figure 11-10. It can be seen that when the maximum eigenvalue of the W matrix is greater than 1, the matrix multiplication will make the result larger and larger.

```
val = [W]
for i in range(10): # Matrix multiplication n times
    val.append([val[-1]@W])
# Calculate L2 norm
norm = list(map(lambda x:tf.norm(x).numpy(),val))
```

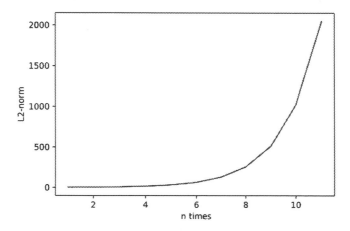

Figure 11-10. *Matrix multiplication when the largest eigenvalue is greater than 1*

Consider the case when the maximum eigenvalue is less than 1.

```
In [14]:
W = tf.ones([2,2])*0.4 # Create a matrix
eigenvalues = tf.linalg.eigh(W)[0] # Calculate eigenvalues
print(eigenvalues)
Out[14]:
tf.Tensor([0.  0.8], shape=(2,), dtype=float32)
```

It can be seen that the maximum eigenvalue of the W matrix at this time is 0.8. In the same way, consider the results of multiple multiplications of the W matrix as follows:

```
val = [W]
for i in range(10):
    val.append([val[-1]@W])
# Calculate the L2 norm
norm = list(map(lambda x:tf.norm(x).numpy(),val))
plt.plot(range(1,12),norm)
```

Its L2-norm curve is shown in Figure 11-11. It can be seen that when the maximum eigenvalue of the W matrix is less than 1, the matrix multiplication will make the result smaller and smaller, close to 0.

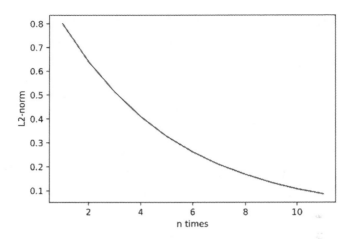

Figure 11-11. *Matrix multiplication when the largest eigenvalue is less than 1*

We call the phenomenon where the gradient value is close to 0 gradient vanishing and the phenomenon where the gradient value is far greater than 1 gradient exploding. Details about the gradient propagation mechanism can be found in Chapter 7. Gradient vanishing and gradient exploding are two situations that appear in the process of neural network optimization, and they are also not conducive to network training.

Consider the gradient descent algorithm:

$$\theta' = \theta - \eta \nabla_\theta L$$

When gradient vanishing occurs, $\nabla_\theta L \approx 0$, at this time $\theta' \approx \theta$, which means that the parameters remain unchanged after each gradient update, and the parameters of the neural network cannot be updated for a long time. The specific performance is that L has almost no change, other evaluation indicators, such as accuracy, also remain the same. When the gradient exploding occurs, $\nabla_\theta L \gg 1$, the update step size of the gradient $\eta \nabla_\theta L$ is very large, so that the updated θ' and θ are very different, and the network L has a sudden change, and even oscillates back and forth with non-convergence.

By deriving the gradient propagation formula of the recurrent neural network, we found that the recurrent neural network is prone to gradient vanishing and gradient exploding. So how to solve these two problems?

11.6.1 Gradient Clipping

Gradient exploding can be solved to a certain extent by gradient clipping. Gradient clipping is very similar to tensor limiting. It also limits the value or norm of the gradient tensor to a small interval, thereby reducing the gradient value far greater than 1 and avoiding gradient exploding.

In deep learning, there are three commonly used gradient clipping methods.

- Limit the value of the tensor directly so that all the elements of the tensor W are $w_{ij} \in [min, max]$. In TensorFlow, it can be achieved through the tf.clip_by_value() function. For example:

```
In [15]:
a=tf.random.uniform([2,2])
tf.clip_by_value(a,0.4,0.6) # Gradient value clipping
Out[15]:
<tf.Tensor: id=1262, shape=(2, 2), dtype=float32, numpy=
array([[0.5410726, 0.6      ],
       [0.4      , 0.6      ]], dtype=float32)>
```

- Limit the norm of the gradient tensor W. For example, the L2 norm of W – $\|W\|_2$ is constrained between [0,max]. If $\|W\|_2$ is greater than the max value, use:

$$W' = \frac{W}{\|W\|_2} \cdot max$$

to restrict $\|W\|_2$ to max. This can be done through the tf.clip_by_norm function. For example:

```
In [16]:
a=tf.random.uniform([2,2]) * 5
# Clip by norm
b = tf.clip_by_norm(a, 5)
# Norm before and after clipping
tf.norm(a),tf.norm(b)
Out[16]:
(<tf.Tensor: id=1338, shape=(), dtype=float32, numpy=5.380655>,
 <tf.Tensor: id=1343, shape=(), dtype=float32, numpy=5.0>)
```

It can be seen that for tensors with L2 norm greater than max, the norm value is reduced to 5 after clipping.

- The update direction of the neural network is represented by the gradient tensor W of all parameters. The first two methods only consider a single gradient tensor, and so the update direction of the network may change. If the norm of the gradient W of all parameters can be considered, and equal scaling can be achieved, then the gradient value of the network can be well restricted without changing the update direction of the network. This is the third method of gradient clipping: global norm clipping. In TensorFlow, the norm of the overall network gradient W can be quickly scaled through the tf.clip_by_global_norm function.

Let $W^{(i)}$ denote the i-th gradient tensor of the network parameters. Use the following formula to calculate the global norm of the network.

$$global\_norm = \sqrt{\sum_i \|W^{(i)}\|_2^{\,2}}$$

For the i-th parameter $W^{(i)}$, use the following formula to clip.

$$W^{(i)} = \frac{W^{(i)} \cdot max\_norm}{max(global\_norm, \, max\_norm)}$$

where max_norm is the global maximum norm value specified by the user. For example:

```
In [17]:
w1=tf.random.normal([3,3]) # Create gradient tensor 1
w2=tf.random.normal([3,3]) # Create gradient tensor 2
# Calculate global norm
global_norm=tf.math.sqrt(tf.norm(w1)**2+tf.norm(w2)**2)
# Clip by global norm and max norm=2
(ww1,ww2),global_norm=tf.clip_by_global_norm([w1,w2],2)
# Calcualte global norm after clipping
global_norm2 = tf.math.sqrt(tf.norm(ww1)**2+tf.norm(ww2)**2)
# Print the global norm before cropping and the global norm
after cropping
print(global_norm, global_norm2)
Out[17]:
tf.Tensor(4.1547523, shape=(), dtype=float32)
tf.Tensor(2.0, shape=(), dtype=float32)
```

It can be seen that after clipping, the global norm of the gradient group of the network parameters is reduced to max_norm=2. It should be noted that tf.clip_by_global_norm returns two objects of the clipped tensor – list and global_norm, where global_norm represents the global norm sum of the gradient before clipping.

Through gradient clipping, the gradient exploding phenomenon can be suppressed. As shown in Figure 11-12, the error value J of the $J(w, b)$ function represented by the surface in the figure under different network parameters w and b. There is a region where the gradient of the $J(w, b)$

function changes greatly. When parameters enter this area, gradient exploding are prone to occur, which makes the network state deteriorate rapidly. Figure 11-12 on the right shows the optimized trajectory after adding gradient clipping. Since the gradient is effectively restricted, the step size of each update is effectively controlled, thereby preventing the network from suddenly deteriorating.

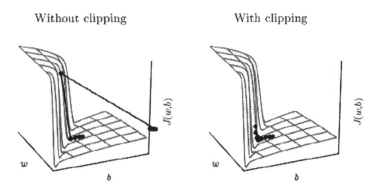

Figure 11-12. *Diagram of the optimized trajectory of gradient clipping [1]*

During network training, gradient clipping is generally performed after the gradient is calculated and before the gradient is updated. For example:

```
with tf.GradientTape() as tape:
  logits = model(x) # Forward calculation
  loss = criteon(y, logits) # Calculate error
# Calcualte gradients
grads = tape.gradient(loss, model.trainable_variables)
grads, _ = tf.clip_by_global_norm(grads, 25) # Global norm
clipping
# Update parameters using clipped gradient
optimizer.apply_gradients(zip(grads, model.trainable_
variables))
```

11.6.2 Gradient Vanishing

The gradient vanishing phenomenon can be suppressed by a series of measures such as increasing the learning rate, reducing the network depth, and adding Skip Connection.

Increasing the learning rate η can prevent gradient vanishing to a certain extent. When gradient vanishing occurs, the gradient of the network $\nabla_\theta L$ is close to 0. At this time, if the learning rate η is also small, such as $\eta=1e-5$, the gradient update step is even smaller. By increasing the learning rate, such as letting $\eta = 1e - 2$, it is possible to quickly update the state of the network and escape the gradient vanishing area.

For deep neural networks, the gradient gradually propagates from the last layer to the first layer, and gradient vanishing is generally more likely to appear in the first few layers of the network. Before the emergence of deep residual networks, it was very difficult to train deep networks with dozens or hundreds of layers. The gradients of the previous layers of the network were very prone to gradient vanishing, which made the network parameters not updated for a long time. The deep residual network better overcomes the gradient vanishing phenomenon, so that the number of neural network layers can reach hundreds or thousands. Generally speaking, reducing the network depth can reduce the gradient vanishing phenomenon, but after the number of network layers is reduced, the network expression ability will be weaker.

11.7 RNN Short-Term Memory

In addition to the training difficulty of recurrent neural networks, there is a more serious problem, that is, short-term memory. Consider a long sentence:

Today's weather is so beautiful, even though an unpleasant thing happened on the road..., I immediately adjusted my state and happily prepared for a beautiful day.

According to our understanding, the reason why we "happily prepared for a beautiful day" is that "Today's weather is so beautiful" which is mentioned at the beginning of the sentence. It can be seen that humans can understand long sentences well, but recurrent neural networks are not necessary. Researchers have found that when recurrent neural networks process long sentences, they can only understand information within a limited length, while useful information in a longer range cannot be used well. We call this phenomenon short-term memory.

So, can this short-term memory be prolonged so that the recurrent neural network can effectively use the training data in a longer range, thereby improving model performance? In 1997, Swiss artificial intelligence scientist Jürgen Schmidhuber proposed the Long Short-Term Memory (LSTM) model. Compared with the basic RNN network, LSTM has longer memory and is better at processing longer sequence data. After LSTM was proposed, it has been widely used in tasks such as sequence prediction and natural language processing, almost replacing the basic RNN model .

Next, we will introduce the more popular and powerful LSTM network.

11.8 LSTM Principle

The basic RNN network structure is shown in Figure 11-13. After the state vector h_{t-1} of the previous time stamp and the input x_t of the current time stamp are linearly transformed, the new state vector h_t is obtained through the activation function tanh. Compared with the basic RNN network which has only one state vector h_t, LSTM adds a new state vector C_t, and at the same time introduces a gate control mechanism, which controls the forgetting and updating of information through the gate control unit, as shown in Figure 11-14.

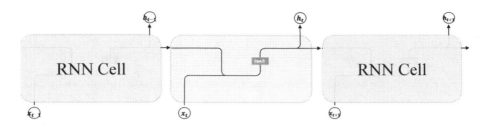

Figure 11-13. *Basic RNN structure*

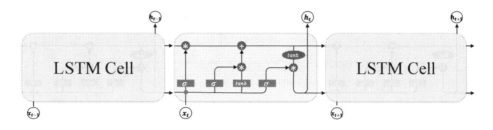

Figure 11-14. *LSTM structure*

In LSTM, there are two state vectors c and h, where c is the internal state vector of LSTM, which can be understood as the memory state vector of LSTM, and h represents the output vector of LSTM. Compared with the basic RNN, LSTM separates the internal memory and output into two variables and uses three gates, input gate, forget gate, and output gate, to control the internal information flow.

The gate mechanism can be understood as a way of controlling the data flow, analogous to a water valve: when the water valve is fully opened, water flows unimpeded; when the water valve is fully closed, the water flow is completely blocked. In LSTM, the valve opening degree are represented by the gate control value vector g, as shown in Figure 11-15, the gate control is compressed to the interval between $[0,1]$ through the $\sigma(g)$ activation function. When $\sigma(g) = 0$, all gates are closed, and output is $o = 0$. When $\sigma(g) = 1$, all gates are open, and output is $o = x$. Through the gate mechanism, the data flow can be better controlled.

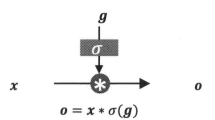

$$o = x * \sigma(g)$$

Figure 11-15. *Gate mechanism*

In the following, we respectively introduce the principles and functions of the three gates.

11.8.1 Forget Gate

The forget gate acts on the LSTM state vector c to control the impact of the memory c_{t-1} of the previous time stamp on the current time stamp. As shown in Figure 11-16, the control variable g_f of the forget gate is determined by:

$$g_f = \sigma\left(W_f\left[h_{t-1}, x_t\right] + b_f\right)$$

where W_f and b_f are the parameter tensors of the forget gate, which can be automatically optimized by the backpropagation algorithm. σ is the activation function, and the Sigmoid function is generally used. When $g_f = 1$, the forget gates are all open, and LSTM accepts all the information of the previous state c_{t-1}. When the gating $g_f = 0$, the forget gate is closed, and LSTM directly ignores c_{t-1}, and the output is a vector of 0. This is why it's called the forget gate.

After passing through the forget gate, the state vector of LSTM becomes $g_f c_{t-1}$.

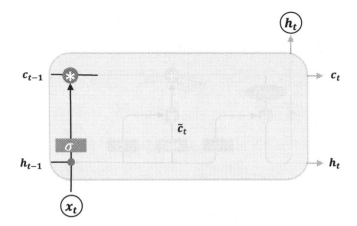

Figure 11-16. *Forget gate*

11.8.2 Input Gate

The input gate is used to control the degree to which the LSTM receives input. First, the new input vector $\tilde{c}_t$ is obtained by nonlinear transformation of the input x_t of the current time stamp and the output h_{t-1} of the previous time stamp:

$$\tilde{c}_t = tanh\ tanh\left(W_c\left[h_{t-1},x_t\right]+b_c\right)$$

where W_c and b_c are the parameters of the input gate, which need to be automatically optimized by the back propagation algorithm, and Tanh is the activation function, which is used to normalize the input to [-1,1]. $\tilde{c}_t$ does not completely refresh the memory that enters the LSTM but controls the amount of input received through the input gate. The control variables of the input gate also come from the input x_t and the output h_{t-1}:

$$g_i = \sigma\left(W_i\left[h_{t-1},x_t\right]+b_i\right)$$

where W_i and b_i are the parameters of the input gate, which need to be automatically optimized by the back propagation algorithm, and σ is the activation function, and the Sigmoid function is generally used. The input gate control variable g_i determines how LSTM accepts the new input $\tilde{c}_t$ of the current time stamp: when $g_i = 0$, LSTM does not accept any new input $\tilde{c}_t$; when $g_i = 1$, LSTM accepts all new input $\tilde{c}_t$, As shown in Figure 11-17.

After passing through the input gate, the vector to be written into Memory is $g_i\tilde{c}_t$.

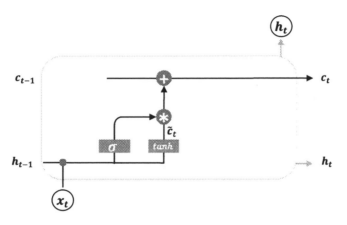

Figure 11-17. *Input gate*

11.8.3 Update Memory

Under the control of the forget gate and the input gate, LSTM selectively reads the memory c_{t-1} of the previous time stamp and the new input $\tilde{c}_t$ of the current time stamp. The refresh mode of the state vector c_t is:

$$c_t = g_i\tilde{c}_t + g_f c_{t-1}$$

The new state vector c_t obtained is the state vector of the current time stamp, as shown in Figure 11-17.

11.8.4 Output Gate

The internal state vector c_t of LSTM is not directly used for output, which is different from the basic RNN. The state vector h of the basic RNN network is used for both memory and output, so the basic RNN can be understood as the state vector c and the output vector h are the same object. In LSTM, the state vector is not totally outputted, but selectively under the action of the output gate. The gate variable g_o of the output gate is:

$$g_o = \sigma\left(W_o\left[h_{t-1}, x_t\right] + b_o\right)$$

where W_o and b_o are the parameters of the output gate, which also need to be automatically optimized by the back propagation algorithm. σ is the activation function, and the Sigmoid function is generally used. When the output gate $g_o = 0$, the output is closed, and the internal memory of LSTM is completely blocked and cannot be used as an output. At this time, the output is a vector of 0; when the output gate $g_o = 1$, the output is fully open, and the LSTM state vector c_t is all used for output. The output of LSTM is composed of:

$$h_t = g_o \cdot tanh\,tanh\left(c_t\right)$$

That is, the memory vector c_t interacts with the input gate after passing the Tanh activation function to obtain the output of the LSTM. Since $g_o \in [0, 1]$ and $tanh\,tanh\left(c_t\right) \in [-1, 1]$, the output of LSTM is $h_t \in [-1, 1]$.

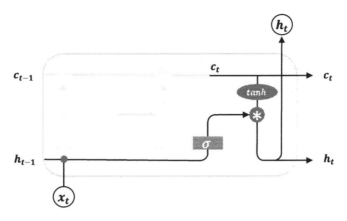

Figure 11-18. *Output gate*

11.8.5 Summary

Although LSTM has a large number of state vectors and gates, the calculation process is relatively complicated. But since each gate control function is clear, the role of each state is also easier to understand. Here, the typical gating behavior is listed and the LSTM behavior of the code is explained, as shown in Table 11-1.

Table 11-1. *Typical behavior of input gate and forget gate*

Input gating	Forget Gating	LSTM behavior
0	1	Only use memory
1	1	Integrated input and memory
0	0	Clear memory
1	0	Input overwrites memory

11.9 How to Use the LSTM Layer

In TensorFlow, there are also two ways to implement LSTM networks. Either LSTMCell can be used to manually complete the cyclic operation on the time stamp, or the forward operation can be completed in one step through the LSTM layer.

11.9.1 LSTMCell

The usage of LSTMCell is basically the same as SimpleRNNCell. The difference is that there are two state variables – list for LSTM, namely, $[h_t, c_t]$, which need to be initialized separately. The first element of list is h_t and the second element is c_t. When the cell is called to complete the forward operation, two elements are returned. The first element is the output of the cell, which is h_t, and the second element is the updated state list of the cell: $[h_t, c_t]$. First create a new LSTMCell with a state vector length of $h = 64$, where the length of the state vector c_t and the output vector h_t are both h. The code is as follows:

```
In [18]:
x = tf.random.normal([2,80,100])
xt = x[:,0,:] # Get a timestamp input
cell = layers.LSTMCell(64) # Create LSTM Cell
# Initialization state and output List,[h,c]
state = [tf.zeros([2,64]),tf.zeros([2,64])]
out, state = cell(xt, state) # Forward calculation
# View the id of the returned element
id(out),id(state[0]),id(state[1])
Out[18]: (1537587122408, 1537587122408, 1537587122728)
```

It can be seen that the returned output out is the same as the id of the first element h_t of the list, which is consistent with the original intention of the basic RNN and is for the unification of the format.

By unrolling the loop operation on the timestamp, the forward propagation of a layer can be completed, and the writing method is the same as the basic RNN. For example:

```
# Untie it in the sequence length dimension, and send it to the
LSTM Cell unit in a loop
for xt in tf.unstack(x, axis=1):
    # Forward calculation
    out, state = cell(xt, state)
```

The output can use only the output on the last time stamp, or it can aggregate the output vectors on all time stamps.

11.9.2 LSTM layer

Through the layers.LSTM layer, the operation of the entire sequence can be conveniently completed at one time. First create a new LSTM network layer, for example:

```
# Create an LSTM layer with a memory vector length of 64
layer = layers.LSTM(64)
# The sequence passes through the LSTM layer and returns the
output h of the last time stamp by default
out = layer(x)
```

After forward propagation through the LSTM layer, only the output of the last timestamp will be returned by default. If you need to return the output above each timestamp, you need to set the return_sequences=True. For example:

```
# When creating the LSTM layer, set to return the output on
each timestamp
layer = layers.LSTM(64, return_sequences=True)
```

```
# Forward calculation, the output on each timestamp is
automatically concated to form a tensor
out = layer(x)
```

The out returned at this time contains the status output above all timestamps, and its shape is [2,80,64], where 80 represents 80 timestamps.

For multilayer neural networks, you can wrap multiple LSTM layers with Sequential containers, and set all non-final layer networks return_sequences=True, because the non-final LSTM layer needs the output of all timestamps of the previous layer as input. For example:

```
# Like the CNN network, LSTM can also be simply stacked layer
by layer
net = keras.Sequential([
    layers.LSTM(64, return_sequences=True), # The non-final
    layer needs to return all timestamp output
    layers.LSTM(64)
])
# Once through the network model, you can get the output of the
last layer and the last time stamp
out = net(x)
```

11.10 GRU Introduction

LSTM has a longer memory capacity and has achieved better performance than the basic RNN model on most sequence tasks. More importantly, LSTM is not prone to gradient vanishing. However, the LSTM structure is relatively complex, the calculation cost is high, and the model parameters are large. Therefore, scientists try to simplify the calculation process inside LSTM, especially to reduce the number of gates. Studies found that the forget gate is the most important gate control in LSTM [2], and even found that the simplified version of the network with only the forget gate is better than the standard LSTM network on multiple benchmark data sets. Among many simplified versions of LSTM, Gated Recurrent Unit (GRU) is one of the most widely used RNN variants. GRU merges the internal state vector and output vector into a state vector h, and the number of gates is also reduced to two, reset gate and update gate, as shown in Figure 11-19.

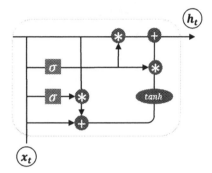

Figure 11-19. *GRU network structure*

Let's introduce the principle and function of reset gate and update gate respectively.

11.10.1 Reset Door

The reset gate is used to control the amount of the state h_{t-1} of the last time stamp into the GRU. The gating vector g_r is obtained by transforming the current time stamp input x_t and the last time stamp state h_{t-1}, the relationship is as follows:

$$g_r = \sigma\left(W_r\left[h_{t-1}, x_t\right] + b_r\right)$$

where W_r and b_r are the parameters of the reset gate, which are automatically optimized by the back propagation algorithm, σ is the activation function, and the Sigmoid function is generally used. The gating vector g_r only controls the state h_{t-1}, but not the input x_t:

$$\tilde{h}_t = tanh\ tanh\left(W_h\left[g_r h_{t-1}, x_t\right] + b_h\right)$$

When $g_r = 0$, the new input $\tilde{h}_t$ all comes from the input x_t, and h_{t-1} is not accepted, which is equivalent to resetting h_{t-1}. When $g_r = 1$, h_{t-1} and input x_t jointly generate a new input $\tilde{h}_t$, as shown in Figure 11-20.

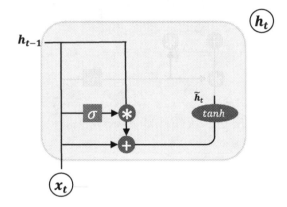

Figure 11-20. *Reset gate*

11.10.2 Update Gate

The update gate controls the degree of influence of the last time stamp state h_{t-1} and the new input $\tilde{h}_t$ on the new state vector h_t. Update the gating vector g_z by:

$$g_z = \sigma\left(W_z\left[h_{t-1}, x_t\right] + b_z\right)$$

where W_z and b_z are the parameters of the update gate, which are automatically optimized by the back propagation algorithm, σ is the activation function, and the Sigmoid function is generally used. g_z is used to control the new input $\tilde{h}_t$ signal, and $1 - g_z$ is used to control the state h_{t-1} signal:

$$h_t = (1 - g_z)h_{t-1} + g_z\tilde{h}_t$$

It can be seen that the updates of $\tilde{h}_t$ and h_{t-1} to h_t are in a state of competing with each other. When the update gate $g_z = 0$, all h_t comes from the last time stamp state h_{t-1}; when the update gate $g_z = 1$, all h_t comes from the new input $\tilde{h}_t$.

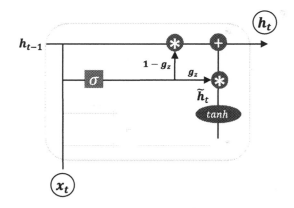

Figure 11-21. *Update gate*

11.10.3 How to Use GRU

Similarly, in TensorFlow, there are also cell and layer methods to implement GRU networks. The usage of GRUCell and GRU layer is very similar to the previous SimpleRNNCell, LSTMCell, SimpleRNN and LSTM. First, use GRUCell to create a GRU cell object, and cyclically unroll operations on the time axis. For example:

```
In [19]:
# Initialize the state vector, there is only one GRU
h = [tf.zeros([2,64])]
```

```
cell = layers.GRUCell(64) # New GRU Cell, vector length is 64
# Untie in the timestamp dimension, loop through the cell
for xt in tf.unstack(x, axis=1):
    out, h = cell(xt, h)
# Out shape
out.shape
Out[19]:TensorShape([2, 64])
```

You can easily create a GRU network layer through the layers.GRU class, and stack a network of multiple GRU layers through the Sequential container. For example:

```
net = keras.Sequential([
    layers.GRU(64, return_sequences=True),
    layers.GRU(64)
])
out = net(x)
```

11.11 Hands-On LSTM/GRU Sentiment Classification

Earlier we introduced the sentiment classification problem and used the SimpleRNN model to solve the problem. After introducing the more powerful LSTM and GRU networks, we upgraded the network model. Thanks to the unified format of TensorFlow's recurrent neural network related interfaces, only a few modifications on the original code can be perfectly upgraded to the LSTM or GRU model.

11.11.1 LSTM Model

First, let's use the cell method. There are two state lists of the LSTM network, and the *h* and *c* vectors of each layer need to be initialized respectively. For example:

```
self.state0 = [tf.zeros([batchsz, units]),tf.
zeros([batchsz, units])]
self.state1 = [tf.zeros([batchsz, units]),tf.
zeros([batchsz, units])]
```

Modify the model to LSTMCell model as in the following:

```
self.rnn_cell0 = layers.LSTMCell(units, dropout=0.5)
self.rnn_cell1 = layers.LSTMCell(units, dropout=0.5)
```

Other codes can run without modification. For the layer method, only one part of the network model needs to be modified, as follows:

```
# Build RNN, replace with LSTM class
self.rnn = keras.Sequential([
    layers.LSTM(units, dropout=0.5, return_
    sequences=True),
    layers.LSTM(units, dropout=0.5)
])
```

11.11.2 GRU model

For the cell method, there is only one GRU state list. Like the basic RNN, you only need to modify the type of cell created. The code is as follows:

```
# Create 2 Cells
self.rnn_cell0 = layers.GRUCell(units, dropout=0.5)
self.rnn_cell1 = layers.GRUCell(units, dropout=0.5)
```

For the layer method, just modify the network layer type as follows:

```
# Create RNN
self.rnn = keras.Sequential([
    layers.GRU(units, dropout=0.5, return_
    sequences=True),
    layers.GRU(units, dropout=0.5)
])
```

11.12 Pre-trained Word Vectors

In the sentiment classification task, the Embedding layer is trained from scratch. In fact, for text processing tasks, most of the domain knowledge is shared, so we can use the word vectors trained on other tasks to initialize the Embedding layer to complete the domain knowledge transfer. Start training based on the pre-trained Embedding layer, and good results can be achieved with a small number of samples.

We take the pre-trained GloVe word vector as an example to demonstrate how to use the pre-trained word vector model to improve task performance. First, download the pre-trained GloVe word vector table from the official website. We choose the file glove.6B.100d.txt with a feature length of 100, and each word is represented by a vector of length 100, which can be decompressed after downloading.

Name	Date modified	Type	Size
glove.6B.50d.txt	1/3/2018 9:04 PM	Text Document	167,335 KB
glove.6B.100d.txt	8/5/2014 6:14 AM	Text Document	338,982 KB

Figure 11-22. *GloVe word vector model file*

Use the Python file IO code to read the word encoding vector table and store it in the Numpy array. code show as in the following:

```
print('Indexing word vectors.')
embeddings_index = {} # Extract words and their vectors and
save them in a dictionary
# Word vector model file storage path
GLOVE_DIR = r'C:\Users\z390\Downloads\glove6b50dtxt'
with open(os.path.join(GLOVE_DIR, 'glove.6B.100d.
txt'),encoding='utf-8') as f:
    for line in f:
        values = line.split()
        word = values[0]
        coefs = np.asarray(values[1:], dtype='float32')
        embeddings_index[word] = coefs
print('Found %s word vectors.' % len(embeddings_index))
```

The GloVe.6B version stores a vector table of 400,000 words in total. We only considered up to 10,000 common words. We obtained the word vectors from the GloVe model according to the number code table of the words and wrote them into the corresponding positions as in the following:

```
num_words = min(total_words, len(word_index))
embedding_matrix = np.zeros((num_words, embedding_len)) # Word
vector table
for word, i in word_index.items():
    if i >= MAX_NUM_WORDS:
        continue # Filter out other words
    embedding_vector = embeddings_index.get(word) # Query word
    vector from GloVe
```

```
if embedding_vector is not None:
    # words not found in embedding index will be all-zeros.
    embedding_matrix[i] = embedding_vector # Write the
    corresponding location
print(applied_vec_count, embedding_matrix.shape)
```

After obtaining the vocabulary data, use the vocabulary to initialize the Embedding layer, and set the Embedding layer not to participate in gradient optimization as in the following:

```
# Create Embedding layer
self.embedding = layers.Embedding(total_words,
embedding_len, input_length=max_review_len,
trainable=False)# Does not participate in
gradient updates
self.embedding.build(input_shape=(None, max_
review_len))
# Initialize the Embedding layer using the GloVe model
self.embedding.set_weights([embedding_matrix])#
initialization
```

The other parts are consistent. We can simply compare the training results of the Embedding layer initialized by the pre-trained GloVe model with the training results of the randomly initialized Embedding layer. After training 50 Epochs, the accuracy of the pre-training model reached 84.7%, an increase of approximately 2%.

11.13 Pre-trained Word Vectors

In this chapter, we introduced the recurrent neural network (RNN) that is appropriate to handle sequence related problems such as speech and stock market signals. Several sequence representation methods were discussed including one-hot encoding and word embedding. Then we introduced the motivation of developing the RNN structure along with examples of the SimpleRNNCell network. Hands-on sentiment classification was implemented using RNN to help us get familiar with using RNN to solve real world problems. Gradient vanishing and exploding are common issues during the RNN training process. Fortunately, the gradient clipping method can be used to overcome the gradient exploding issue. And different variants of RNN such as LSTM and GRU can be used to avoid the gradient vanishing issue. The sentiment classification example shows the better performance of using LSTM and GRU models because their ability of avoiding gradient exploding issue.

11.14 References

[1].   I. Goodfellow, Y. Bengio and A. Courville, Deep Learning, MIT Press, 2016.

[2].   J. Westhuizen and J. Lasenby, "The unreasonable effectiveness of the forget gate," *CoRR*, abs/1804.04849, 2018.

CHAPTER 12

Autoencoder

Suppose machine learning is a cake, reinforcement learning is the cherry on the cake, supervised learning is the icing on the outside, and unsupervised learning is the cake itself.

—Yann LeCun

Earlier we introduced the neural network learning algorithm given the sample and its corresponding label. This type of algorithm actually learns the conditional probability $P(y|x)$ given the sample x. With the booming social network today, it is relatively easy to obtain massive sample data x, such as photos, voices, and texts, but the difficulty is to obtain the label information corresponding to these data. For example, in addition to collecting source language text, the target language text data to be translated is also required for machine translation. Data labeling is mainly based on human prior knowledge. For example, Amazon's Mechanical Turk system is responsible for data labeling, recruiting part-time staff from all over the world to complete customer data labeling tasks. The scale of data required for deep learning is generally very large. This method of relying heavily on manual data annotation is expensive and inevitably introduces the subjective prior bias of the annotator.

For massive unlabeled data, is there a way to learn the data distribution $P(x)$ from it? This is the unsupervised learning algorithm that we will introduce in this chapter. In particular, if the algorithm learns x as a

© Liangqu Long and Xiangming Zeng 2022
L. Long and X. Zeng, *Beginning Deep Learning with TensorFlow*,
https://doi.org/10.1007/978-1-4842-7915-1_12

supervised signal, this type of algorithm is called self-supervised learning, and the autoencoder algorithm introduced in this chapter is one type of self-supervised learning algorithms.

12.1 Principle of Autoencoder

Let us consider the function of neural networks in supervised learning:

$$o = f_\theta(x), x \in R^{d_{in}}, o \in R^{d_{out}}$$

d_{in} is the length of the input feature vector, and d_{out} is the length of the network output vector. For classification problems, the network model transforms the input feature vector x of length d_{in} to the output vector o of length d_{out}. This process can be considered as a feature reduction process, transforming the original high-dimensional input vector x to a low-dimensional variable o. Dimensionality reduction has a wide range of applications in machine learning, such as file compression and data preprocessing. The most common dimension reduction algorithm is principal component analysis (PCA), which obtains the main components of the data by eigen-decomposing the covariance matrix, but PCA is essentially a linear transformation, and the ability to extract features is limited.

So can we use the powerful nonlinear expression capabilities of neural networks to learn low-dimensional data representation? The key to the problem is that training neural networks generally requires an explicit label data (or supervised signal), but unsupervised data has no additional labeling information, only the data x itself.

Therefore, we try to use the data x itself as a supervision signal to guide the training of the network, that is, we hope that the neural network can learn the mapping $f_\theta : x \rightarrow x$. We divide the network f_θ into two parts. The first sub-network tries to learn the mapping relationship: $g_{\theta_1} : x \rightarrow z$, and the latter sub-network tries to learn the mapping relationship $h_{\theta_2} : z \rightarrow x$, as shown in Figure 12-1. We consider g_{θ_1} as a process of data encoding which encodes the high-dimensional input x into a low-dimensional hidden variable z (latent variable or hidden variable), which is called an encoder network. h_{θ_2} is considered as the process of data decoding, which decodes the encoded input z into high-dimensional x, which is called a decoder network.

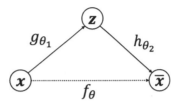

Figure 12-1. *Autoencoder model*

The encoder and decoder jointly complete the encoding and decoding process of the input data x. We call the entire network model f_θ an autoencoder for short. If a deep neural network is used to parameterize g_{θ_1} and h_{θ_2} functions, it is called deep autoencoder, as shown in Figure 12-2.

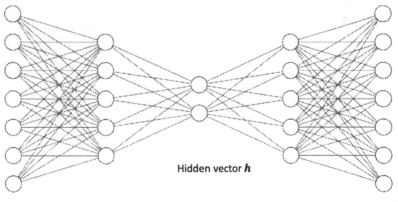
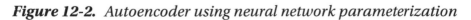

Figure 12-2. *Autoencoder using neural network parameterization*

The self-encoder can transform the input to the hidden vector z, and reconstruct $\underline{x}$ through the decoder. We hope that the output of the decoder can perfectly or approximately recover the original input, that is $\underline{x} \approx x$, then the optimization goal of the autoencoder can be written as:

$$min\ L = dist(x, \underline{x})$$

$$\underline{x} = h_{\theta_2}\left(g_{\theta_1}(x)\right)$$

where $dist(x, \underline{x})$ represents the distance measurement between x and $\underline{x}$, which is called the reconstruction error function. The most common measurement method is the square of the Euclidean distance. The calculation method is as follows:

$$L = \sum_i (x_i - \underline{x}_i)^2$$

It is equivalent in principle to the mean square error. There is no essential difference between the autoencoder network and the ordinary neural network, except that the trained supervision signal has changed from the label y to its own x. With the help of the nonlinear feature extraction capability of deep neural networks, the autoencoder can obtain good data representation, for example, smaller size and dimension data representation than the original input data. This is very useful for data and information compression. Compared with linear methods such as PCA, the autoencoder has better performance and can even recover the input x more perfectly.

In Figure 12-3(a), the first row is a real MNIST handwritten digit picture randomly sampled from the test set, and the second, third, and fourth rows are reconstructed using a hidden vector of length 30, using autoencoder, logistic PCA, and standard PCA, respectively. In Figure 12-3(b), the first row is a real portrait image, and the second and third rows are based on a hidden vector of length 30, which is recovered using the autoencoder and the standard PCA algorithm. It can be seen that the image reconstructed by the autoencoder is relatively clear and has a high degree of restoration, while the image reconstructed by the PCA algorithm is blurry.

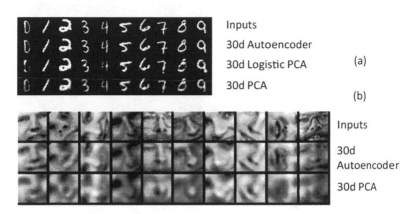

Inputs
30d Autoencoder
30d Logistic PCA   (a)
30d PCA
(b)

Inputs
30d
Autoencoder
30d PCA

Figure 12-3. *Autoencoder vs. PCA [1]*

12.2 Hands-On Fashion MNIST Image Reconstruction

The principle of the autoencoder algorithm is very simple, easy to implement, and stable in training. Compared with the PCA algorithm, the powerful expression ability of the neural network can learn the high-level abstract hidden feature vector z of the input, and it can also reconstruct the input based on z. Here we perform actual picture reconstruction based on the Fashion MNIST dataset.

12.2.1 Fashion MNIST Dataset

Fashion MNIST is a dataset that is a slightly more complicated problem than MNIST image recognition. Its settings are almost the same as MNIST. It contains ten types of grayscale images of different types of clothes, shoes, and bags, and the size of the image is 28 × 28, with a total of 70,000 pictures, of which 60,000 are used for the training set and 10,000 are used for the test set, as shown in Figure 12-4. Each row is a category of pictures. As you can see, Fashion MNIST has the same settings except that the picture content is different from MNIST. In most cases, the original algorithm code based on MNIST can be directly replaced without additional modification. Since Fashion MNIST image recognition is more difficult than MNIST, it can be used to test the performance of a slightly more complex algorithm.

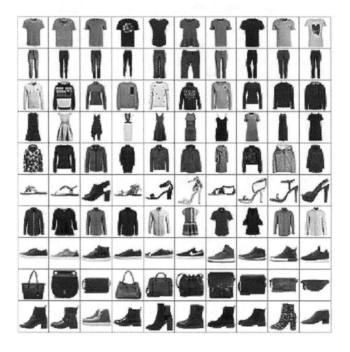

Figure 12-4. *Fashion MNIST Dataset*

In TensorFlow, it is also very convenient to load the Fashion MNIST dataset, which can be downloaded, managed, and loaded online using the keras.datasets.fashion_mnist.load_data() function as in the following:

```
# Load Fashion MNIST data set
(x_train, y_train), (x_test, y_test) = keras.datasets.fashion_
mnist.load_data()
# Normalize
```

```
x_train, x_test = x_train.astype(np.float32) / 255., x_test.
astype(np.float32) / 255.
# Only need to use image data to build data set objects, no
tags required
train_db = tf.data.Dataset.from_tensor_slices(x_train)
train_db = train_db.shuffle(batchsz * 5).batch(batchsz)
#  Build test set objects
test_db = tf.data.Dataset.from_tensor_slices(x_test)
test_db = test_db.batch(batchsz)
```

12.2.2 Encoder

We use the encoder to reduce the dimensionality of the input picture
$x \in R^{784}$ to a lower-dimensional hidden vector, $h \in R^{20}$, and use the decoder
to reconstruct the picture based on the hidden vector h. The autoencoder
model is shown in Figure 12-5. The decoder is composed of a 3-layer fully
connected network with output nodes of 256, 128, and 20, respectively.
The decoder is also composed of a three-layer fully connected network
with output nodes of 128, 256, and 784, respectively.

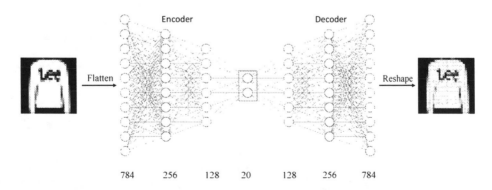

Figure 12-5. *Fashion MNIST autoencoder network architecture*

The first is the realization of the encoder sub-network. A three-layer neural network is used to reduce the dimensionality of the image vector from 784 to 256, 128, and finally to h_dim. Each layer uses the ReLU activation function, and the last layer does not use any activation function.

```
# Create Encoders network, implemented in the
initialization function of the autoencoder class
self.encoder = Sequential([
    layers.Dense(256, activation=tf.nn.relu),
    layers.Dense(128, activation=tf.nn.relu),
    layers.Dense(h_dim)
])
```

12.2.3 Decoder

Let's create the decoder sub-network. Here, the hidden vector h_dim is upgraded to the length of 128, 256, and 784 in turn. Except for the last layer, the ReLU activation function are used. The output of the decoder is a vector of length 784, which represents a 28 × 28 size picture after being flattened, and can be restored to a picture matrix through the reshape operation as in the following:

```
# Create Decoders network
self.decoder = Sequential([
    layers.Dense(128, activation=tf.nn.relu),
    layers.Dense(256, activation=tf.nn.relu),
    layers.Dense(784)
])
```

12.2.4 Autoencoder

The preceding two sub-networks of encoder and decoder are implemented in the autoencoder class AE, and we create these two sub-networks in the initialization function at the same time.

```
class AE(keras.Model):
    # Self-encoder model class, including Encoder and Decoder
    2 subnets
    def __init__(self):
        super(AE, self).__init__()
        #  Create Encoders network
        self.encoder = Sequential([
            layers.Dense(256, activation=tf.nn.relu),
            layers.Dense(128, activation=tf.nn.relu),
            layers.Dense(h_dim)
        ])
        #  Create Decoders network
        self.decoder = Sequential([
            layers.Dense(128, activation=tf.nn.relu),
            layers.Dense(256, activation=tf.nn.relu),
            layers.Dense(784)
        ])
```

Next, the forward propagation process is implemented in the call function. The input image first obtains the hidden vector h through the encoder sub-network and then obtains the reconstructed image through the decoder. Just call the forward propagation function of the encoder and decoder in turn as follows:

```
    def call(self, inputs, training=None):
        # Forward propagation function
        # Encoding to obtain hidden vector h,[b, 784]
        => [b, 20]
```

```
        h = self.encoder(inputs)
        # Decode to get reconstructed picture, [b, 20] =>
        [b, 784]
        x_hat = self.decoder(h)

        return x_hat
```

12.2.5 Network Training

The training process of the autoencoder is basically the same as that of a classifier. The distance between the reconstructed vector $\underline{x}$ and the original input vector x is calculated through the error function, and then the gradients of the encoder and decoder are simultaneously calculated using the automatic derivation mechanism of TensorFlow.

First, create an instance of the autoencoder and optimizer, and set an appropriate learning rate. For example:

```
#  Create network objects
model = AE()
#  Specify input size
model.build(input_shape=(4, 784))
#  Print network information
model.summary()
#  Create an optimizer and set the learning rate
optimizer = optimizers.Adam(lr=lr)
```

Here 100 Epochs are trained, and the reconstructed image vector is obtained through forward calculation each time, and the tf.nn.sigmoid_cross_entropy_with_logits loss function is used to calculate the direct error between the reconstructed image and the original image. In fact, it is also feasible to use the MSE error function as in the following:

```
for epoch in range(100): # Train 100 Epoch
    for step, x in enumerate(train_db): #  Traverse the
    training set
        # Flatten, [b, 28, 28] => [b, 784]
        x = tf.reshape(x,    [-1, 784])
        # Build a gradient recorder
        with tf.GradientTape() as tape:
            # Forward calculation to obtain the
            reconstructed picture
            x_rec_logits = model(x)
            # Calculate the loss function between the
            reconstructed picture and the input
            rec_loss = tf.nn.sigmoid_cross_entropy_with_
            logits(labels=x, logits=x_rec_logits)
            # Calculate the mean
            rec_loss = tf.reduce_mean(rec_loss)
        # Automatic derivation, including the gradient of 2
        sub-networks
        grads = tape.gradient(rec_loss, model.trainable_
        variables)
        # Automatic update, update 2 subnets at the same time
        optimizer.apply_gradients(zip(grads, model.trainable_
        variables))
        if step % 100 ==0:
            # Interval print training error
            print(epoch, step, float(rec_loss))
```

12.2.6 Image Reconstruction

Different from the classification problem, the model performance of
the autoencoder is generally not easy to quantify. Although the L value
can represent the learning effect of the network to a certain extent, we
ultimately hope to obtain reconstruction samples with a higher degree of
reduction and richer styles. Therefore, it is generally necessary to discuss
the learning effect of the autoencoder according to specific issues. For
image reconstruction, it generally depends on the quality of artificial
subjective evaluation of the image generation, or the use of certain image
fidelity calculation methods such as Inception Score and Frechet Inception
Distance.

In order to test the effect of image reconstruction, we divide the dataset
into a training set and a test set, where the test set does not participate in
training. We randomly sample the test picture $x \in D^{test}$ from the test set,
calculate the reconstructed picture through the autoencoder, and then
save the real picture and the reconstructed picture as a picture array and
visualize it for easy comparison as in the following:

```
# Reconstruct pictures, sample a batch of pictures from the
test set
x = next(iter(test_db))
logits = model(tf.reshape(x, [-1, 784])) # Flatten and send
to autoencoder
x_hat = tf.sigmoid(logits) # Convert the output to pixel
values, using the sigmoid function
# Recover to 28x28,[b, 784] => [b, 28, 28]
x_hat = tf.reshape(x_hat, [-1, 28, 28])

# The first 50 input + the first 50 reconstructed pictures
merged, [b, 28, 28] => [2b, 28, 28]
x_concat = tf.concat([x[:50], x_hat[:50]], axis=0)
x_concat = x_concat.numpy() * 255. #  Revert to 0~255 range
```

```
    x_concat = x_concat.astype(np.uint8)  # Convert to integer
save_images(x_concat, 'ae_images/rec_epoch_%d.png'%epoch)
# Save picture
```

The effect of image reconstruction is shown in Figure 12-6, Figure 12-7, and Figure 12-8. The five columns on the left of each picture are real pictures, and the five columns on the right are the corresponding reconstructed pictures. It can be seen that in the first Epoch, the picture reconstruction effect is poor, the picture is very blurry, and the fidelity is poor. As the training progresses, the edges of the reconstructed picture become clearer and clearer. At the 100th Epoch, the reconstructed picture effect is already closer to the real picture.

Figure 12-6. *First Epoch*

Figure 12-7. *Tenth Epoch*

531

Figure 12-8. *Hundredth Epoch*

The save_images function here is responsible for merging multiple pictures and saving them as a big picture. This is done using the PIL picture library. The code is as follows:

```
def save_images(imgs, name):
    #  Create 280x280 size image array
    new_im = Image.new('L', (280, 280))
    index = 0
    for i in range(0, 280, 28): # 10-row image array
        for j in range(0, 280, 28): # 10-column picture array
            im = imgs[index]
            im = Image.fromarray(im, mode='L')
            new_im.paste(im, (i, j)) # Write the corresponding
            location
            index += 1
    # Save picture array
    new_im.save(name)
```

12.3 Autoencoder Variants

Generally speaking, the training of the autoencoder network is relatively stable, but because the loss function directly measures the distance between the reconstructed sample and the underlying features of the real sample, rather than evaluating abstract indicators such as the fidelity and diversity of the reconstructed sample, the effect on some tasks is mediocre, such as image reconstruction where the edges of the reconstructed image are prone to be blurred, and the fidelity is not good compared to the real image. In order to learn the true distribution of the data, a series of autoencoder variant networks were produced: denoising autoencoder.

In order to prevent the neural network from memorizing the underlying features of the input data, denoising autoencoders adds random noise disturbances to the input data, such as adding noise ε sampled from the Gaussian distribution to the input x:

$$\tilde{x} = x + \varepsilon, \varepsilon \sim N(0, var)$$

After adding noise, the network needs to learn the real hidden variable $\mathbf{z}$ of the data from x, and restore the original input x, as shown in Figure 12-9. The optimization goals of the model are:

$$\theta^* = argmin_{\smile\theta} dist(h_{\theta_2}(g_{\theta_1}(\tilde{x})), x)$$

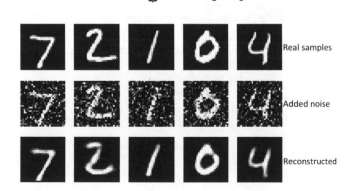

Real samples

Added noise

Reconstructed

Figure 12-9. *Denoising autoencoder diagram*

12.3.1 Dropout Autoencoder

The autoencoder network also faces the risk of overfitting. Dropout autoencoder reduces the expressive power of the network by randomly disconnecting the network and prevents overfitting. The implementation of dropout autoencoder is very simple. Random disconnection of the network connection can be achieved by inserting the Dropout layer in the network layer.

12.3.2 Adversarial Autoencoder

In order to be able to conveniently sample the hidden variable z from a known prior distribution $p(z)$, it is convenient to use $p(z)$ to reconstruct the input, and the adversarial autoencoder uses an additional discriminator network (discriminator, referred to as D network) to determine whether the hidden variable z for dimensionality reduction is sampled from the prior distribution $p(z)$, as shown in Figure 12-10. The output of the discriminator network is a variable belonging to the interval $[0,1]$, which represents whether the hidden vector is sampled from the prior distribution $p(z)$: all samples from the prior distribution $p(z)$ are marked as true, and those generated from the conditional probability $q(z|x)$ are marked as false. In this way, in addition to reconstructing samples, the conditional probability distribution $q(x)$ can also be constrained to approximate the prior distribution $p(z)$.

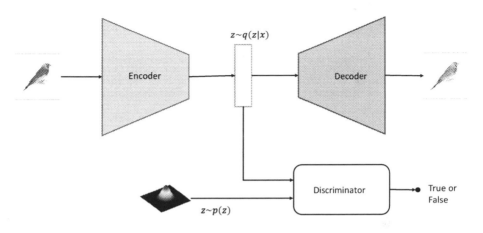

Figure 12-10. *Adversarial autoencoder*

The adversarial autoencoder is derived from the generative adversarial network algorithm introduced in the next chapter. After learning the adversarial generative network, you can deepen your understanding of the adversarial autoencoder.

12.4 Variational Autoencoder

The basic autoencoder essentially learns the mapping relationship between the input x and the hidden variable z. It is a discriminative model, not a generative model. So can the autoencoder be adjusted to a generative model to easily generate samples?

Given the distribution of hidden variables $P(z)$, if the conditional probability distribution $P(z)$ can be learned, then we can sample the joint probability distribution $P(x, z) = P(z)P(z)$ to generate different samples. Variational autoencoders (VAE) can achieve this goal, as shown in Figure 12-11. If you understand it from the perspective of neural networks, VAE is the same as the previous autoencoders, which is very intuitive and easy to understand; but the theoretical derivation of VAE is a little more complicated. Next, we will first explain VAE from the perspective of neural networks, and then derive VAE from the perspective of probability.

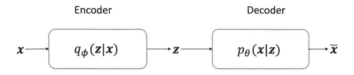

Figure 12-11. *VAE model structure*

From the point of view of neural network, VAE also has two sub-networks of encoder and decoder compared to the self-encoder model. The decoder accepts the input x, and the output is the latent variable z; the decoder is responsible for decoding the latent variable z into the reconstructed x ‾. The difference is that the VAE model has explicit constraints on the distribution of the hidden variable z, and hopes that the hidden variable z conforms to the preset prior distribution P(z). Therefore, in the design of the loss function, in addition to the original reconstruction error term, a constraint term for the z distribution of the hidden variable is added.

12.4.1 Principle of VAE

From a probability point of view, we assume that any dataset is sampled from a certain distribution $p(x|z)$; z is a hidden variable and represents a certain internal feature, such as a picture of handwritten digits x; z can represent font size, writing style, bold, italic, and other settings, which conform to a certain prior distribution $p(z)$. Given a specific hidden variable z, we can sample a series of samples from the learned distribution $p(x|z)$. These samples all have the commonality represented by z.

It is usually assumed that $p(z)$ follows a known distribution, such as $N(0, 1)$. Under the condition that $p(z)$ is known, our goal is to learn to a generative probability model $p(x|z)$. The maximum likelihood estimation method can be used here: a good model should have a high probability of generating a real sample $x \in D$. If our generative model $p(x|z)$ is parameterized with θ, then the optimization goal of our neural network is:

$$p(x) = \int_z p(x|z)p(z)dz$$

Unfortunately, since z is a continuous variable, the preceding integral cannot be converted into a discrete form, which makes it difficult to optimize directly.

Another way of thinking is using the idea of variational inference, we approximate $p(z|x)$ through the distribution $q_\phi(x)$, that is, we need to minimize the distance between $q_\phi(x)$ and $p(z|x)$:

$$\min_\phi D_{KL}\big(q_\phi(x)\|p(x)\big)$$

The KL divergence D_{KL} is a measure of the gap between the distribution q and p, defined as:

$$D_{KL}(q\|p) = \int_x q(x)\log \log \frac{q(x)}{p(x)} dx$$

Strictly speaking, the distance is generally symmetric, while the KL divergence is asymmetric. Expand the KL divergence to:

$$D_{KL}\big(q_\phi(x)\|p(x)\big) = \int_z q_\phi(x)\log \log \frac{q_\phi(x)}{p(x)} dz$$

Use

$$p(z|x) \cdot p(x) = p(x,z)$$

Get

$$D_{KL}\left(q_\phi\left(x\right)\|p\left(x\right)\right)=\int_z q_\phi\left(x\right)log\ log\ \frac{q_\phi\left(x\right)p\left(x\right)}{p\left(x,z\right)}\ dz$$

$$=\int_z q_\phi\left(x\right)log\ log\ \frac{q_\phi\left(x\right)}{p\left(x,z\right)}\ dz+\int_z q_\phi\left(x\right)log\ log\ p\left(x\right)dz$$

$$=-\left(-\int_z q_\phi(x)\ log\ log\ \frac{q_\phi(x)}{p(x,z)}\ dz\right)_{\underbrace{}_{L(\phi,\theta)}}+log\ log\ p\ (x)$$

We define $-\int_z q_\phi\left(x\right)log\ log\ \frac{q_\phi\left(x\right)}{p\left(x,z\right)}\ dz$ as $L(\phi,\theta)$, so the preceding equation becomes:

$$D_{KL}\left(q_\phi\left(x\right)\|p\left(x\right)\right)=-L\left(\phi,\theta\right)+log\ log\ p\left(x\right)$$

where

$$L(\phi,\theta)=-\int_z q_\phi\left(x\right)log\ log\ \frac{q_\phi\left(x\right)}{p\left(x,z\right)}\ dz$$

Consider

$$D_{KL}\left(q_\phi\left(x\right)\|p\left(x\right)\right)\geq 0$$

We have

$$L(\phi,\theta)\leq log\ log\ p\left(x\right)$$

In other words, $L(\phi,\theta)$ is the lower bound of $loglog\ p\ (x)$, and the optimization objective $L(\phi,\theta)$ is called evidence lower bound objective (ELBO). Our goal is to maximize the likelihood probability $p(x)$, or to maximize $loglog\ p\ (x)$, which can be achieved by maximizing its lower bound $L(\phi,\theta)$.

Now let's analyze how to maximize the $L(\phi, \theta)$ function, and expand it to get:

$$L(\theta, \phi) = \int_z q_\phi(x) log\ log\ \frac{p_\theta(x,z)}{q_\phi(x)}$$

$$= \int_z q_\phi(x) log\ log\ \frac{p(z)p_\theta(z)}{q_\phi(x)}$$

$$= \int_z q_\phi(x) log\ log\ \frac{p(z)}{q_\phi(x)} + \int_z q_\phi(x) log\ log\ p_\theta(z)$$

$$= -\int_z q_\phi(x) log\ log\ \frac{q_\phi(x)}{p(z)} + E_{z-q}\Big[log\ log\ p_\theta(z) \Big]$$

$$= -D_{KL}\big(q_\phi(x)\|p(z)\big) + E_{z-q}\Big[log\ log\ p_\theta(z) \Big]$$

So,

$$L(\theta, \phi) = -D_{KL}\big(q_\phi(x)\|p(z)\big) + E_{z-q}\Big[log\ log\ p_\theta(z) \Big] \qquad (12\text{-}1)$$

You can use the encoder network to parameterize the $q_\phi(x)$ function, and the decoder network to parameterize the $p_\theta(z)$ function. The target function $L(\theta, \phi)$ can be optimized by calculating KL divergence between the output distribution of the decoder $q_\phi(x)$ and the prior distribution $p(z)$, and the likelihood probability $loglog\ p_\theta(z)$ of the decoder.

In particular, when both $q_\phi(x)$ and $p(z)$ are assumed to be normally distributed, the calculation of $D_{KL}(q_\phi(x)\|p(z))$ can be simplified to:

$$D_{KL}\big(q_\phi(x)\|p(z)\big) = log\ log\ \frac{\sigma_2}{\sigma_1} + \frac{\sigma_1^2 + (\mu_1 - \mu_2)^2}{2\sigma_2^2} - \frac{1}{2}$$

More specifically, when $q_\phi(x)$ is the normal distribution $N(\mu_1, \sigma_1)$ and $p(z)$ is the normal distribution $N(0, 1)$, that is, $\mu_2 = 0$, $\sigma_2 = 1$, at this time:

$$D_{KL}\left(q_\phi(x)\|p(z)\right) = -\log\sigma_1 + 0.5\sigma_1^2 + 0.5\mu_1^2 - 0.5 \qquad (12\text{-}2)$$

The preceding process makes the $D_{KL}(q_\phi(x)\|p(z))$ term in $L(\theta, \phi)$ easier to calculate, while $E_{z\sim q}[\log\log p_\theta(z)]$ can also be implemented based on the reconstruction error function in the autoencoder.

Therefore, the optimization objective of the VAE model is transformed from maximizing the $L(\phi, \theta)$ function to:

$$min\ D_{KL}\left(q_\phi(x)\|p(z)\right)$$

and

$$max\ E_{z\sim q}\left[\log\log p_\theta(z)\right]$$

The first optimization goal can be understood as constraining the distribution of latent variable z, and the second optimization goal can be understood as improving the reconstruction effect of the network. It can be seen that after our derivation, the VAE model is also very intuitive and easy to understand.

12.4.2 Reparameterization Trick

Now consider a serious problem encountered in the implementation of the above-mentioned VAE model. The hidden variable z is sampled from the output $q_\phi(x)$ of the encoder, as shown on the left in Figure 12-12. When both $q_\phi(x)$ and $p(z)$ are assumed to be normally distributed, the encoder outputs the mean μ and variance σ^2 of the normal distribution, and the decoder's input is sampled from $N(\mu, \sigma^2)$. Due to the existence of the sampling operation, the gradient propagation is discontinuous, and the VAE network cannot be trained end-to-end through the gradient descent algorithm.

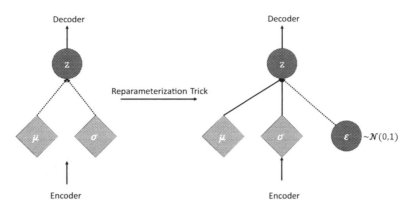

Figure 12-12. *Reparameterization trick diagram*

The paper [2] proposed a continuous and derivable solution called reparameterization trick. It samples the hidden variable z through

$z = \mu + \sigma \odot \varepsilon$, where $\dfrac{\partial z}{\partial \mu}$ and $\dfrac{\partial z}{\partial \sigma}$ are both continuous and differentiable, thus connecting the gradient propagation. As shown on the right of Figure 12-12, the ε variable is sampled from the standard normal distribution $N(0, I)$, and μ and σ are generated by the encoder network. The hidden variable after sampling can be obtained through $z = \mu + \sigma \odot \varepsilon$, which ensures that the gradient propagation is continuous.

The VAE network model is shown in Figure 12-13, the input x is calculated through the encoder network $q_\phi(x)$ to obtain the mean and variance of the hidden variable z, and the hidden variable z is obtained by sampling through the reparameterization trick method, and sent to the decoder network to obtain the distribution (z), and calculate the error and optimize the parameters by formula (12 1).

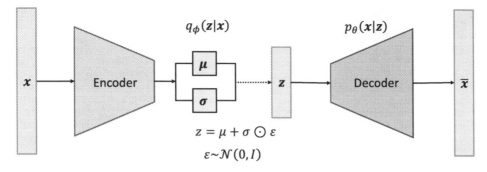

Figure 12-13. *VAE model architecture*

12.5 Hands-On VAE Image Reconstruction

In this section, we'll work on the reconstruction and generation of Fashion MNIST pictures based on the VAE model. As shown in Figure 12-13, the input is the Fashion MNIST picture vector. After three fully connected layers, the mean and variance of the hidden vector z are obtained, which are represented by two fully connected layers with 20 output nodes. The 20 output nodes of FC2 represent the mean vector μ of the 20 feature distributions, and the 20 output nodes of FC3 represent the log variance vectors of the 20 feature distributions. The hidden vector z with a length of 20 is obtained through reparameterization trick sampling, and the sample picture is reconstructed through FC4 and FC5.

As a generative model, VAE can not only reconstruct the input samples but also use the decoder alone to generate samples. The hidden vector z is obtained by directly sampling from the prior distribution $p(z)$, and the generated samples can be generated after decoding.

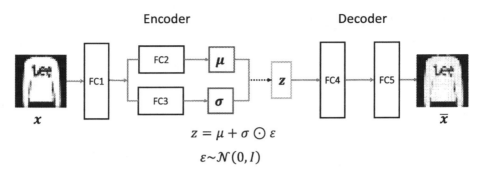

Figure 12-14. *VAE model architecture*

12.5.1 VAE model

We implement encoder and decoder sub-networks in the VAE category. In the initialization function, we create the network layers required by encoder and decoder, respectively, as in the following:

```python
class VAE(keras.Model):
    # Variational Encoder
    def __init__(self):
        super(VAE, self).__init__()

        # Encoder
        self.fc1 = layers.Dense(128)
        self.fc2 = layers.Dense(z_dim) # output mean
        self.fc3 = layers.Dense(z_dim) # output variance

        # Decoder
        self.fc4 = layers.Dense(128)
        self.fc5 = layers.Dense(784)
```

The input of the encoder first passes through the shared layer FC1, and then through the FC2 and FC3 networks, respectively, to obtain the log vector value of the mean vector and variance of the hidden vector distribution.

```
def encoder(self, x):
    # Get mean and variance
    h = tf.nn.relu(self.fc1(x))
    # Mean vector
    mu = self.fc2(h)
    # Log of variance
    log_var = self.fc3(h)

    return mu, log_var
```

Decoder accepts the hidden vector z after sampling, and decodes it into picture output.

```
def decoder(self, z):
    #  Generate image data based on hidden variable z
    out = tf.nn.relu(self.fc4(z))
    out = self.fc5(out)
    #  Return image data, 784 vector
    return out
```

In the forward calculation process of VAE, the distribution of the input latent vector z is first obtained by the encoder, and then the latent vector z is obtained by sampling the reparameterize function implemented by reparameterization trick, and finally the reconstructed picture vector can be restored by the decoder. The implementation is as follows:

```
def call(self, inputs, training=None):
    # Forward calculation
    # Encoder [b, 784] => [b, z_dim], [b, z_dim]
    mu, log_var = self.encoder(inputs)
    # Sampling - reparameterization trick
    z = self.reparameterize(mu, log_var)
    # Decoder
```

```
x_hat = self.decoder(z)
# Return sample, mean and log variance
return x_hat, mu, log_var
```

12.5.2 Reparameterization Trick

The reparameterize function accepts the mean and variance parameters and obtains ε by sampling from the normal distribution $N(0, I)$, and returns the sampled hidden vector by $z = \mu + \sigma \odot \varepsilon$.

```
def reparameterize(self, mu, log_var):
    # reparameterize trick
    eps = tf.random.normal(log_var.shape)
    # calculate standard variance
    std = tf.exp(log_var)**0.5
    # reparameterize trick
    z = mu + std * eps
    return z
```

12.5.3 Network Training

The network is trained for 100 Epochs, and the reconstruction samples are obtained from the forward calculation of the VAE model each time. The reconstruction error term $E_{z \sim q}[log \, log \, p_\theta(z)\,]$ is calculated based on cross-entropy loss function. The error term $D_{KL}(q_\phi(x)\|p(z))$ is calculated based on equation (12-2).

```
# Create network objects
model = VAE()
model.build(input_shape=(4, 784))
# Optimizer
optimizer = optimizers.Adam(lr)
```

```python
for epoch in range(100): # Train 100 Epochs
    for step, x in enumerate(train_db): #  Traverse the
    training set
        # Flatten, [b, 28, 28] => [b, 784]
        x = tf.reshape(x, [-1, 784])
        #  Build a gradient recorder
        with tf.GradientTape() as tape:
            # Forward calculation
            x_rec_logits, mu, log_var = model(x)
            #  Reconstruction loss calculation
            rec_loss = tf.nn.sigmoid_cross_entropy_with_
            logits(labels=x, logits=x_rec_logits)
            rec_loss = tf.reduce_sum(rec_loss) / x.shape[0]
            # Calculate KL convergence N(mu, var) VS N(0, 1)
            # Refernece:https://stats.stackexchange.com/
            questions/7440/kl-divergence-between-two-
            univariate-gaussians
            kl_div = -0.5 * (log_var + 1 - mu**2 -
            tf.exp(log_var))
            kl_div = tf.reduce_sum(kl_div) / x.shape[0]
            # Combine error
            loss = rec_loss + 1. * kl_div
        # Calculate gradients
        grads = tape.gradient(loss, model.trainable_variables)
        # Update parameters
        optimizer.apply_gradients(zip(grads, model.trainable_
        variables))

        if step % 100 == 0:
            # Print error
            print(epoch, step, 'kl div:', float(kl_div), 'rec
            loss:', float(rec_loss))
```

12.5.4 Image Generation

Picture generation only uses the decoder network. First, the hidden vector is sampled from the prior distribution $N(0, I)$, and then the picture vector is obtained through the decoder, and finally is reshaped to picture matrix. For example:

```
#  Test generation effect, randomly sample z from normal
distribution
z = tf.random.normal((batchsz, z_dim))
logits = model.decoder(z) #  Generate pictures only
by decoder
x_hat = tf.sigmoid(logits) #  Convert to pixel range
x_hat = tf.reshape(x_hat, [-1, 28, 28]).numpy() *255.
x_hat = x_hat.astype(np.uint8)
save_images(x_hat, 'vae_images/epoch_%d_sampled.png'%epoch)
# Save pictures

# Reconstruct the picture, sample pictures from the
test set
x = next(iter(test_db))
logits, _, _ = model(tf.reshape(x, [-1, 784])) # Flatten
and send to autoencoder
x_hat = tf.sigmoid(logits) #  Convert output to pixel value
# Restore to 28x28,[b, 784] => [b, 28, 28]
x_hat = tf.reshape(x_hat, [-1, 28, 28])
# The first 50 input + the first 50 reconstructed pictures
merged, [b, 28, 28] => [2b, 28, 28]
x_concat = tf.concat([[x[:50], x_hat[:50]], axis=0)
x_concat = x_concat.numpy() * 255.
x_concat = x_concat.astype(np.uint8)
save_images(x_concat, 'vae_images/epoch_%d_rec.png'%epoch)
```

The effect of picture reconstruction is shown in Figure 12-15, Figure 12-16, and Figure 12-17, which show the reconstruction effect obtained by inputting the pictures of the test set at the first, tenth, and 100th Epoch respectively. The left five columns of each picture are real pictures, and the five columns on the right are the corresponding reconstruction effects. The effect of picture generation is shown in Figure 12-18, Figure 12-19, and Figure 12-20, respectively showing the effect of the image generation at the first, tenth, and 100th Epoch.

Figure 12-15. *Picture reconstruction:epoch=1*

Figure 12-16. *Picture reconstruction:epoch=10*

Figure 12-17. *Picture reconstruction:epoch=100*

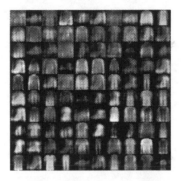

Figure 12-18. *Picture generation:epoch=1*

Figure 12-19. *Picture generation:epoch=10*

Figure 12-20. *Picture generation:epoch=100*

It can be seen that the effect of image reconstruction is slightly better than that of image generation, which also shows that image generation is a more complex task. Although the VAE model has the ability to generate images, the generated effect is still not good enough, and the human eye can still distinguish the difference between machine-generated and real picture samples. The generative confrontation network that will be introduced in the next chapter performs better in image generation.

12.6 Summary

In this chapter, we introduced the powerful self-supervised learning algorithm – the autoencoder and its variants. We started with the principle of autoencoder in order to understand its mathematical mechanism and then we walked through the actual implementation of Autoencoder through the Fashion MNIST image reconstruction exercise. Following similar steps, the VAE model was discussed and applied to the Fashion MNIST image dataset to demonstrate the image generation process. While developing machine learning or deep learning models, one common challenge is the high dimensionality of input data. Compared to traditional dimension reduction methods (e.g., PCA), the autoencoder and its variants usually have better performance in terms of generating data representation in lower dimensions and size.

12.7 References

[1]. G. E. Hinton, "Reducing the Dimensionality of Data with Neural," 2008.

[2]. D. P. Kingma and M. Welling, "Auto-Encoding Variational Bayes,"2nd International Conference on Learning Representations, ICLR 2014, Banff, AB, Canada, April 14-16, 2014, Conference Track Proceedings, 2014.

CHAPTER 13

Generative Adversarial Networks

What I cannot create, I have not yet fully understood.

—Richard Feynman

Before the invention of the generative adversarial network (GAN), the variational autoencoder was considered to be theoretically complete and simple to implement. It is very stable when trained using neural networks, and the resulting images are more approximate, but the human eyes can still easily distinguish real pictures and machine-generated pictures.

In 2014, Ian Goodfellow, a student of Yoshua Bengio (the winner of the Turing Award in 2018) at the Université de Montréal, proposed the GAN [1], which opened up one of the hottest research directions in deep learning. From 2014 to 2019, GAN research has been steadily advancing, and research successes have been reported frequently. The effect of the latest GAN algorithm on image generation has reached a level that is difficult to distinguish with the naked eyes, which is really exciting. Due to the invention of GAN, Ian Goodfellow was awarded the title of Father of GAN, and was granted the 35 Innovators Under 35 award by the Massachusetts Institute of Technology Review in 2017. Figure 13-1

© Liangqu Long and Xiangming Zeng 2022
L. Long and X. Zeng, *Beginning Deep Learning with TensorFlow*,
https://doi.org/10.1007/978-1-4842-7915-1_13

shows that from 2014 to 2018, the GAN model achieved the effect of book generation. It can be seen that both the size of the picture and the fidelity of the picture have been greatly improved. [1]

Figure 13-1. *GAN generated image effect from 2014 to 2018*

Next, we will start from the example of game learning in life, step by step, to introduce the design ideas and model structure of the GAN algorithm.

13.1 Examples of Game Learning

We use the growth trajectory of a cartoonist to vividly introduce the idea of GAN. Consider a pair of twin brothers, called G and D. G learns how to draw cartoons, and D learns how to appreciate paintings. The two brothers at young ages only learned how to use brushes and papers. G drew an unknown painting, as shown in Figure 13-2(a). At this time, D's discriminating ability is not high, so D thinks G's work is OK, but the main character is not clear enough. Under D's guidance and encouragement, G began to learn how to draw the outline of the subject and use simple color combinations.

---

[1] Image source: https://twitter.com/goodfellow_ian/status/1084973596236144640?lang=en

A year later, G improved the basic skills of painting, and D also initially mastered the ability to identify works by analyzing masterpieces and the works of G. At this time, D feels that G's work has the main character, as shown in Figure 13-2(b), but the use of color is not mature enough. A few years later, G's basic painting skills have been very solid, and he can easily draw paintings with bright subjects, appropriate color matching, and high fidelity, as shown in Figure 13-2(c), but D also observes the differences between G and other masterpieces, and improved the ability to distinguish paintings. At this time, D felt that G's painting skills have matured, but his observation of life is not enough. G's work does not convey the expression and some details are not perfect. After a few more years, G's painting skills have reached the point of perfection. The details of the paintings are perfect, the styles are very different and vivid, just like a master level, as shown in Figure 13-2(d). Even at this time, D's discrimination skills are quite excellent. It is also difficult for D to distinguish G from other masterpieces.

The growth process of the above-mentioned painters is actually a common learning process in life, through the game of learning between the two sides and mutual improvement, and finally reaches a balance point. The GAN network draws on the idea of game learning and sets up two sub-networks: a generator G responsible for generating samples and a discriminator D responsible for authenticating. The discriminator D learns how to distinguish between true and false by observing the difference between the real sample and the sample produced by the generator G, where the real sample is true and the sample produced by the generator G is false. The generator G is also learning. It hopes that the generated samples can be recognized by the discriminator D as true. Therefore, the generator G tries to make the samples it generates be considered as true by discriminant D. The generator G and the discriminator D play a game with each other and improve together until they reach an equilibrium point. At this time, the samples generated by the generator G are very realistic, making the discriminator D difficult to distinguish between true and false.

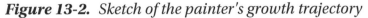

Figure 13-2. *Sketch of the painter's growth trajectory*

In the original GAN paper, Ian Goodfellow used another vivid metaphor to introduce the GAN model: The function of the generator network G is to generate a series of very realistic counterfeit banknotes to try to deceive the discriminator D, and the discriminator D learns the difference between the real money and the counterfeit banknotes generated by generator G to master the banknote identification method. These two networks are synchronized in the process of mutual games, until the counterfeit banknotes produced by the generator G are very real, and even the discriminator D can barely distinguish.

This idea of game learning makes the network structure and training process of GAN slightly different from the previous network model. Let's introduce the network structure and algorithm principle of GAN in detail in the following.

13.2 GAN Principle

Now we will formally introduce the network structure and training methods of GAN.

13.2.1 Network Structure

GAN contains two sub-networks: the generator network (referred to as G) and the discriminator network (referred to as D). The generator network G is responsible for learning the true distribution of samples, and the discriminator network D is responsible for distinguish the samples generated by the generator network from the real samples.

Generator $G(z)$ The generator network G is similar to the function of decoder of the autoencoder. The hidden variables $z \sim p_z(\cdot)$ are sampled from the prior distribution $p_z(\cdot)$. The generated sample $x \sim p_g(x|z)$ is obtained by the parameterized distribution $p_g(x|z)$ of the generator network G, as shown in Figure 13-3. The prior distribution $p_z(\cdot)$ of the hidden variable z can be assumed to be a known distribution, such as a multivariate uniform distribution $z \sim Uniform(-1, 1)$.

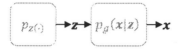

Figure 13-3. *Generator G*

$p_g(x|z)$ can be parameterized by a deep neural network. As shown in Figure 13-4, the hidden variable z is sampled from the uniform distribution $p_z(\cdot)$, and then sample x_f is obtained from the $p_g(x|z)$ distribution. From the perspective of input and output, the function of the generator G is to convert the hidden vector z into a sample vector x_f through a neural network, and the subscript f represents fake samples.

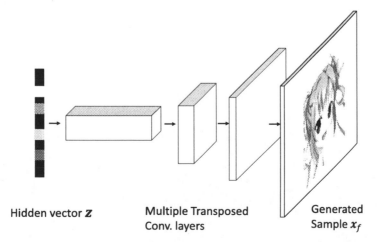

Hidden vector **z**   Multiple Transposed   Generated
                          Conv. layers   Sample x_f

Figure 13-4. *Generator network composed of transposed convolution*

Discriminator $D(x)$ The function of the discriminator network is similar to that of the ordinary binary classification network. It accepts a dataset of input sample x, including samples $x_r \sim p_r(\cdot)$ sampled from the real data distribution $p_r(\cdot)$, and also includes fake samples sampled from the generator network $x_f \sim p_g(x|z)$. x_r and x_f together form the training data set of the discriminator network. The output of the discriminator network is the probability of x belonging to the real sample $P(x\ is\ real\ |x)$. We label all the real samples x_r as true (1), and all the samples x_f generated by the generator network are labeled as false (0). The error between the predicted value of the discriminator network D and the label is used to optimize the discriminator network parameters as shown in Figure 13-5.

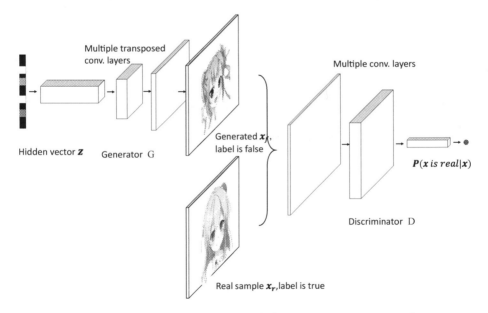

Figure 13-5. *Generator network and discriminator network*

13.2.2 Network Training

The idea of GAN game learning is reflected in its training method. Since the optimization goals of generator G and discriminator D are different, they cannot be the same as the previous network model training, and only one loss function is used. Let us introduce how to train the generator G and the discriminator D respectively.

For the discriminator network D, its goal is to be able to distinguish the real sample x_r from the fake sample x_f. Taking picture generation as an example, its goal is to minimize the cross-entropy loss function between the predicted value and the true value of the picture:

$$L = CE\left(D_\theta\left(x_r\right), y_r, D_\theta\left(x_f\right), y_f\right)$$

where $D_\theta(x_r)$ represents the output of the real sample x_r in the discriminant network D_θ, θ is the parameter set of the discriminator network, $D_\theta(x_f)$ is the output of the generated sample x_f in the discriminator network, and y is the label of x_r. Because the real sample is labeled as true, So $y_r = 1$. y_f is the label of x_f of the generated sample. Since the generated sample is labeled as false, $y_f = 0$. The CE function represents the cross-entropy loss function CrossEntropy. The cross-entropy loss function of the two classification problem is defined as:

$$L = -\sum_{x_r \sim p_r(\cdot)} logD_\theta\left(x_r\right) - \sum_{x_f \sim p_g(\cdot)} log\left(1 - D_\theta\left(x_f\right)\right)$$

Therefore, the optimization goal of the discriminator network D is:

$$\theta^* = -\sum_{x_r \sim p_r(\cdot)} logD_\theta\left(x_r\right) - \sum_{x_f \sim p_g(\cdot)} log\left(1 - D_\theta\left(x_f\right)\right)$$

Convert L to $-L$, and write it in the expectation form:

$$\theta^* = E_{x_r \sim p_r(\cdot)} \, logD_\theta\left(x_r\right) + E_{x_f \sim p_g(\cdot)} \, log\left(1 - D_\theta\left(x_f\right)\right)$$

For the generator network $G(z)$, we hope that $x_f = G(z)$ can deceive the discriminator network D well, and the output of the fake sample x_f is as close to the real label as possible. That is to say, when training the generator network, it is hoped that the output $D(G(z))$ of the discriminator network is as close to 1 as possible, and the cross-entropy loss function between $D(G(z))$ and 1 is minimized:

$$L = CE\left(D\left(G_\phi\left(z\right)\right),1\right) = -logD\left(G_\phi\left(z\right)\right)$$

Convert L to $-L$, and write it in the expectation form:

$$\phi^* = E_{z \sim p_z(\cdot)} logD\left(G_\phi\left(z\right)\right)$$

It can be equivalently transformed into:

$$\phi^* = L = E_{z \sim p_z(\cdot)} log\left[1 - D\left(G_\phi\left(z\right)\right)\right]$$

where ϕ is the parameter set of the generator network G, and the gradient descent algorithm can be used to optimize the parameters ϕ.

13.2.3 Unified Objective Function

We can merge the objective functions of the generator and discriminator networks and write it in the form of a min-max game:

$$\min_\phi \max_\theta L\left(D,G\right) = E_{x_r \sim p_r(\cdot)} \, logD_\theta\left(x_r\right) + E_{x_f \sim p_g(\cdot)} \, log\left(1 - D_\theta\left(x_f\right)\right)$$

$$= E_{x \sim p_r(\cdot)} logD_\theta\left(x\right) + E_{z \sim p_z(\cdot)} log\left(1 - D_\theta\left(G_\phi\left(z\right)\right)\right) \tag{13-1}$$

The algorithm is as follows:

Algorithm 1:GAN training algorithm

Randomly initialize parameters θ *and* ϕ

repeat

  for k times **do**

    Randomly sample hidden vectors $z \sim p_z(\cdot)$

    Randomly sample of real samples $x_r \sim p_r(\cdot)$

    Update the D network according to the gradient descent algorithm:

$$\nabla_\theta E_{x_r \sim p_r(\cdot)} \, logD_\theta\left(x_r\right) + E_{x_f \sim p_g(\cdot)} \, log\left(1 - D_\theta\left(x_f\right)\right)$$

    Randomly sample hidden vectors $z \sim p_z(\cdot)$

    Update the G network according to the gradient descent algorithm:

$$\nabla_\phi E_{z \sim p_z(\cdot)} log\left(1 - D_\theta\left(G_\phi\left(z\right)\right)\right)$$

  end for

until the number of training rounds meets the requirements

output:Trained generator G_ϕ

13.3 Hands-On DCGAN

In this section, we will complete the actual generation of cartoon avatar images. Refer to the network structure of DCGAN [2], where the discriminator D is implemented by a common convolutional layer, and the generator G is implemented by a transposed convolutional layer, as shown in Figure 13-6.

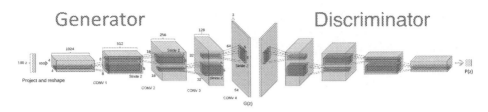

Figure 13-6. *DCGAN Network structure*

13.3.1 Cartoon Avatar Dataset

Here we use a dataset of cartoon avatars, a total of 51,223 pictures, without annotation information. The main body of the pictures have been cropped, aligned, and uniformly scaled to a size of 96 × 96. Some samples are shown in Figure 13-7.

Figure 13-7. *Cartoon avatar dataset*

For customized datasets, you need to complete the data loading and preprocessing work by yourself. We focus here on the GAN algorithm itself. The subsequent chapter on customized datasets will introduce in detail how to load your own datasets. Here the processed dataset is obtained directly through the pre-written make_anime_dataset function.

```
# Dataset path. URL: https://drive.google.com/file/
d/1lRPATrjePnX_n8laDNmPkKCtkf8j_dMD/view?usp=sharing
img_path = glob.glob(r'C:\Users\z390\Downloads\
faces\*.jpg')
# Create dataset object, return Dataset class and size
dataset, img_shape, _ = make_anime_dataset(img_path, batch_
size, resize=64)
```

The dataset object is an instance of the tf.data.Dataset class. Operations such as random dispersal, preprocessing, and batching have been completed, and sample batches can be obtained directly, and img_shape is the preprocessed image size.

13.3.2 Generator

The generator network G is formed by stacking five transposed convolutional layers in order to realize the layer-by-layer enlargement of the height and width of the feature map and the layer-by-layer reduction of the number of feature map channels. First, the hidden vector z with a length of 100 is adjusted to a four-dimensional tensor of $[b, 1, 1, 100]$ through the reshape operation, and the convolutional layer is transposed in order to enlarge the height and width dimensions, reduce the number of channels, and finally get the color picture with a width of 64 and a channel number of 3. A BN layer is inserted between each convolutional layer to improve training stability, and the convolutional layer chooses not to use a bias vector. The generator class code is implemented as follows:

```
class Generator(keras.Model):
    # Generator class
    def __init__(self):
        super(Generator, self).__init__()
        filter = 64
```

```
# Transposed convolutional layer 1, output channel
is filter*8, kernel is 4, stride is 1, no padding,
no bias.
self.conv1 = layers.Conv2DTranspose(filter*8, 4,1,
'valid', use_bias=False)
self.bn1 = layers.BatchNormalization()
# Transposed convolutional layer 2
self.conv2 = layers.Conv2DTranspose(filter*4, 4,2,
'same', use_bias=False)
self.bn2 = layers.BatchNormalization()
# Transposed convolutional layer 3
self.conv3 = layers.Conv2DTranspose(filter*2, 4,2,
'same', use_bias=False)
self.bn3 = layers.BatchNormalization()
# Transposed convolutional layer 4
self.conv4 = layers.Conv2DTranspose(filter*1, 4,2,
'same', use_bias=False)
self.bn4 = layers.BatchNormalization()
# Transposed convolutional layer 5
self.conv5 = layers.Conv2DTranspose(3, 4,2, 'same',
use_bias=False)
```

The forward propagation of generator network G is implemented as follow:

```
def call(self, inputs, training=None):
    x = inputs # [z, 100]
    # Reshape to 4D tensor:(b, 1, 1, 100)
    x = tf.reshape(x, (x.shape[0], 1, 1, x.shape[1]))
    x = tf.nn.relu(x) # activation function
    # Transposed convolutional layer-BN-activation
    function:(b, 4, 4, 512)
```

```
x = tf.nn.relu(self.bn1(self.conv1(x),
training=training))
# Transposed convolutional layer-BN-activation
function:(b, 8, 8, 256)
x = tf.nn.relu(self.bn2(self.conv2(x),
training=training))
# Transposed convolutional layer-BN-activation
function:(b, 16, 16, 128)
x = tf.nn.relu(self.bn3(self.conv3(x),
training=training))
# Transposed convolutional layer-BN-activation
function:(b, 32, 32, 64)
x = tf.nn.relu(self.bn4(self.conv4(x),
training=training))
# Transposed convolutional layer-BN-activation
function:(b, 64, 64, 3)
x = self.conv5(x)
x = tf.tanh(x) # output x range -1~1

return x
```

The output size of the generated network is $[b, 64, 64, 3]$, and the value range is $-1\sim1$.

13.3.3 Discriminator

The discriminator network D is the same as the ordinary classification network. It accepts image tensors of size [b,64,64,3] and continuously extracts features through five convolutional layers. The final output size of the convolutional layer is [b ,2,2,1024], and then convert the feature size to [b,1024] through the pooling layer GlobalAveragePooling2D, and finally obtain the probability of the binary classification task through a

fully connected layer. The code for the discriminator network class D is
implemented as follows:

```
class Discriminator(keras.Model):
    # Discriminator class
    def __init__(self):
        super(Discriminator, self).__init__()
        filter = 64
        # Convolutional layer 1
        self.conv1 = layers.Conv2D(filter, 4, 2, 'valid', use_
        bias=False)
        self.bn1 = layers.BatchNormalization()
        # Convolutional layer 2
        self.conv2 = layers.Conv2D(filter*2, 4, 2, 'valid',
        use_bias=False)
        self.bn2 = layers.BatchNormalization()
        # Convolutional layer 3
        self.conv3 = layers.Conv2D(filter*4, 4, 2, 'valid',
        use_bias=False)
        self.bn3 = layers.BatchNormalization()
        # Convolutional layer 4
        self.conv4 = layers.Conv2D(filter*8, 3, 1, 'valid',
        use_bias=False)
        self.bn4 = layers.BatchNormalization()
        # Convolutional layer 5
        self.conv5 = layers.Conv2D(filter*16, 3, 1, 'valid',
        use_bias=False)
        self.bn5 = layers.BatchNormalization()
        # Global pooling layer
        self.pool = layers.GlobalAveragePooling2D()
        # Flatten feature layer
        self.flatten = layers.Flatten()
```

```
# Binary classification layer
self.fc = layers.Dense(1)
```

The forward calculation process of the discriminator D is implemented as follows:

```
def call(self, inputs, training=None):
    # Convolutional layer-BN-activation function:
    (4, 31, 31, 64)
    x = tf.nn.leaky_relu(self.bn1(self.conv1(inputs),
    training=training))
    # Convolutional layer-BN-activation function:
    (4, 14, 14, 128)
    x = tf.nn.leaky_relu(self.bn2(self.conv2(x),
    training=training))
    # Convolutional layer-BN-activation function:
    (4, 6, 6, 256)
    x = tf.nn.leaky_relu(self.bn3(self.conv3(x),
    training=training))
    # Convolutional layer-BN-activation function:
    (4, 4, 4, 512)
    x = tf.nn.leaky_relu(self.bn4(self.conv4(x),
    training=training))
    # Convolutional layer-BN-activation function:
    (4, 2, 2, 1024)
    x = tf.nn.leaky_relu(self.bn5(self.conv5(x),
    training=training))
    # Convolutional layer-BN-activation function:(4, 1024)
    x = self.pool(x)
    # Flatten
```

```
x = self.flatten(x)
# Output, [b, 1024] => [b, 1]
logits = self.fc(x)

return logits
```

The output size of the discriminator is [b,1]. The Sigmoid activation function is not used inside the class, and the probability that b samples belong to the real samples can be obtained through the Sigmoid activation function.

13.3.4 Training and Visualization

Discriminator According to formula (13-1), the goal of the discriminator network is to maximize the function $L(D, G)$, so that the probability of true sample prediction is close to 1, and the probability of generated sample prediction is close to 0. We implement the error function of the discriminator in the d_loss_fn function, label all real samples as 1, and label all generated samples as 0, and maximize the function L(D,G) by minimizing the corresponding cross-entropy loss function. The d_loss_fn function is implemented as follows:

```
def d_loss_fn(generator, discriminator, batch_z, batch_x, is_
training):
    # Loss function for discriminator
    # Generate images from generator
    fake_image = generator(batch_z, is_training)
    # Distinguish images
    d_fake_logits = discriminator(fake_image, is_training)
    # Determine whether the image is real or not
    d_real_logits = discriminator(batch_x, is_training)
    # The error between real image and 1
    d_loss_real = celoss_ones(d_real_logits)
```

```
# The error between generated image and 0
d_loss_fake = celoss_zeros(d_fake_logits)
# Combine loss
loss = d_loss_fake + d_loss_real

return loss
```

The celoss_ones function calculates the cross-entropy loss between the current predicted probability and label 1. The code is as follows:

```
def celoss_ones(logits):
    # Calculate the cross entropy belonging to and label 1
    y = tf.ones_like(logits)
    loss = keras.losses.binary_crossentropy(y, logits, from_
    logits=True)
    return tf.reduce_mean(loss)
```

The celoss_zeros function calculates the cross entropy loss between the current predicted probability and label 0. The code is as follows:

```
def celoss_zeros(logits):
    # Calculate the cross entropy that belongs to and the
    note is 0
    y = tf.zeros_like(logits)
    loss = keras.losses.binary_crossentropy(y, logits, from_
    logits=True)
      return tf.reduce_mean(loss)
```

Generator The training goal of generator network is to minimize the $L(D, G)$ objective function. Since the real sample has nothing to do with the generator, the error function only needs to minimize $E_{z-p_z(\cdot)} log\left(1 - D_\theta\left(G_\phi(z)\right)\right)$. The cross-entropy error at this time can be minimized by marking the generated sample as 1. It should be noted that in the process of back propagating errors, the discriminator also participates in the construction of the calculation graph, but at this stage only the generator network parameters need to be updated. The error function of the generator is as follows:

```
def g_loss_fn(generator, discriminator, batch_z, is_training):
    # Generate images
    fake_image = generator(batch_z, is_training)
    #  When training the generator network, it is necessary to
    force the generated image to be judged as true
    d_fake_logits = discriminator(fake_image, is_training)
    # Calculate error between generated images and 1
    loss = celoss_ones(d_fake_logits)

    return loss
```

Network training In each Epoch, first randomly sample the hidden vector from the prior distribution $p_z(\cdot)$, randomly sample the real pictures from the true data set, calculate the loss of the discriminator network through the generator and the discriminator, and optimize the discriminator network parameters θ. When training the generator, the discriminator is needed to calculate the error, but only the gradient information of the generator is calculated and ϕ is updated. Here set the discriminator training times $k = 5$, and set the generator training time as one.

First, create the generator network and the discriminator network, and create the corresponding optimizers, respectively, as in the following:

```
generator = Generator() #  Create generator
generator.build(input_shape = (4, z_dim))
discriminator = Discriminator() #  Create discriminator
discriminator.build(input_shape=(4, 64, 64, 3))
# Create optimizers for generator and discriminator
respectively
```

```
g_optimizer = keras.optimizers.Adam(learning_rate=learning_
rate, beta_1=0.5)
d_optimizer = keras.optimizers.Adam(learning_rate=learning_
rate, beta_1=0.5)
```

The main training part of the code is implemented as follows:

```
for epoch in range(epochs): #  Train epochs times
    # 1. Train discriminator
    for _ in range(5):
        # Sample hidden vectors
        batch_z = tf.random.normal([batch_size, z_dim])
        batch_x = next(db_iter) # Sample real images
        # Forward calculation - discriminator
        with tf.GradientTape() as tape:
            d_loss = d_loss_fn(generator, discriminator,
            batch_z, batch_x, is_training)
        grads = tape.gradient(d_loss, discriminator.
        trainable_variables)
        d_optimizer.apply_gradients(zip(grads,
        discriminator.trainable_variables))
    # 2. Train generator
    # Sample hidden vectors
    batch_z = tf.random.normal([batch_size, z_dim])
    batch_x = next(db_iter) # Sample real images
    # Forward calculation - generator
    with tf.GradientTape() as tape:
        g_loss = g_loss_fn(generator, discriminator,
        batch_z, is_training)
    grads = tape.gradient(g_loss, generator.trainable_
    variables)
    g_optimizer.apply_gradients(zip(grads, generator.
    trainable_variables))
```

Every 100 Epochs, a picture generation test is performed. The hidden vector is randomly sampled from the prior distribution, sent to the generator to obtain the generated picture which is saved as a file.

As shown in Figure 13-8, it shows a sample of generated pictures saved by the DCGAN model during the training process. It can be observed that most of the pictures have clear subjects, vivid colors, rich picture diversity, and the generated pictures are close to the real pictures in the data set. At the same time, it can be found that a small amount of generated pictures are still damaged, and the main body of the pictures cannot be recognized by human eyes. To obtain the image generation effect shown in Figure 13-8, it is necessary to carefully design the network model structure and fine-tune the network hyperparameters.

Figure 13-8. *DCGAN image generation effect*

13.4 GAN Variants

In the original GAN paper, Ian Goodfellow analyzed the convergence of the GAN network from a theoretical level and tested the effect of image generation on multiple classic image data sets, as shown in Figure 13-9, where Figure 13-9 (a) is the MNIST dataset, Figure 13-9 (b) is the Toronto Face dataset, and Figure 13-9 (c) and Figure 13-9 (d) are the CIFAR10 dataset.

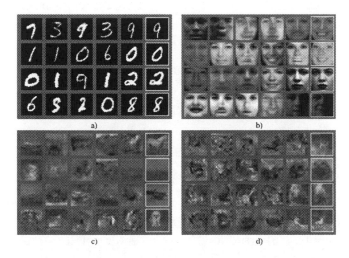

Figure 13-9. *Original GAN image generation effect [1]*

It can be seen that the original GAN model is not outstanding in terms of image generation effect, and the difference from VAE is not obvious. At this time, it does not show its powerful distribution approximation ability. However, because GAN is relatively new in theory, there are many areas for improvement, which greatly stimulated the research interest of the academic community. In the next few years, GAN research is in full swing, and substantial progress has also been made. Next we will introduce several significant GAN variants.

13.4.1 DCGAN

The initial GAN network is mainly based on the fully connected layer to realize the generator G and the discriminator D. Due to the high dimensionality of the picture and the huge amount of network parameters, the training effect is not excellent. DCGAN [2] proposed a generator network implemented using transposed convolutional layers, and a discriminator network implemented by ordinary convolutional layers, which greatly reduces the amount of network parameters and greatly improves the effect of image generation, showing that the GAN model has the potential of outperforming the VAE model in image generation. In addition, the author of DCGAN also proposed a series of empirical GAN network training techniques, which were proved to be beneficial to the stable training of the GAN network. We have used the DCGAN model to complete the actual picture generation of the animation avatars.

13.4.2 InfoGAN

InfoGAN [3] tried to use an unsupervised way to learn the interpretable representation of the interpretable hidden vector z of the input x, that is, it is hoped that the hidden vector z can correspond to the semantic features of the data. For example, for MNIST handwritten digital pictures, we can consider the category, font size, and writing style of the digits to be hidden variables of the picture. We hope that the model can learn these disentangled interpretable feature representation methods, so that the hidden variables can be controlled artificially to generate a sample of the specified content. For the CelebA celebrity photo dataset, it is hoped that the model can separate features such as hairstyles, glasses wearing conditions, and facial expressions, to generate face images of specified shapes.

What are the benefits of disentangled interpretable features? It can make the neural network more interpretable. For example, z contains some separate interpretable features, then we can obtain generated data with different semantics by only changing the features at this position. As shown in Figure 13-10, subtracting the hidden vectors of "men with glasses" and "men without glasses" and adding them to the hidden vectors of "women without glasses" can generate a picture of "women with glasses".

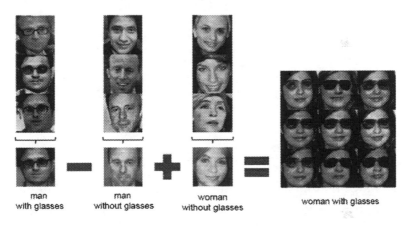

Figure 13-10. Schematic diagram of separated features [3]

13.4.3 CycleGAN

CycleGAN [4] is an unsupervised algorithm for image style conversion proposed by Zhu Junyan. Because the algorithm is clear and simple, and the results are better, this work has received a lot of praise. The basic assumption of CycleGAN is that if you switch from picture A to picture B, and then from picture B to A', then A' should be the same picture as A. Therefore, in addition to setting up the standard GAN loss item, CycleGAN also adds cycle consistency loss to ensure that A' is as close to A as possible. The conversion effect of CycleGAN pictures is shown in Figure 13-11.

Figure 13-11. *Image conversion effect [4]*

13.4.4 WGAN

The training problem of GAN has been criticized all the time, and it
is prone to the phenomenon of training non-convergence and mode
collapse. WGAN [5] analyzed the flaws of the original GAN using JS
divergence from a theoretical level and proposed that the Wasserstein
distance can be used to solve this problem. In WGAN-GP [6], the author
proposed that by adding a gradient penalty term, the WGAN algorithm
was well realized from the engineering level, and the advantages of WGAN
training stability were confirmed.

13.4.5 Equal GAN

From the birth of GAN to the end of 2017, GAN Zoo has collected more
than 214 GAN network variants. These GAN variants have more or less
proposed some innovations, but several researchers from Google Brain
provided another point in a paper [7]: There is no evidence that the GAN
variant algorithms we tested have been consistently better than the

original GAN paper. In that paper, these GAN variants are compared fairly and comprehensively. With sufficient computing resources, it is found that almost all GAN variants can achieve similar performance (FID score). This work reminds the industry whether these GAN variants are essentially innovative.

13.4.6 Self-Attention GAN

The attention mechanism has been widely used in natural language processing (NLP). Self-Attention GAN (SAGAN) [8] borrowed from the attention mechanism and proposed a variant of GAN based on the self-attention mechanism. SAGAN improved the fidelity index of the picture: Inception score from the 36.8 to 52.52, and Frechet inception distance from 27.62 to 18.65. From the effect of image generation perspective, SAGAN's breakthrough is very significant, and it also inspired the industry's attention to the self-attention mechanism.

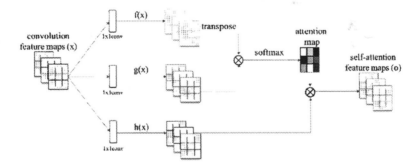

Figure 13-12. Attention mechanism in SAGAN [8]

13.4.7 BigGAN

On the basis of SAGAN, BigGAN [9] attempts to extend the training of GAN to a large scale, using techniques such as orthogonal regularization to ensure the stability of the training process. The significance of BigGAN is to inspire people that the training of GAN networks can also benefit from big data and large computing power. The effect of BigGAN image generation has reached an unprecedented height: the inception score record has increased to 166.5 (an increase of 52.52); Frechet inception distance has dropped to 7.4, which has been reduced by 18.65. As shown in Figure 13-13, the image resolution can reach 512×512, and the image details are extremely realistic.

Figure 13-13. *BigGAN generated images*

13.5 Nash Equilibrium

Now we analyze from the theoretical level, through the training method of game learning, what equilibrium state the generator G and the discriminator D will reach. Specifically, we will explore the following two questions:

- Fix G, what optimal state D^* will D converge to?

- After D reaches the optimal state D^*, what state will G converge to?

First, we give an intuitive explanation through the example of one-dimensional normal distribution $x_r \sim p_r(\cdot)$. As shown in Figure 13-14, the black dashed curve represents the real data distribution $p_r(\cdot)$, which is a normal distribution $N(\mu, \sigma^2)$, and the green solid line represents the distribution $x_f \sim p_g(\cdot)$ learned by the generator network. The blue dotted line represents the decision boundary curve of the discriminator. Figure 13-14 (a), (b), (c), and (d) represents the learning trajectory of the generator network, respectively. In the initial state, as shown in Figure 13-14(a), the distribution of $p_g(\cdot)$ is quite different from $p_r(\cdot)$, and the discriminator can easily learn a clear decision boundary, which is the blue dotted line in Figure 13-14(a), which sets the sampling point from $p_g(\cdot)$ as 0 and the sampling point in $p_r(\cdot)$ as 1. As the distribution $p_g(\cdot)$ of the generator network approaches the true distribution $p_r(\cdot)$, it becomes more and more difficult for the discriminator to distinguish between true and false samples, as shown in Figures 13.14(b)(c). Finally, when the distribution $p_g(\cdot) = p_r(\cdot)$ learned by the generator network, the samples extracted from the generator network are very realistic, and the discriminator cannot distinguish the difference, that is, the probability of determining the true and false samples is equal, as shown in Figure 13-14(d).

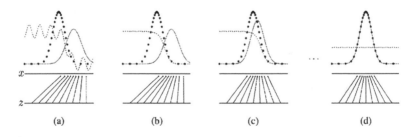

Figure 13-14. *Nash Equilibrium [1]*

This example intuitively explains the training process of the GAN network.

13.5.1 Discriminator State

Now let's derive the first question. Review the loss function of GAN:

$$L(G,D) = \int_x p_r(x) log\ log\ (D(x))\,dx + \int_z p_z(z) log\ log\ (1 - D(g(z)))\,dz$$

$$= \int_x p_r(x) log\ log\ (D(x)) + p_g(x) log\ log\ (1 - D(x))\,dx$$

For the discriminator D, the optimization goal is to maximize the $L(G, D)$ function, and the maximum value of the following function needs to be found:

$$f_\theta = p_r(x) log\ log\ (D(x)) + p_g(x) log\ log\ (1 - D(x))$$

where θ is the network parameter of the discriminator D.

Let us consider the maximum value of the more general function of f_θ:

$$f(x) = A\ log\ log\ x + B\ log\ log\ (1 - x)$$

The maximum value of the function $f(x)$ is required. Consider the derivative of $f(x)$:

$$\frac{df(x)}{dx} = A \frac{1}{ln\ ln\ 10} \frac{1}{x} - B \frac{1}{ln\ ln\ 10} \frac{1}{1-x}$$

$$= \frac{1}{ln\ ln\ 10} \left(\frac{A}{x} - \frac{B}{1-x} \right)$$

$$= \frac{1}{ln\ ln\ 10} \frac{A - (A+B)x}{x(1-x)}$$

Let $\frac{df(x)}{dx} = 0$, we can find the extreme points of the $f(x)$ function:

$$x = \frac{A}{A+B}$$

Therefore, it can be known that the extreme points of the f_θ function are also:

$$D_\theta = \frac{p_r(x)}{p_r(x) + p_g(x)}$$

That is to say, when the discriminator network D_θ is in the D_{θ^*} state, the f_θ function takes the maximum value, and the $L(G, D)$ function also takes the maximum value.

Now back to the problem of maximizing $L(G, D)$, the maximum point of $L(G, D)$ is obtained at:

$$D^* = \frac{A}{A+B} = \frac{p_r(x)}{p_r(x) + p_g(x)}$$

which is also the optimal state D^* of D_θ.

13.5.2 Generator State

Before deriving the second question, we first introduce another distribution distance metric similar to KL divergence: JS divergence, which is defined as a combination of KL divergence:

$$D_{KL}(p\|q) = \int_x p(x) log\, log\, \frac{p(x)}{q(x)} dx$$

$$D_{JS}(p\|q) = \frac{1}{2}D_{KL}\left(p\Big\|\frac{p+q}{2}\right) + \frac{1}{2}D_{KL}\left(q\Big\|\frac{p+q}{2}\right)$$

JS divergence overcomes the asymmetry of KL divergence.

When D reaches the optimal state D^*, let us consider the JS divergence of p_r and p_g at this time:

$$D_{JS}\left(p_r\|p_g\right)=\frac{1}{2}D_{KL}\left(p_r\|\frac{p_r+p_g}{2}\right)+\frac{1}{2}D_{KL}\left(p_g\|\frac{p_r+p_g}{2}\right)$$

According to the definition of KL divergence:

$$D_{JS}\left(p_r\|p_g\right)=\frac{1}{2}\left(\log\log 2+\int_x p_r(x)\log\log\frac{p_r(x)}{p_r+p_g(x)}dx\right)$$

$$+\frac{1}{2}\left(\log\log 2+\int_x p_g(x)\log\log\frac{p_g(x)}{p_r+p_g(x)}dx\right)$$

Combining the constant terms, we can get:

$$D_{JS}\left(p_r\|p_g\right)=\frac{1}{2}(\log\log 2+\log\log 2\ )$$

$$+\frac{1}{2}\left(\int_x p_r(x)\log\log\frac{p_r(x)}{p_r+p_g(x)}dx+\int_x p_g(x)\log\log\frac{p_g(x)}{p_r+p_g(x)}dx\right)$$

That is:

$$D_{JS}\left(p_r\|p_g\right)=\frac{1}{2}(\log\log 4)$$

$$+\frac{1}{2}\left(\int_x p_r(x)\log\log\frac{p_r(x)}{p_r+p_g(x)}dx+\int_x p_g(x)\log\log\frac{p_g(x)}{p_r+p_g(x)}dx\right)$$

Consider when the network reaches D^*, the loss function at this time is:

$$L\left(G,D^*\right)=\int_x p_r(x)\log\log\left(D^*(x)\right)+p_g(x)\log\log\left(1-D^*(x)\right)dx$$

$$=\int_x p_r(x)\log\log\frac{p_r(x)}{p_r+p_g(x)}dx+\int_x p_g(x)\log\log\frac{p_g(x)}{p_r+p_g(x)}dx$$

Therefore, when the discriminator network reaches D^*, $D_{JS}(p_r\|p_g)$ and $L(G, D^*)$ satisfy the relationship:

$$D_{JS}\left(p_r\|p_g\right) = \frac{1}{2}\left(log\ log\ 4 + L\left(G,D^*\right)\right)$$

That is:

$$L\left(G,D^*\right) = 2D_{JS}\left(p_r\|p_g\right) - 2\ log\ log\ 2$$

For the generator network G, the training target is $L(G, D)$, considering the nature of the JS divergence:

$$D_{JS}\left(p_r\|p_g\right) \geq 0$$

Therefore, $L(G, D^*)$ obtains the minimum value only when $D_{JS}(p_r\|p_g) = 0$ (at this time $p_g = p_r$), $L(G, D^*)$ obtains the minimum value:

$$L\left(G^*, D^*\right) = -2\ log\ log\ 2$$

At this time, the state of the generator network G^* is:

$$p_g = p_r$$

That is, the learned distribution p_g of G^* is consistent with the real distribution p_r, and the network reaches a balance point. At this time:

$$D^* = \frac{p_r(x)}{p_r(x) + p_g(x)} = 0.5$$

13.5.3 Nash Equilibrium Point

Through the preceding derivation, we can conclude that the generation network G will eventually converge to the true distribution, namely:$p_g = p_r$

At this time, the generated sample and the real sample come from the same distribution, and it is difficult to distinguish between true and false. The discriminator has the same probability to judge as true or false, that is:

$$D(\cdot) = 0.5$$

At this time, the loss function is

$$L(G^*, D^*) = -2\ log\ log\ 2$$

13.6 GAN Training Difficulty

Although the GAN network can learn the true distribution of data from the theoretical level, the problem of difficulty in GAN network training often arises in engineering implementation, which is mainly reflected in that the GAN model is more sensitive to hyperparameters, and it is necessary to carefully select the hyperparameters that can make the model work. Hyperparameter settings are also prone to mode collapse.

13.6.1 Hyperparameter Sensitivity

Hyperparameter sensitivity means that the network's structure setting, learning rate, initialization state and other hyper-parameters have a greater impact on the training process of the network. A small amount of hyperparameter adjustment may lead to completely different network training results. Figure 13-15 (a) shows the generated samples obtained from good training of the GAN model. The network in Figure 13-15 (b) does not use the batch normalization layer and other settings, resulting in unstable GAN network training and failure to converge. The generated samples are different from each other. The real sample gap is very large.

(a)   (b)

Figure 13-15.   *Hyperparameter sensitive example [5]*

In order to train the GAN network well, the author of the DCGAN paper proposes not to use the pooling layer, not to use the fully connected layer, to use the batch normalization layer more, and the activation function in the generated network should use ReLU. The activation function of the last layer should be Tanh, and the activation function of the discriminator network should use a series of empirical training techniques such as LeakyLeLU. However, these techniques can only avoid the phenomenon of training instability to a certain extent and do not explain from the theoretical level why there is training difficulty and how to solve the problem of training instability.

13.6.2 Model Collapse

Mode collapse refers to the phenomenon that the sample generated by the model is single and the diversity is poor. Since the discriminator can only identify whether a single sample is sampled from the true distribution and does not impose explicit constraints on the sample diversity, the generative model may tend to generate a small number of high-quality samples in a partial interval of the true distribution, without learning all the true distributions. The phenomenon of model collapse is more common in GAN, as shown in Figure 13-16. During the training process, it can be observed by visualizing the samples of the generator network that the types of pictures generated are very single, and the generator network always tends to generate samples of a certain single style to fool the discriminator.

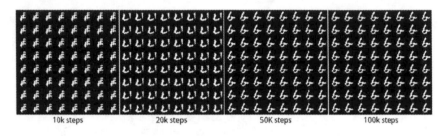

Figure 13-16. *Image generation – model collapsed [10]*

Another example of intuitive understanding of mode collapse is shown in Figure 13-17. The first row is the training process of the generator network without mode collapse, and the last column is the real distribution, that is, the 2D Gaussian mixture model. The second row shows the training process of generator network with model collapse. The last column is the true distribution. It can be seen that the real distribution is a mixture of eight Gaussian models. After model collapse occurs, the generator network always tends to approach a narrow interval of the real distribution, as shown in the first six columns of the second row in Figure 13-17. The samples from this interval of can often be judged as real samples with a higher probability in the discriminator, thus deceiving the discriminator. But this phenomenon is not what we want to see. We hope that the generator network can approximate the real distribution, rather than a certain part of the real distribution.

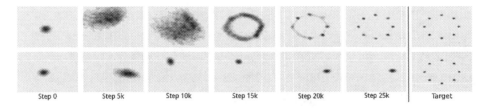

Figure 13-17. *Schematic diagram of model collapse [10]*

So how to solve the problem of GAN training so that GAN can be trained more stably like ordinary neural networks? The WGAN model provides a solution.

13.7 WGAN Principle

The WGAN algorithm analyzes the reasons for the instability of GAN training from a theoretical level, and proposes an effective solution. So what makes GAN training so unstable? WGAN proposed that the gradient surface of the JS divergence on the non-overlapping distributions p and q is always 0. As shown in Figure 13-18, when the distributions p and q do not overlap, the gradient value of the JS divergence is always 0, which leads to the gradient vanishing phenomenon; therefore, the parameters cannot be updated for a long time, and the network cannot converge.

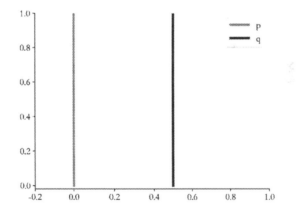

Figure 13-18. *Schematic diagram of distribution p and q*

Next we will elaborate on the defects of JS divergence and how to solve this defect.

13.7.1 JS Divergence Disadvantage

In order to avoid too much theoretical derivation, we use a simple distribution example to explain the defects of JS divergence. Consider two distributions p and q that are completely non-overlapping ($\theta \neq 0$), where the distribution p is:

$$\forall (x,y) \in p, x = 0, y \sim U(0,1)$$

And the distribution of q is:

$$\forall (x,y) \in q, x = \theta, y \sim U(0,1)$$

where $\theta \in R$, when $\theta = 0$, the distributions p and q overlap, and the two are equal; when $\theta \neq 0$, the distributions p and q do not overlap.

Let us analyze the variation of the JS divergence between the preceding distributions p and q with θ. According to the definition of KL divergence and JS divergence, calculate the JS divergence $D_{JS}(p\|q)$ when $\theta = 0$:

$$D_{KL}(p\|q) = \sum_{x=0,y\sim U(0,1)} 1 \cdot log\ log\ \frac{1}{0} = +\infty$$

$$D_{KL}(q\|p) = \sum_{x=\theta,y\sim U(0,1)} 1 \cdot log\ log\ \frac{1}{0} = +\infty$$

$$D_{JS}(p\|q) = \frac{1}{2}\left(\sum_{x=0,y\sim U(0,1)} 1 \cdot log\ log\ \frac{1}{1/2} + \sum_{x=0,y\sim U(0,1)} 1 \cdot log\ log\ \frac{1}{1/2} \right) = log\ log\ 2$$

When $\theta = 0$, the two distributions completely overlap. At this time, the JS divergence and KL divergence both achieve the minimum value, which is 0:

$$D_{KL}(p\|q) = D_{KL}(q\|p) = D_{JS}(p\|q) = 0$$

From the preceding derivation, we can get the trend of $D_{JS}(p\|q)$ with θ:

$$D_{JS}\left(p\|q\right)=\{log\;log\;2\quad \theta\neq 0\;0\;\theta =0$$

In other words, when the two distributions do not overlap at all, regardless of the distance between the distributions, the JS divergence is a constant value $log\;log\;2$, then the JS divergence will not be able to produce effective gradient information. When the two distributions overlap, the JS divergence changes smoothly and produces effective gradient information. When the two distributions completely coincide, the JS divergence takes the minimum value of 0. As shown in Figure 13-19, the red curve divides the two normal distributions. Since the two distributions do not overlap, the gradient value at the generated sample position is always 0, and the parameters of the generator network cannot be updated, resulting in difficulty in network training.

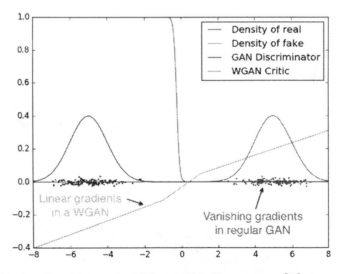

Figure 13-19.   *Gradient vanishing of JS divergence [5]*

Therefore, the JS divergence cannot smoothly measure the distance between the distributions when the distributions p and q do not overlap. As a result, effective gradient information cannot be generated at this position, and the GAN training is unstable. To solve this problem, we need to use a better distribution distance measurement, so that it can smoothly reflect the true distance change between the distributions even when the distributions p and q do not overlap.

13.7.2 EM Distance

The WGAN paper found that JS divergence leads to the instability of GAN training and introduced a new distribution distance measurement method: Wasserstein distance, also called earth mover's distance (EM distance), which represents the minimum cost of transforming a distribution to another distribution. It's defined as:

$$W(p,q) = E_{(x,y)\sim\gamma}\left[\|x-y\|\right]$$

where $\prod(p, q)$ is the set of all possible joint distributions combined by the distributions p and q. For each possible joint distribution $\gamma \sim \prod(p,q)$, calculate the expectation distance $E_{(x,y)\sim\gamma}[\|x-y\|]$ of $\|x-y\|$, where (x,y) is sampled from the joint distribution γ. Different joint distributions γ have different expectations $E_{(x,y)\sim\gamma}[\|x-y\|]$, and the infimum of these expectations is defined as the Wasserstein distance of distributions p and q, where inf{·} represents the infimum of the set, for example, the infimum of $\{x|1 < x < 3, x \in R\}$ is 1.

Continuing to consider the example in Figure 13-18, we directly give the expression of the EM distance between the distributions p and q:

$$W(p,q) = |\theta|$$

Draw the curves of JS divergence and EM distance, as shown in Figure 13-20. It can be seen that the JS divergence is not continuous at $\theta = 0$, the other position derivatives are all 0, and the EM distance can always produce effective derivative information. Therefore, EM distance is more suitable for guiding the training of GAN network than JS divergence.

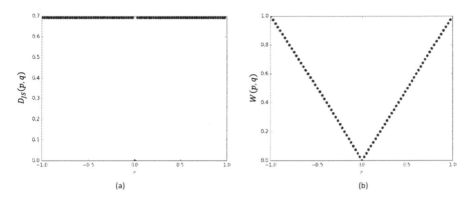

Figure 13-20. *JS divergence and EM distance change curve with* θ *WGAN-GP*

Considering that it is almost impossible to traverse all the joint distributions γ to calculate the distance expectation $E_{(x,y) \sim \gamma}[\|x - y\|]$ of $\|x - y\|$, so it's not realistic to calculate the distance between the distribution p_g of the generator network and $W(p_r, p_g)$. Based on the Kantorovich-Rubinstein duality, the WGAN author converts the direct calculation of $W(p_r, p_g)$ into:

$$W\left(p_r, p_g\right) = \frac{1}{K} E_{x \sim p_r} \left[f(x)\right] - E_{x \sim p_g}\left[f(x)\right]$$

where $sup\{\cdot\}$ represents the supremum of the set, $\|f\|_L \leq K$ represents the function $f: R \to R$ which satisfies the K-order Lipschitz continuity, that is,

$$\left|f(x_1) - f(x_2)\right| \leq K \cdot \left|x_1 - x_2\right|$$

Therefore, we use the discriminant network $D_\theta(x)$ to parameterize the $f(x)$ function, under the condition that D_θ satisfies the 1-Lipschitz constraint, that is, $K = 1$, at this time:

$$W\left(p_r, p_g\right) = E_{x \sim p_r}\left[D_\theta(x)\right] - E_{x \sim p_g}\left[D_\theta(x)\right]$$

Therefore, the problem of solving $W(p_r, p_g)$ can be transformed into:

$$E_{x\sim p_r}\left[D_\theta(x)\right]-E_{x\sim p_g}\left[D_\theta(x)\right]$$

This is the optimization goal of the discriminator D. The discriminant network function $D_\theta(x)$ needs to satisfy the 1-Lipschitz constraint:

$$\nabla_{\hat{x}}D(\hat{x})\leq I$$

In the WGAN-GP paper, the author proposes to increase the gradient penalty method to force the discriminator network to meet the first-order-Lipschitz function constraint, and the author found that the engineering effect is better when the gradient value is constrained around 1, so the gradient penalty term is defined as:

$$GP \triangleq E_{\hat{x}\sim P_{\hat{x}}}\left[\left(\left\|\nabla_{\hat{x}}D(\hat{x})\right\|_2-1\right)^2\right]$$

Therefore, the training objective of WGAN discriminator D is:

$$max_\theta L(G,D) = E_{x_r\sim p_r}[D(x_r)]-E_{x_f\sim p_g}[D(x_f)]\underbrace{\quad}_{EM\ distance}$$
$$-\lambda E_{\hat{x}\sim P_{\hat{x}}}[(\|\nabla_{\hat{x}}D(\hat{x})\|_2-1)^2]\underbrace{\quad}_{GP\ penalty\ term}$$

where $\hat{x}$ comes from the linear difference between x_r and x_f:

$$\hat{x}=tx_r+(1-t)x_f, t\in[0,1]$$

The goal of the discriminator D is to minimize the above-mentioned error $L(G,D)$, that is, to force the EM distance $E_{x_r\sim p_r}\left[D(x_r)\right]-E_{x_f\sim p_g}\left[D(x_f)\right]$ as large as possible, and $\nabla_{\hat{x}}D(\hat{x})_2$ close to 1.

The training objectives of WGAN generator G are:

$$\min_\phi L(G,D) = E_{x_r\sim p_r}[D(x_r)]-E_{x_f\sim p_g}[D(x_f)]\underbrace{\quad}_{EM\ distance}$$

That is, the EM distance between the generator's distribution p_g and the real distribution p_r is as small as possible. Considering that $E_{x_r \sim p_r} \left[D(x_r) \right]$ has nothing to do with the generator, the training objective of the generator is abbreviated as:

$$\min_{\phi} L(G,D) = -E_{x_f \sim p_g} \left[D(x_f) \right]$$

$$= -E_{z \sim p_z(\cdot)} \left[D(G(z)) \right]$$

From the implementation point of view, the output of the discriminator network D does not need to add a Sigmoid activation function. This is because the original version of the discriminator is a binary classification network, the Sigmoid function is added to obtain the probability of belonging to a certain category; while the discriminator in WGAN is used to measure the EM distance between the distribution p_g of the generator network and the real distribution p_r. It belongs to the real number space, so there is no need to add a Sigmoid activation function. When calculating the error function, WGAN also does not have a log function. When training WGAN, WGAN authors recommend using RMSProp or SGD and other optimizers without momentum.

WGAN discovered the reason why the original GAN is prone to training instability from the theoretical level and gave a new distance metric and engineering implementation solution, which achieved good results. WGAN also alleviates the problem of model collapse to a certain extent, and the model using WGAN is not prone to model collapse. It should be noted that WGAN generally does not improve the generation effect of the model but only ensures the stability of model training. Of course, the training stability is also a prerequisite for good model performance. As shown in Figure 13-21, the original version of DCGAN showed unstable training when the BN layer and other settings were not used. Under the same settings, using WGAN to train the discriminator can avoid this phenomenon, as shown in Figure 13-22.

Figure 13-21. *DCGAN generator effect without BN layer [5]*

Figure 13-22. *WGAN generator effect without BN layer [5]*

13.8 Hands-On WGAN-GP

The WGAN-GP model can be modified slightly on the basis of the original GAN implementation. The output of the discriminator D of the WGAN-GP model is no longer the probability of the sample category, and the output does not need to add the Sigmoid activation function. At the same time, we need to add a gradient penalty term as follows:

```python
def gradient_penalty(discriminator, batch_x, fake_image):
    # Gradient penalty term calculation function
    batchsz = batch_x.shape[0]

    # Each sample is randomly sampled at t for interpolation
    t = tf.random.uniform([batchsz, 1, 1, 1])
    # Automatically expand to the shape of x, [b, 1, 1, 1] =>
    [b, h, w, c]
    t = tf.broadcast_to(t, batch_x.shape)
    # Perform linear interpolation between true and false
    pictures
    interplate = t * batch_x + (1 - t) * fake_image
    # Calculate the gradient of D to interpolated samples in a
    gradient environment
    with tf.GradientTape() as tape:
        tape.watch([interplate]) # Add to the gradient
        watch list
        d_interplote_logits = discriminator(interplate)
    grads = tape.gradient(d_interplote_logits, interplate)

    # Calculate the norm of the gradient of each sample:[b, h,
    w, c] => [b, -1]
    grads = tf.reshape(grads, [grads.shape[0], -1])
    gp = tf.norm(grads, axis=1) #[b]
```

```
# Calculate the gradient penalty
gp = tf.reduce_mean( (gp-1.)**2 )

return gp
```

The loss function calculation of WGAN discriminator is different from GAN. WGAN directly maximizes the output value of real samples and minimizes the output value of generated samples. There is no cross-entropy calculation process. The code is implemented as follows:

```
def d_loss_fn(generator, discriminator, batch_z, batch_x, is_
training):
    # Calculate loss function for D
    fake_image = generator(batch_z, is_training) #
    Generated sample
    d_fake_logits = discriminator(fake_image, is_training)
    # Output of generated sample
    d_real_logits = discriminator(batch_x, is_training)
    # Output of real sample
    # Calculate gradient penalty term
    gp = gradient_penalty(discriminator, batch_x, fake_image)
    # WGAN-GP loss function of D. Here is not to calculate the
      cross entropy, but to directly maximize the output of the
      positive sample
    # Minimize the output of false samples and the gradient
      penalty term
    loss = tf.reduce_mean(d_fake_logits) - tf.reduce_mean
    (d_real_logits) + 10. * gp

    return loss, gp
```

The loss function of the WGAN generator G only needs to maximize the output value of the generated sample in the discriminator D, and there is also no cross-entropy calculation step. The code is implemented as follows:

```
def g_loss_fn(generator, discriminator, batch_z, is_training):
    # Generator loss function
    fake_image = generator(batch_z, is_training)
    d_fake_logits = discriminator(fake_image, is_training)
    # WGAN-GP G loss function. Maximize the output value of
    false samples
    loss = - tf.reduce_mean(d_fake_logits)

    return loss
```

Comparing with the original GAN, the main training logic of WGAN is basically the same. The role of the discriminator D for WGAN is a measure of EM distance. Therefore, the more accurate the discriminator is, the more beneficial it is to the generator. The discriminator D can be trained multiple times for a step, and the generator G can be trained once to obtain a more accurate EM distance estimation.

13.9 References

[1].   I. Goodfellow, J. Pouget-Abadie, M. Mirza, B. Xu, D. Warde-Farley, S. Ozair, A. Courville and Y. Bengio, "Generative Adversarial Nets," *Advances in Neural Information Processing Systems 27*, Z. Ghahramani, M. Welling, C. Cortes, N. D. Lawrence and K. Q. Weinberger, Curran Associates, Inc., 2014, pp. 2672-2680.

[2].   A. Radford, L. Metz and S. Chintala, Unsupervised Representation Learning with Deep Convolutional Generative Adversarial Networks, 2015.

[3].   X. Chen, Y. Duan, R. Houthooft, J. Schulman,
       I. Sutskever and P. Abbeel, "InfoGAN: Interpretable
       Representation Learning by Information
       Maximizing Generative Adversarial Nets,"*Advances
       in Neural Information Processing Systems 29*,
       D. D. Lee, M. Sugiyama, U. V. Luxburg, I. Guyon
       and R. Garnett, Curran Associates, Inc., 2016,
       pp. 2172-2180.

[4].   J.-Y. Zhu, T. Park, P. Isola and A. A. Efros, "Unpaired
       Image-to-Image Translation using Cycle-Consistent
       Adversarial Networks,"*Computer Vision (ICCV),
       2017 IEEE International Conference on*, 2017.

[5].   M. Arjovsky, S. Chintala and L. Bottou, "Wasserstein
       Generative Adversarial Networks," *Proceedings
       of the 34th International Conference on Machine
       Learning*, International Convention Centre, Sydney,
       Australia, 2017.

[6].   I. Gulrajani, F. Ahmed, M. Arjovsky, V. Dumoulin
       and A. C. Courville, "Improved Training of
       Wasserstein GANs,"*Advances in Neural Information
       Processing Systems 30*, I. Guyon, U. V. Luxburg,
       S. Bengio, H. Wallach, R. Fergus, S. Vishwanathan
       and R. Garnett, Curran Associates, Inc., 2017,
       pp. 5767-5777.

[7].   M. Lucic, K. Kurach, M. Michalski, O. Bousquet
       and S. Gelly, "Are GANs Created Equal? A Large-
       scale Study," *Proceedings of the 32Nd International
       Conference on Neural Information Processing
       Systems*, USA, 2018.

[8].   H. Zhang, I. Goodfellow, D. Metaxas and A. Odena,
       "Self-Attention Generative Adversarial Networks,"
       *Proceedings of the 36th International Conference
       on Machine Learning*, Long Beach, California,
       USA, 2019.

[9].   A. Brock, J. Donahue and K. Simonyan, "Large
       Scale GAN Training for High Fidelity Natural Image
       Synthesis," *International Conference on Learning
       Representations*, 2019.

[10]. L. Metz, B. Poole, D. Pfau and J. Sohl-Dickstein,
       "Unrolled Generative Adversarial Networks," *CoRR*,
       abs/1611.02163, 2016.

CHAPTER 14

Reinforcement Learning

> Artificial intelligence = deep learning + reinforcement learning
>
> —David Silver

Reinforcement learning is another field of machine learning besides supervised learning and unsupervised learning. It mainly uses agents to interact with the environment in order to learn strategies that can achieve good results. Different from supervised learning, the action of reinforcement learning does not have clear label information. It only has the reward information from the feedback of the environment. It usually has a certain lag and is used to reflect the "good and bad" of the action.

With the rise of deep neural networks, the field of reinforcement learning has also developed vigorously. In 2015, the British company DeepMind proposed a deep neural network-based reinforcement learning algorithm DQN, which achieved a human level performance in 49 Atari games such as space invaders, bricks, and table tennis [1]. In 2017, the AlphaGo program proposed by DeepMind defeated Ke Jie, the no. 1 Go player at the time by a score of 3:0. In the same year, the new version of AlphaGo, AlphaGo Zero, used self-play training without any human knowledge defeated AlphaGo at 100:0 [3]. In 2019, the OpenAI Five

© Liangqu Long and Xiangming Zeng 2022
L. Long and X. Zeng, *Beginning Deep Learning with TensorFlow*,
https://doi.org/10.1007/978-1-4842-7915-1_14

program defeated the Dota2 world champion OG team 2:0. Although the game rules of this game are restricted, it requires a super individual intelligence level for Dota2. With a good teamwork game, this victory undoubtedly strengthened the belief of mankind in AGI.

In this chapter, we will introduce the mainstream algorithms in reinforcement learning, including the DQN algorithm for achieving human-like level in games such as Space Invaders, and the PPO algorithm for winning Dota2.

14.1 See It Soon

The design of reinforcement learning algorithm is different from traditional supervised learning and contains a large number of new mathematical formula derivations. Before entering the learning process of reinforcement learning algorithms, let us first experience the charm of reinforcement learning algorithms through a simple example.

In this section, you don't need to master every detail but should focus on intuitive experience and get the first impression.

14.1.1 Balance Bar Game

The balance bar game system contains three objects: sliding rail, trolley and pole. As shown in Figure 14-1, the trolley can move freely on the slide rail, and one side of the rod is fixed on the trolley through a bearing. In the initial state, the trolley is located in the center of the slide rail and the rod stands on the trolley. The agent controls the balance of the rod by controlling the left and right movement of the trolley. When the angle between the rod and the vertical is greater than a certain angle or the trolley deviates from the center of the slide rail after a certain distance, the game is deemed to be over. The longer the game time, the more rewards the game will give, and the higher the control level of the agent.

In order to simplify the representation of the environment, we directly take the high-level environment feature vector s as the input of the agent. It contains a total of four high-level features, namely, car position, car speed, rod angle, and rod speed. The output action a of the agent is to move to the left or to the right. The action applied to the balance bar system will generate a new state, and the system will also return a reward value. This reward value can be simply recorded as 1, which is instantaneously adding 1 unit time. At each time stamp t, the agent generates an action a_t by observing the environment state s_t. After the environment receives the action, the state changes to s_{t+1} and returns the reward

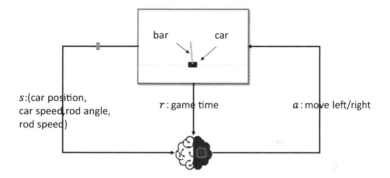

Figure 14-1. *Balance bar game system*

14.1.2 Gym Platform

In reinforcement learning, the robot can directly interact with the real environment, and the updated environment state and rewards can be obtained through sensors. However, considering the complexity of the real environment and the cost of experiments, it is generally preferred to test algorithms in a virtual software environment, and then consider migrating to the real environment.

Reinforcement learning algorithms can be tested through a large number of virtual game environments. In order to facilitate researchers to debug and evaluate algorithm models, OpenAI has developed a gym game interactive platform. Users can use Python language to complete game creation and interaction with only a small amount of code. It's very convenient.

The OpenAI Gym environment includes many simple and classic control games, such as balance bar and roller coaster (Figure 14-2). It can also call the Atari game environment and the complex MuJoCo physical environment simulator (Figure 14-4). In the Atari game environment, there are familiar mini-games, such as Space Invaders, Brick Breaker (Figure 14-3), and racing. Although these games are small in scale, they require high decision-making capabilities and are very suitable for evaluating the intelligence of algorithms.

Figure 14-2. Roller coaster

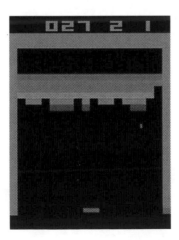

Figure 14-3. Brick Breaker

Figure 14-4. *Walking robot*

At present, you may encounter some problems when installing the Gym environment on the Windows platform, because some of the software libraries are not friendly to the Windows platform. It is recommended that you use a Linux system for installation. The balance bar game environment used in this chapter can be used perfectly on the Windows platform, but other complex game environments are not necessarily.

Running the pip install gym command will only install the basic library of the Gym environment, and the balance bar game is already included in the basic library. If you need to use Atari or MuJoCo emulators, additional installation steps are required. Let's take the installation of the Atari emulator as an example:

```
git clone https://github.com/openai/gym.git # Pull the code
cd gym # Go to directory
pip install -e '.[all]' # Install Gym
```

Generally speaking, creating a game and interacting in the Gym environment mainly consists of five steps:

[1]. Create a game. Through gym.make(name), you can create a game with the specified name and return the game object env.

[2].   Reset the game state. Generally, the game environment has an initial state. You can reset the game state by calling env.reset() and return to the initial state observation of the game.

[3].   Display the game screen. The game screen of each time stamp can be displayed by calling env.render(), which is generally used for testing. Rendering images during training will introduce a certain computational cost, so images may not be displayed during training.

[4].   Interact with the game environment. The action can be executed through env.step(action), and the system can return the new state observation, current reward, the game ending flag done and the additional information carrier. By looping this step, you can continue to interact with the environment until the end of the game.

[5].   Destroy the game. Just call env.close().

The following demonstrates a piece of interactive code for the balance bar game CartPole-v1. During each interaction, an action is randomly sampled in the action space: {left, right}, interact with the environment until the end of the game.

```
import gym # Import gym library
env = gym.make("CartPole-v1") # Create game environment
observation = env.reset() # Reset game state
for _ in range(1000): # Loop 1000 times
  env.render() # Render game image
  action = env.action_space.sample() # Randomly select
  an action
```

```
# Interact with the environment, return new status, reward,
end flag, other information
observation, reward, done, info = env.step(action)
if done:# End of game round, reset state
   observation = env.reset()
env.close() # End game environment
```

14.1.3 Policy Network

Let's discuss the most critical link in reinforcement learning: how to judge and make decisions? We call judgment and decision-making policy. The input of the policy is the state s, and the output is a specific action a or the distribution of the action $\pi_\theta(a|s)$, where θ is the parameter of the strategy function π, and the π_θ function can be parameterized using neural networks, as shown in Figure 14-5. The input of the neural network π_θ is the state s of the balance bar system, that is, a vector of length 4, and the output is the probability of all actions $\pi_\theta(a|s)$: the probability to the left $P(to\ left|s)$ and the probability to the right $P(to\ right|s)$. The sum of all action probabilities is 1:

$$\sum_{a \in A} \pi_\theta(a|s) = 1$$

where A is the set of all actions. The π_θ network represents the policy of the agent and is called the policy network. Naturally, we can embody the policy function as a neural network with four input nodes, multiple fully connected hidden layers in the middle, and two output nodes in the output layer, which represents the probability distribution of these two actions. When interacting, choose the action with the highest probability:

$$a_t = \pi_\theta(s_t)$$

As a result of the decision, it acts in the environment and gets a new state s_{t+1} and reward r_t, and so on, until the end of the game.

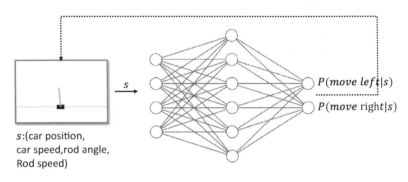

s:(car position,
car speed,rod angle,
Rod speed)

Figure 14-5. *Strategy network*

We implement the policy network as a two-layer fully connected network. The first layer converts a vector of length 4 to a vector of length 128, and the second layer converts a vector of 128 to a vector of 2, which is the probability distribution of actions. Just like the creation process of a normal neural network, the code is as follows:

```
class Policy(keras.Model):
    # Policy network, generating probability distribution
    of actions
    def __init__(self):
        super(Policy, self).__init__()
        self.data = [] # Store track
        # The input is a vector of length 4, and the output
        is two actions - left and right, specifying the
        initialization scheme of the W tensor
        self.fc1 = layers.Dense(128, kernel_initializer='
        he_normal')
        self.fc2 = layers.Dense(2, kernel_initializer='he_normal')
        # Network optimizer
        self.optimizer = optimizers.Adam(lr=learning_rate)
```

```
def call(self, inputs, training=None):
    # The shape of the state input s is a vector:[4]
    x = tf.nn.relu(self.fc1(inputs))
    x = tf.nn.softmax(self.fc2(x), axis=1) # Get the
    probability distribution of the action
    return x
```

During the interaction, we record the state input s_t at each timestamp, the action distribution output a_t, the environment reward r_t, and the new state s_{t+1} as a four-tuple item for training the policy network.

```
def put_data(self, item):
    # Record r,log_P(a|s)
    self.data.append(item)
```

14.1.4 Gradient Update

If you need to use the gradient descent algorithm to optimize the network, you need to know the label information a_t of each input s_t and ensure that the loss value is continuously differentiable from the input to the loss. However, reinforcement learning is not the same as traditional supervised learning, which is mainly reflected in the fact that the action a_t of reinforcement learning at each timestamp t does not have a clear standard for good and bad. The reward r_t can reflect the quality of the action to a certain extent, but it cannot directly determine the quality of the action. Even some game interaction processes only have a final reward r_t signal representing the game result, such as Go. So is it feasible to define an optimal action a_t^* for each state as the label of the neural network input s_t? The first is that the total number of states in the game is usually huge. For example, the total number of states in Go is about 10^{170}. Furthermore, it is difficult to define an optimal action for each state. Although some actions have low short-term returns, long-term returns are better, and sometimes even humans do not know which action is the best.

Therefore, the optimization goal of the strategy should not be to make the output of the input s_t as close as possible to the labeling action, but to maximize the expected value of the total return. The total reward can be defined as the sum of incentives $\sum r_t$ from the beginning of the game to the end of the game. A good strategy should be able to obtain the highest expected value of total return $J(\pi_\theta)$ in the environment. According to the principle of the gradient ascent algorithm, if we can find $\dfrac{\partial J(\theta)}{\partial\theta}$, then the policy network only needs to follow:

$$\theta' = \theta + \eta \cdot \frac{\partial J(\theta)}{\partial\theta}$$

to update the network parameters in order to maximize the expectation reward.

Unfortunately, the total return expectation $J(\pi_\theta)$ is given by the game environment. If the environment model is not known, then $\dfrac{\partial J(\theta)}{\partial\theta}$ cannot be calculated by automatic differentiation. So even if the expression of $J(\pi_\theta)$ is unknown, can the partial derivative $\dfrac{\partial J(\theta)}{\partial\theta}$ be solved directly?

The answer is yes. We directly give the derivation result of $\dfrac{\partial J(\theta)}{\partial\theta}$ here. The specific derivation process will be introduced in detail in 14.3:

$$\frac{\partial J(\theta)}{\partial\theta} = E_{\tau \sim p_\theta(\tau)}\left[\left(\sum_{t=1}^{T}\frac{\partial}{\partial\theta}log\ log\ \pi_\theta\left(s_t\right)\right)R(\tau)\right]$$

Using the preceding formula, you only need to calculate $\dfrac{\partial}{\partial\theta}log\ log\ \pi_\theta\left(s_t\right)$, and multiply it by $R(\tau)$ to update and calculate $\dfrac{\partial J(\theta)}{\partial\theta}$. According to $\theta' = \theta - \eta \cdot \dfrac{\partial L(\theta)}{\partial\theta}$, the policy network can be updated to maximize the $J(\theta)$ function, where $R(\tau)$ is the total return of a certain interaction; τ is the interaction trajectory $s_1, a_1, r_1, s_2, a_2, r_2, \cdots, s_T$; T is the number of timestamps or steps of the interaction; and $log\ log\ \pi_\theta\ (s_t)$ is the log function of the probability value of the a_t action in the output of the policy network. $\frac{\partial}{\partial\theta}log\ log\ \pi_\theta\ (s_t)$ can be solved by TensorFlow automatic differentiation. The code of the loss function is implemented as:

```
for r, log_prob in self.data[::-1]:# Get trajectory
data in reverse order
    R = r + gamma * R # Accumulate the return on each
    time stamp
    # The gradient is calculated once for each timestamp
    # grad_R=-log_P*R*grad_theta
    loss = -log_prob * R
```

The whole training and updating code is as follows:

```
def train_net(self, tape):
    # Calculate the gradient and update the policy network
    parameters. tape is a gradient recorder
    R = 0 # The initial return of the end state is 0
    for r, log_prob in self.data[::-1]:# Reverse order
        R = r + gamma * R # Accumulate the return on each
        time stamp
        # The gradient is calculated once for each timestamp
        # grad_R=-log_P*R*grad_theta
        loss = -log_prob * R
        with tape.stop_recording():
            # Optimize strategy network
            grads = tape.gradient(loss, self.trainable_
            variables)
            # print(grads)
            self.optimizer.apply_gradients(zip(grads,
            self.trainable_variables))
    self.data = [] # Clear track
```

14.1.5 Hands-On Balance Bar Game

We train for a total of 400 rounds. At the beginning of the round, we reset the game state, sample actions by sending input states, interact with the environment, and record the information of each time stamp until the end of the game.

The interactive and training part of the code is as follows:

```
for n_epi in range(10000):
    s = env.reset() # Back to the initial state of the
    game, return to s0
    with tf.GradientTape(persistent=True) as tape:
        for t in range(501): # CartPole-v1 forced to
        terminates at 500 step.
            # Send the state vector to get the strategy
            s = tf.constant(s,dtype=tf.float32)
            # s: [4] => [1,4]
            s = tf.expand_dims(s, axis=0)
            prob = pi(s) # Action distribution: [1,2]
            # Sample 1 action from the category
            distribution, shape: [1]
            a = tf.random.categorical(tf.math.
            log(prob), 1)[0]
            a = int(a) # Tensor to integer
            s_prime, r, done, info = env.step(a) # Interact
            with the environment
            # Record action a and the reward r generated by
            the action
            # prob shape:[1,2]
            pi.put_data((r, tf.math.log(prob[0][a])))
            s = s_prime # Refresh status
            score += r # Cumulative reward
```

```
    if done:  # The current episode is terminated
        break
    # After the episode is terminated, train the
    network once
    pi.train_net(tape)
del tape
```

The training process of the model is shown in Figure 14-6. The horizontal axis is the number of training rounds, and the vertical axis is the average return value of the rounds. It can be seen that as the training progresses, the average return obtained by the network is getting higher and higher, and the strategy is getting better and better. In fact, reinforcement learning algorithms are extremely sensitive to parameters, and modifying the random seed will result in completely different performance. In the process of implementation, it is necessary to carefully select parameters to realize the potential of the algorithm.

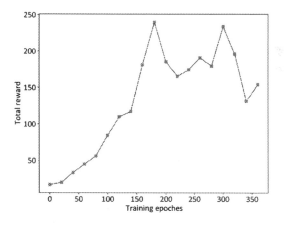

Figure 14-6. *Balance bar game training process*

Through this example, we have a preliminary impression and understanding of the interaction process between reinforcement learning algorithms and reinforcement learning, and then we will formally describe the reinforcement learning problem.

14.2 Reinforcement Learning Problems

In the reinforcement learning problem, the object with perception and decision-making capabilities is called an agent, which can be a piece of algorithm code, or a robotic software and hardware system with a mechanical structure. The agent completes a certain task by interacting with the external environment. The environment here refers to the sum of the external environment that can be affected by the action of the agent and gives corresponding feedback. For the agent, it generates decision-making actions (action) by sensing the state of the environment (state). For the environment, it starts from an initial state s_1, and dynamically changes its state by accepting the actions of the agent, and give the corresponding reward signal (Reward).

We describe the reinforcement learning process from a probabilistic perspective. It contains the following five basic objects:

- State s reflects the state characteristics of the environment. The state on the time stamp t is marked as s_t. It can be the original visual image, voice waveform, and other signals, or it can be the features after high-level abstraction, such as the speed and position of the car. All (finite) states constitute the state space S.

- Action a is the action taken by the agent. The state on the timestamp t is recorded as a_t, which can be discrete actions such as leftward and rightward, or continuous actions such as strength and position. All (finite) actions constitute action space A.

- Policy $\pi(a|s)$ represents the decision model of the agent. It accepts the input as the state s and gives the probability distribution $p(a|s)$ of the action executed after the decision, which satisfies:

$$\sum_{a \in A} \pi(s) = 1$$

This kind of action probability output with a certain randomness is called a stochastic policy. In particular, when the policy model always outputs a certain action with a probability of 1 and others at 0, this kind of policy model is called a deterministic policy, namely:

$$a = \pi(s)$$

- Reward $r(s, a)$ expresses the feedback signal given by the environment after accepting action a in state s. It is generally a scalar value, which reflects the good or bad of the action to a certain extent. The reward obtained at the timestamp t is recorded as r_t (in some materials, it is recorded as r_{t+1}, because the reward often has a certain hysteresis)

- The state transition probability $p(s'|s, a)$ expresses the changing law of the state of the environment model, that is, after the environment of the current state s accepts the action a, the probability distribution that the state changes to s' satisfies:

$$\sum_{s' \in S} p(s'|s, a) = 1$$

The interaction process between the agent and the environment can be represented by Figure 14-7.

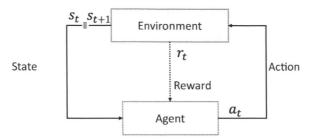

Figure 14-7. *The interaction process between the agent and the environment*

14.2.1 Markov Decision Process

The agent starts from the initial state s_1 of the environment and executes a specific action a_1 through the policy model $\pi(a|s)$. The environment is affected by the action a_1, and the state s_1 changes to s_2 according to the internal state transition model $p(s'|s, a)$. In the meantime, it gives the feedback signal of the agent: the reward r_1, which is generated by the reward function $r(s_1, a_1)$. This cycle of interaction continues until the game reaches termination state s_T. This process produces a series of ordered data:

$$\tau = s_1, a_1, r_1, s_2, a_2, r_2, \cdots, s_T$$

This sequence represents an exchange process between the agent and the environment, called trajectory, denoted as τ. An interaction process is called an episode, and T represents the timestamp (or number of steps). Some environments have a clear terminal state. For example, the game ends when a small plane in the space invaders is hit, while some environments do not have a clear termination mark. For example, some games can be played indefinitely as long as they remain healthy. At this time, T represents ∞.

The conditional probability $P(s_{t+1}|s_1, s_2, ..., s_t)$ is very important, but it requires multiple historical state, which is very complicated to calculate. For simplicity, we assume that the state s_{t+1} on the next time stamp is only affected by the current time stamp s_t, and has nothing to do with other historical states $s_1, s_2, ..., s_{t-1}$, that is :

$$P(s_1, s_2, ..., s_t) = P(s_t)$$

The property that next state s_{t+1} is only related to the current state s_t is called Markov property, and the sequence $s_1, s_2, ..., s_T$ with Markov property is called Markov process.

If the action a is also taken into consideration of the state transition probability, the Markov hypothesis is also applied: the state s_{t+1} of the next time stamp is only related to the current state s_t and the action a_t performed on the current state, then the condition probability becomes:

$$P(s_1, a_1, ..., s_t, a_t) = P(s_t, a_t)$$

We call the sequence of states and actions $s_1, a_1, ..., s_T$ the Markov decision process (MDP). In some scenarios, the agent can only observe part of the state of the environment, which is called partially observable Markov decision process (POMDP). Although the Markovian hypothesis does not necessarily correspond to the actual situation, it is the cornerstone of a large number of theoretical derivations in reinforcement learning. We will see the application of Markovianness in subsequent derivations.

Now let's consider a certain trajectory:

$$\tau = s_1, a_1, r_1, s_2, a_2, r_2, \cdots, s_T$$

It's probability of occurrence $P(\tau)$:

$$P(\tau) = P(s_1, a_1, s_2, a_2, \cdots, s_T)$$

$$= P(s_1)\pi(s_1)P(s_1, a_1)\pi(s_2)P(s_1, a_1, s_2, a_2)\cdots$$

$$= P(s_1)\prod_{t=1}^{T-1}\pi(s_t)p(s_1, a_1, \ldots, s_t, a_t)$$

After applying Markovianity, we simplify the preceding expression to:

$$P(\tau) = P(s_1)\prod_{t=1}^{T-1}\pi(s_t)p(s_t, a_t)$$

The diagram of Markov decision process is shown in Figure 14-8.

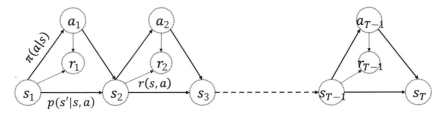

Figure 14-8. *Markov decision process*

If the state transition probability $p(s'|s, a)$ and the reward function $r(s, a)$ of the environment can be obtained, the value function can be directly calculated iteratively. This method of known environmental models is collectively called model-based reinforcement learning. However, environmental models in the real world are mostly complex and unknown. Such methods with unknown models are collectively called model-free reinforcement learning. Next, we will mainly introduce model-free reinforcement learning algorithms.

14.2.2 Objective Function

Each time the agent interacts with the environment, it will get a (lagging) reward signal:

$$r_t = r(s_t, a_t)$$

The cumulative reward of one interaction trajectory τ is called total return:

$$R(\tau) = \sum_{t=1}^{T-1} r_t$$

where T is the number of steps in the trajectory. If we only consider the cumulative return of s_t, s_{t+1}, ..., s_T starting from the intermediate state s_t of the trajectory, it can be recorded as:

$$R(s_t) = \sum_{k=1}^{T-t-1} r_{t+k}$$

In some environments, the stimulus signal is very sparse, such as Go, the stimulus of the previous move is 0, and only at the end of the game will there be a reward signal representing the win or loss.

Therefore, in order to weigh the importance of short-term and long-term rewards, discounted returns that decay over time (Discounted Return) can be used:

$$R(\tau) = \sum_{t=1}^{T-1} \gamma^{t-1} r_t$$

where $\gamma \in [0, 1]$ is called the discount rate. It can be seen that the recent incentive r_1 is all used for total return, while the long-term incentive r_{T-1} can be used to contribute to the total return $R(\tau)$ after attenuating

γ^{T-2}. When $\gamma \approx 1$, the short-term and long-term reward weights are approximately the same, and the algorithm is more forward-looking; when $\gamma \approx 0$, the later long-term reward decays close to 0, short-term reward becomes more important. For an environment with no termination state, that is, $T = \infty$, the discounted return becomes very important, because $\sum_{t=1}^{\infty} \gamma^{t-1} r_t$ may increase to infinity, and the discounted return can be approximately ignored for long-term rewards to facilitate algorithm implementation.

We hope to find a policy $\pi(a|s)$ model so that the higher the total return $R(\tau)$ of the trajectory τ generated by the interaction between the agent and the environment under the control of the policy $\pi(a|s)$, the better. Due to the randomness of environment state transition and policy, the same policy model acting on the same environment with the same initial state may also produce completely different trajectory sequence τ. Therefore, the goal of reinforcement learning is to maximize the expected return:

$$J\left(\pi_\theta\right) = E_{\tau \sim p(\tau)}\left[R\left(\tau\right)\right] = E_{\tau \sim p(\tau)}\left[\sum_{t=1}^{T-1} \gamma^{t-1} r_t\right]$$

The goal of training is to find a policy network π_θ represented by a set of parameters θ, so that $J(\pi_\theta)$ is the largest:

$$\theta^* = E_{\tau \sim p(\tau)}\left[R\left(\tau\right)\right]$$

where $p(\tau)$ represents the distribution of trajectory τ, which is jointly determined by the state transition probability $p(s'|s, a)$ and the strategy $\pi(a|s)$. The quality of strategy π can be measured by $J(\pi_\theta)$. The greater the expected return, the better the policy; otherwise, the worse the strategy.

14.3 Policy Gradient Method

Since the goal of reinforcement learning is to find an optimal policy $\pi_\theta(s)$ that maximizes the expected return $J(\theta)$, this type of optimization problem is similar to supervised learning. It is necessary to solve the partial derivative of the expected return with the network parameters $\dfrac{\partial J}{\partial \theta}$, and use gradient ascent algorithm to update network parameters:

$$\theta' = \theta + \eta \cdot \frac{\partial J}{\partial \theta}$$

That is, where η is the learning rate.

The policy model $\pi_\theta(s)$ can use a multilayer neural network to parameterize $\pi_\theta(s)$. The input of the network is the state s, and the output is the probability distribution of the action a. This kind of network is called a policy network.

To optimize this network, you only need to obtain the partial derivative of each parameter $\dfrac{\partial J}{\partial \theta}$. Now we come to derive the expression of $\dfrac{\partial J}{\partial \theta}$. First, expand it by trajectory distribution:

$$\frac{\partial J}{\partial \theta} = \frac{\partial}{\partial \theta} \int \pi_\theta(\tau) R(\tau) d\tau$$

Move the derivative symbol to the integral symbol:

$$= \int \left(\frac{\partial}{\partial \theta} \pi_\theta(\tau) \right) R(\tau) d\tau$$

Adding $\pi_\theta(\tau) \cdot \dfrac{1}{\pi_\theta(\tau)}$ does not change the result:

$$= \int \pi_\theta(\tau) \left(\frac{1}{\pi_\theta(\tau)} \frac{\partial}{\partial \theta} \pi_\theta(\tau) \right) R(\tau) d\tau$$

Considering:

$$\frac{dlog\left(f\left(x\right)\right)}{dx}=\frac{1}{f\left(x\right)}\frac{df\left(x\right)}{dx}$$

So:

$$\frac{1}{\pi_{\theta}\left(\tau\right)}\frac{\partial}{\partial\theta}\pi_{\theta}\left(\tau\right)=\frac{\partial}{\partial\theta}log\ log\ \pi_{\theta}\left(\tau\right)$$

We can get:

$$=\int\pi_{\theta}\left(\tau\right)\left(\frac{\partial}{\partial\theta}log\ log\ \pi_{\theta}\left(\tau\right)\right)R\left(\tau\right)d\tau$$

That is:

$$\frac{\partial J}{\partial\theta}=E_{\tau\sim p_{\theta}\left(\tau\right)}\left[\frac{\partial}{\partial\theta}log\ log\ \pi_{\theta}\left(\tau\right)R\left(\tau\right)\right]$$

where $loglog\ \pi_{\theta}\left(\tau\right)$ represents the log probability value of trajectory $\tau = s_1$, $a_1, s_2, a_2, \cdots, s_T$. Considering that $R(\tau)$ can be obtained by sampling, the key becomes to solve $\frac{\partial}{\partial\theta}log\ log\ \pi_{\theta}\left(\tau\right)$, we can decompose $\pi_\theta(\tau)$ to get:

$$\frac{\partial}{\partial\theta}log\ log\ \pi_{\theta}\left(\tau\right)=\frac{\partial}{\partial\theta}log\ log\left(p\left(s_{1}\right)\prod_{t=1}^{T-1}\pi_{\theta}\left(s_{t}\right)p\left(s_{t},a_{t}\right)\right)$$

Convert $log\prod\cdot$ to $\sum log\left(\cdot\right)$:

$$=\frac{\partial}{\partial\theta}\left(log\ log\ p\left(s_{1}\right)+\sum_{t=1}^{T-1}log\ log\ \pi_{\theta}\left(s_{t}\right)+log\ log\ p\left(s_{t},a_{t}\right)\right)$$

Considering that both $log\ log\ p\ (s_t, a_t)$ and $log\ log\ p\ (s_1)$ are not related to θ, the preceding formula becomes:

$$\frac{\partial}{\partial\theta}log\ log\ \pi_\theta\ (\tau) = \sum_{t=1}^{T-1}\frac{\partial}{\partial\theta}log\ log\ \pi_\theta\ (s_t)$$

It can be seen that the partial derivative $\frac{\partial}{\partial\theta}log\ log\ \pi_\theta\ (\tau)$ can finally be converted to $loglog\ \pi_\theta\ (s_t)$ which is the derivative of the policy network output to the network parameter θ. It has nothing to do with the state probability transition $p(s'|s, a)$, that is, it can be solved without knowing the environment model $\frac{\partial}{\partial\theta}log\ log\ p_\theta\ (\tau)$.

Put it into $\frac{\partial J}{\partial\theta}$:

$$\frac{\partial J(\theta)}{\partial\theta} = E_{\tau\sim p_\theta(\tau)}\left[\frac{\partial}{\partial\theta}log\ log\ \pi_\theta\ (\tau)R(\tau)\right]$$

$$= E_{\tau\sim p_\theta(\tau)}\left[\left(\sum_{t=1}^{T-1}\frac{\partial}{\partial\theta}log\ log\ \pi_\theta\ (s_t)\right)R(\tau)\right]$$

Let us intuitively understand the preceding formula. When the total return of a certain round $R(\tau) > 0$, $\frac{\partial J(\theta)}{\partial\theta}$ and $\frac{\partial}{\partial\theta}log\ log\ \pi_\theta\ (\tau)$ are in the same direction. According to the gradient ascent algorithm, the θ parameter is updated toward the direction of increasing $J(\theta)$, and also in the direction of increasing $loglog\ \pi_\theta\ (s_t)$, which encourages the generation of more such trajectories τ. When the total return $R(\tau) < 0$, $\frac{\partial J(\theta)}{\partial\theta}$ and $\frac{\partial}{\partial\theta}log\ log\ \pi_\theta\ (\tau)$ are reversed, so when the θ parameter is updated according to the gradient ascent algorithm. It is updated toward the direction of increasing $J(\theta)$ and decreasing $loglog\ \pi_\theta\ (s_t)$, that is, to avoid generating more such trajectories τ. Through this, it is possible to intuitively understand how the network adjusts itself to achieve greater expected return.

With the preceding expression of $\dfrac{\partial J}{\partial \theta}$, we can easily solve $\dfrac{\partial}{\partial \theta} log\, log\, \pi_\theta\,(s_t)$ through the automatic differentiation tool of TensorFlow to calculate $\dfrac{\partial J}{\partial \theta}$. Finally, we can use the gradient ascent algorithm to update the parameters. The general flow of the policy gradient algorithm is shown in Figure 14-9.

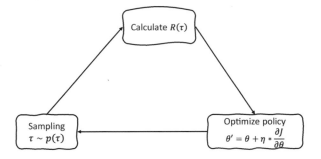

Figure 14-9. *Policy gradient method training process*

14.3.1 Reinforce Algorithm

According to the law of large numbers, write the expectation as the mean value of multiple sampling trajectories τ^n, $n \in [1, N]$:

$$\frac{\partial J(\theta)}{\partial \theta} \approx \frac{1}{N}\sum_{n=1}^{N}\left(\left(\sum_{t=1}^{T-1}\frac{\partial}{\partial \theta}log\,log\,\pi_\theta\left(s_t^{(n)}\right)\right)R\left(\tau^{(n)}\right)\right)$$

where N is the number of trajectories, and $a_t^{(n)}$ and $s_t^{(n)}$ represent the actions and input states of the t-th time stamp of the n-th trajectory τ^n. Then update the θ parameters through gradient ascent algorithm.

This algorithm is called the REINFORCE algorithm [4], which is also the earliest algorithm that uses the policy gradient idea.

Algorithm 1: REINFORCE Algorithm

Randomly initialize θ

repeat

  Interact with environment according to policy (s_t) and generate multiple trajectories $\{\tau^{(n)}\}$

  Calculate $R(\tau^{(n)})$

  Calculate $\dfrac{\partial J(\theta)}{\partial \theta} \approx \dfrac{1}{N}\sum_{n=1}^{N}\left(\left(\sum_{t=1}^{T-1}\dfrac{\partial}{\partial\theta}\log\,\log\,\pi_\theta\left(s_t^{(n)}\right)\right)R\left(\tau^{(n)}\right)\right)$

  Update parameter $\theta' \leftarrow \theta + \eta\cdot\dfrac{\partial J}{\partial \theta}$

until reach certain training times

Output: policy network (s_t)

14.3.2 Improvement of the Original Policy Gradient Method

Because the original REINFORCE algorithm has a large variance between the optimized trajectories, the convergence speed is slow, and the training process is not smooth enough. We can use the idea of variance reduction to make improvements from the perspectives of causality and baseline.

Causality. Considering the partial derivative expression of $\dfrac{\partial J(\theta)}{\partial \theta}$, for the action a_t with a time stamp of t, it has no effect on $\tau_{1:t-1}$, but only has an effect on the subsequent trajectory $\tau_{t:T}$. So for $\pi_\theta(s_t)$, we only consider the cumulative return $R(\tau_{t:T})$ starting from the timestamp t. The expression of $\dfrac{\partial J(\theta)}{\partial \theta}$ is given by

$$\frac{\partial J(\theta)}{\partial \theta} = E_{\tau \sim p_\theta(\tau)} \left[\left(\sum_{t=1}^{T-1} \frac{\partial}{\partial \theta} \log \log \pi_\theta(s_t) \right) R(\tau_{1:T}) \right]$$

It can be written as:

$$\frac{\partial J(\theta)}{\partial \theta} = E_{\tau \sim p_\theta(\tau)} \left[\sum_{t=1}^{T-1} \left(\frac{\partial}{\partial \theta} \log \log \pi_\theta(s_t) R(\tau_{t:T}) \right) \right]$$

$$= E_{\tau \sim p_\theta(\tau)} \left[\sum_{t=1}^{T-1} \left(\frac{\partial}{\partial \theta} \log \log \pi_\theta(s_t) \hat{Q}(s_t, a_t) \right) \right]$$

where $\hat{Q}(s_t, a_t)$ function represents the estimated reward value of π_θ after the a_t action is executed from the state s_t. The definition of the Q function will also be introduced in Section 14.4. Since only the trajectory $\tau_{t:T}$ starting from a_t is considered, the variance of $R(\tau_{t:T})$ becomes smaller.

Bias. The reward r_t in the real environment is not distributed around 0. The rewards of many games are all positive, so that $R(\tau)$ is always greater than 0. The network tends to increase the probability of all sampled actions. The probability of unsampled action is relatively reduced. This is not what we want. We hope that $R(\tau)$ can be distributed around 0, so we introduce a bias variable b, called the baseline, which represents the average level of return $R(\tau)$. The expression of $\frac{\partial J(\theta)}{\partial \theta}$ is converted to:

$$\frac{\partial J(\theta)}{\partial \theta} = E_{\tau \sim p_\theta(\tau)} \left[\sum_{t=1}^{T-1} \frac{\partial}{\partial \theta} \log \log \pi_\theta(s_t) (R(\tau) - b) \right]$$

Considering causality, $\frac{\partial J(\theta)}{\partial \theta}$ can be written as:

$$\frac{\partial J(\theta)}{\partial \theta} = E_{\tau \sim p_\theta(\tau)} \left[\sum_{t=1}^{T-1} \left(\frac{\partial}{\partial \theta} \log \log \pi_\theta(s_t) \left(\hat{Q}(s_t, a_t) - b \right) \right) \right]$$

where $\delta = R(\tau) - b$ is called the advantage function, which represents the advantage of the current action sequence relative to the average return.

After adding bias b, will the value of $\dfrac{\partial J(\theta)}{\partial \theta}$ change? To answer the question, we only need to consider whether $E_{\tau \sim p_\theta(\tau)}\left[\nabla_\theta log\ log\ \pi_\theta(\tau) \cdot b\right]$ can be 0. If it's 0, then the value of $\dfrac{\partial J(\theta)}{\partial \theta}$ will not change.

Expand $E_{\tau \sim p_\theta(\tau)}\left[\nabla_\theta log\ log\ \pi_\theta(\tau) \cdot b\right]$ to :

$$E_{\tau \sim p_\theta(\tau)}\left[\nabla_\theta log\ log\ \pi_\theta(\tau) \cdot b\right] = \int \pi_\theta(\tau)\nabla_\theta log\ log\ \pi_\theta(\tau) \cdot b\ d\tau$$

Because:

$$\pi_\theta(\tau)\nabla_\theta log\ log\ \pi_\theta(\tau) = \nabla_\theta \pi_\theta(\tau)$$

We have:

$$E_{\tau \sim p_\theta(\tau)}\left[\nabla_\theta log\ log\ \pi_\theta(\tau) \cdot b\right] = \int \nabla_\theta \pi_\theta(\tau) b d\tau$$

$$= b\nabla_\theta \int \pi_\theta(\tau) d\tau$$

Consider $\int \pi_\theta(\tau)d\tau = 1$,

$$E_{\tau \sim p_\theta(\tau)}\left[\nabla_\theta log\ log\ \pi_\theta(\tau) \cdot b\right] = b\nabla_\theta 1 = 0$$

Therefore, adding bias b doesn't change the value of $\dfrac{\partial J(\theta)}{\partial \theta}$, but it indeed reduces the variance of $\sum_{t=1}^{T-1}\left(\dfrac{\partial}{\partial \theta} log\ log\ \pi_\theta(s_t)\left(\widehat{Q}(s_t, a_t) - b\right)\right)$.

14.3.3 REINFORCE Algorithm with Bias

Bias b can be estimated using Monte Carlo method:

$$b = \frac{1}{N}\sum_{n=1}^{N}R\left(\tau^{(n)}\right)$$

If causality is considered, then:

$$b = \frac{1}{N} \sum_{n=1}^{N} R\left(\tau_{t:T}^{(n)}\right)$$

Bias b can also be estimated using another neural network, which is also the Actor-Critic method introduced in Section 14.5. In fact, many policy gradient algorithms often use neural networks to estimate bias b. The algorithm can be flexibly adjusted, and it is most important to master the algorithm idea. The REINFORCE algorithm flow with bias is shown in Algorithm 2.

Algorithm 2: REINFORCE algorithm flow with bias

Randomly initialize θ

repeat

  Interact with environment according to policy (s_t)**, generate multiple trajectory** $\{\tau^n\}$

  Calculate $\hat{Q}(s_t, a_t)$

  Estimate bias b **through Monte Carlo method**

  Calculate $\dfrac{\partial J(\theta)}{\partial \theta} \approx \dfrac{1}{N} \sum_{n=1}^{N} \left(\left(\sum_{t=1}^{T-1} \dfrac{\partial}{\partial \theta} log\ log\ \pi_\theta \left(s_t^{(n)}\right) \right) \left(\hat{Q}(s_t, a_t) - b \right) \right)$

  Update parameter $\theta' \leftarrow \theta + \eta \cdot \dfrac{\partial J}{\partial \theta}$

until reach training times

Output: policy network (s_t)

14.3.4 Importance Sampling

After updating the network parameters using the policy gradient method, the policy network $\pi_\theta(s)$ has also changed, and the new policy network must be used for sampling. As a result, the previous historical trajectory data cannot be reused, and the sampling efficiency is very low. How to improve the sampling efficiency and reuse the trajectory data generated by the old policy?

In statistics, importance sampling techniques can estimate the expectation of the original distribution p from another distribution q. Considering that the trajectory τ is sampled from the original distribution p, we hope to estimate the expectation $E_{\tau \sim p}[f(\tau)]$ of the trajectory $\tau \sim p$ function.

$$E_{\tau \sim p}\left[f(\tau)\right] = \int p(\tau) f(\tau) d\tau$$

$$= \int \frac{p(\tau)}{q(\tau)} q(\tau) f(\tau) d\tau$$

$$= E_{\tau \sim q}\left[\frac{p(\tau)}{q(\tau)} f(\tau)\right]$$

Through derivation, we find that the expectation of $f(\tau)$ can be sampled not from the original distribution p, but from another distribution q, which only needs to be multiplied by the ratio $\dfrac{p(\tau)}{q(\tau)}$. This is called importance sampling in statistics.

Let the target policy distribution be $p_\theta(\tau)$, and a certain historical policy distribution is $p_{\bar\theta}(\tau)$, we hope to use the historical sampling trajectory $\tau \sim p_{\bar\theta}(\tau)$ to estimate the expected return of the target policy network:

$$J(\theta) = E_{\tau \sim p_\theta(\tau)}\left[R(\tau)\right]$$

$$= \sum_{t=1}^{T-1} E_{(s_t,a_t)\sim p_\theta(s_t,a_t)}\left[r(s_t,a_t)\right]$$

$$= \sum_{t=1}^{T-1} E_{s_t \sim p_\theta(s_t)} E_{a_t \sim \pi_\theta(s_t)}\left[r(s_t,a_t)\right]$$

Applying importance sampling technique, we can get:

$$J_{\underline\theta}(\theta) = \sum_{t=1}^{T-1} E_{s_t \sim p_{\underline\theta}(s_t)}\left[\frac{p_\theta(s_t)}{p_{\underline\theta}(s_t)} E_{a_t \sim \pi_{\underline\theta}(s_t)}\left[\frac{\pi_\theta(s_t)}{\pi_{\underline\theta}(s_t)} r(s_t,a_t)\right]\right]$$

where $J_{\underline\theta}(\theta)$ represents the value of $J(\theta)$ for the original distribution $p_\theta(\tau)$ estimated through the distribution $p_{\underline\theta}(\tau)$. Under the assumption of approximately ignoring the terms $\dfrac{p_\theta(s_t)}{p_{\underline\theta}(s_t)}$, it is considered that the probability of state s_t appearing under different policies is approximately equal, that is, $\dfrac{p_\theta(s_t)}{p_{\underline\theta}(s_t)} \approx 1$, so:

$$J_{\underline\theta}(\theta) = \sum_{t=1}^{T-1} E_{s_t \sim p_{\underline\theta}(s_t)}\left[E_{a_t \sim \pi_{\underline\theta}(s_t)}\left[\frac{\pi_\theta(s_t)}{\pi_{\underline\theta}(s_t)} r(s_t,a_t)\right]\right]$$

$$= \sum_{t=1}^{T-1} E_{(s_t,a_t)\sim p_{\underline\theta}(s_t,a_t)}\left[\frac{\pi_\theta(s_t)}{\pi_{\underline\theta}(s_t)} r(s_t,a_t)\right]$$

The method in which the sampling policy $p_\theta(\tau)$ and the target policy $p_\theta(\tau)$ to be optimized are not the same is called the off-policy method. Conversely, the method in which the sampling policy and the target policy to be optimized are the same policy is called on-policy method. REINFORCE algorithm belongs to the on-policy method category. The off-policy method can use historical sampling data to optimize the current policy network, which greatly improves data utilization, but also introduces computational complexity. In particular, when importance sampling is implemented by Monte Carlo sampling method, if the difference between the distributions p and q is too large, the expectation estimation will have a large deviation. Therefore, the implementation needs to ensure that the distributions p and q are as similar as possible, such as adding KL divergence constrain to limit the difference between p and q.

We also call the training objective function of the original policy gradient method $L^{PG}(\theta)$:

$$L^{PG}(\theta) = \widehat{E}_t\left[\log \log \pi_\theta(s_t)\widehat{A}_t\right]$$

where PG stands for policy gradient, and $\widehat{E}_t$ and $\widehat{A}_t$ represent empirical estimates. The objective function based on importance sampling is called $L_\theta^{IS}(\theta)$:

$$L_{\underline{\theta}}^{IS}(\theta) = \widehat{E}_t\left[\frac{\pi_\theta(s_t)}{\pi_\theta(s_t)}\widehat{A}_t\right]$$

where IS stands for importance sampling, θ stands for the target policy distribution p_θ, and θ stands for the sampling policy distribution p_θ.

14.3.5 PPO Algorithm

After applying importance sampling, the policy gradient algorithm greatly improves the data utilization rate, which greatly improves the performance and training stability. The more popular off-policy gradient algorithms include TRPO algorithm and PPO algorithm, among which TRPO is the predecessor of PPO algorithm, and PPO algorithm can be regarded as an approximate simplified version of TRPO algorithm.

TRPO algorithm In order to constrain the distance between the target policy $\pi_\theta(s_t)$ and the sampling policy $\pi_{\underline{\theta}}(s_t)$, the TRPO algorithm uses KL divergence to calculate the distance expectation between $\pi_\theta(s_t)$ and $\pi_{\underline{\theta}}(s_t)$. The distance expectation is used as the constraint term of the optimization problem. The implementation of TRPO algorithm is more complicated and computationally expensive. The optimization objective of the TRPO algorithm is:

$$\theta^* = \widehat{E}_t \left[\frac{\pi_\theta(s_t)}{\pi_{\underline{\theta}}(s_t)} \widehat{A}_t \right]$$

$$s.t. \widehat{E}_t \left[D_{KL} \left(\pi_\theta(s_t) \| \pi_{\underline{\theta}}(s_t) \right) \right] \le \delta$$

PPO algorithm. In order to solve the disadvantage of high TRPO calculation cost, the PPO algorithm adds the KL divergence constraint as a penalty item to the loss function. The optimization goal is:

$$\theta^* = \widehat{E}_t \left[\frac{\pi_\theta(s_t)}{\pi_{\underline{\theta}}(s_t)} \widehat{A}_t \right] - \beta \widehat{E}_t \left[D_{KL} \left(\pi_\theta(s_t) \| \pi_{\underline{\theta}}(s_t) \right) \right]$$

where $D_{KL} \left(\pi_\theta(s_t) \| \pi_{\underline{\theta}}(s_t) \right)$ refers to the distance between the policy distribution $\pi_\theta(s_t)$ and $\pi_{\underline{\theta}}(s_t)$, and the hyperparameter β is used to balance the original loss term and the KL divergence penalty term.

Adaptive KL penalty algorithm. The hyperparameter β is dynamically adjusted by setting the threshold KL_{max} of KL divergence. The adjustment rules are as follows: if $\hat{E}_t\left[D_{KL}\left(\pi_\theta\left(s_t\right)\|\pi_\theta\left(s_t\right)\right)\right] > KL_{max}$, increase β; if $\hat{E}_t\left[D_{KL}\left(\pi_\theta\left(s_t\right)\|\pi_\theta\left(s_t\right)\right)\right] < KL_{max}$, then decrease β.

PPO2 algorithm. Based on the PPO algorithm, the PPO2 algorithm adjusts the loss function:

$$L_\theta^{CLIP}\left(\theta\right) = \hat{E}_t\left[\left(\frac{\pi_\theta\left(s_t\right)}{\pi_{\underline{\theta}}\left(s_t\right)}\hat{A}_t, clip\left(\frac{\pi_\theta\left(s_t\right)}{\pi_{\underline{\theta}}\left(s_t\right)}, 1-\epsilon, 1+\epsilon\right)\hat{A}_t\right)\right]$$

The schematic diagram of the error function is shown in Figure 14-10.

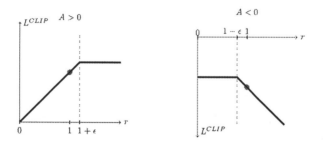

Figure 14-10. *Schematic diagram of PPO2 algorithm error function*

14.3.6 Hands-On PPO

In this section, we implement the PPO algorithm based on importance sampling technology, and test the performance of the PPO algorithm in the balance bar game environment.

Policy network. The policy network is also called the Actor network. The input of the policy network is the state s_t, four input nodes, and the output is the probability distribution $\pi_\theta(s_t)$ of the action a_t, which is implemented by a two-layer fully connected network.

```python
class Actor(keras.Model):
    def __init__(self):
        super(Actor, self).__init__()
        # The policy network is also called the Actor network.
        Output probability p(a|s)
        self.fc1 = layers.Dense(100, kernel_initializer=
        'he_normal')
        self.fc2 = layers.Dense(2, kernel_initializer=
        'he_normal')

    def call(self, inputs):
        # Forward propagation
        x = tf.nn.relu(self.fc1(inputs))
        x = self.fc2(x)
        # Output action probability
        x = tf.nn.softmax(x, axis=1) # Convert to probability
        return x
```

Bias b network Bias b network is also called Critic network, or V-value function network. The input of the network is the state s_t, four input nodes, and the output is the scalar value b. A two-layer fully connected network is used to estimate b. The code is implemented as follows:

```python
class Critic(keras.Model):
    def __init__(self):
        super(Critic, self).__init__()
        # Bias b network is also called Critic network,
        output is v(s)
        self.fc1 = layers.Dense(100, kernel_initializer=
        'he_normal')
        self.fc2 = layers.Dense(1, kernel_initializer=
        'he_normal')
```

```
def call(self, inputs):
    x = tf.nn.relu(self.fc1(inputs))
    x = self.fc2(x)  # Output b's estimate
    return x
```

Next, complete the creation of the strategy network and the value function network and create two optimizers respectively to optimize the parameters of the strategy network and the value function network. We create it in the initialization method of the main class of the PPO algorithm.

```
class PPO():
    # PPO algorithm
    def __init__(self):
        super(PPO, self).__init__()
        self.actor = Actor() # Create Actor network
        self.critic = Critic() # Create Critic network
        self.buffer = [] # Data buffer
        self.actor_optimizer = optimizers.Adam(1e-3) # Actor
        optimizer
        self.critic_optimizer = optimizers.Adam(3e-3) # Critic
        optimizer
```

Action sampling. The select_action function can calculate the action distribution $\pi_\theta(s_t)$ of the current state, and randomly sample actions according to the probability, and return the action and its probability.

```
def select_action(self, s):
    # Send the state vector to get the strategy: [4]
    s = tf.constant(s, dtype=tf.float32)
    # s: [4] => [1,4]
    s = tf.expand_dims(s, axis=0)
    # Get strategy distribution: [1, 2]
    prob = self.actor(s)
```

```
        # Sample 1 action from the category distribution,
        shape: [1]
        a = tf.random.categorical(tf.math.log(prob), 1)[0]
        a = int(a)  # Tensor to integer
        return a, float(prob[0][a]) # Return action and its
        probability
```

Environment interaction. In the main function, interact with the environment for 500 rounds. In each round, the policy is sampled by the select_action function and saved in the buffer pool. The agent.optimizer() function is called to optimize the policy at intervals.

```
def main():
    agent = PPO()
    returns = [] # total return
    total = 0 #  Average return over time
    for i_epoch in range(500): # Number of training rounds
        state = env.reset() # Reset environment
         for t in range(500): # at most 500 rounds
            # Interact with environment with new policy
            action, action_prob = agent.select_action(state)
            next_state, reward, done, _ = env.step(action)
            # Create and store samples
            trans = Transition(state, action, action_prob,
            reward, next_state)
            agent.store_transition(trans)
            state = next_state # Update state
            total += reward # Accumulate rewards
            if done: # Train network
                if len(agent.buffer) >= batch_size:
                    agent.optimize() # Optimize
                break
```

Network optimization. When the buffer pool reaches a certain capacity, the error of the policy network and the error of the value network are constructed through optimizer() function to optimize the parameters of the network. First, the data is converted to the tensor type according to the category, and then the cumulative return $R(\tau_{t:T})$ is calculated by the MC method.

```python
def optimize(self):
    # Optimize the main network function
    # Take sample data from the cache and convert it
    into tensor
    state = tf.constant([t.state for t in self.buffer],
    dtype=tf.float32)
    action = tf.constant([t.action for t in self.buffer],
    dtype=tf.int32)
    action = tf.reshape(action,[-1,1])
    reward = [t.reward for t in self.buffer]
    old_action_log_prob = tf.constant([t.a_log_prob for t
    in self.buffer], dtype=tf.float32)
    old_action_log_prob = tf.reshape(old_action_log_
    prob, [-1,1])
    # Calculate R(st) using MC method
    R = 0
    Rs = []
    for r in reward[::-1]:
        R = r + gamma * R
        Rs.insert(0, R)
    Rs = tf.constant(Rs, dtype=tf.float32)
...
```

Then the data in the buffer pool is taken out according to the batch size. Train the network iteratively ten times. For the policy network, $L_\theta^{CLIP}(\theta)$ is calculated according to the error function of the PPO2 algorithm. For the value network, the distance between the prediction of the value network and $R(\tau_{t:T})$ is calculated through the mean square error, so that the value of the network estimation is getting more and more accurate.

```
    def optimize(self):
...

        # Iterate roughly 10 times on the buffer pool data
        for _ in range(round(10*len(self.buffer)/batch_size)):
            # Randomly sample batch size samples from the
            buffer pool
            index = np.random.choice(np.arange(len(self.
            buffer)), batch_size, replace=False)
            # Build a gradient tracking environment
            with tf.GradientTape() as tape1, tf.GradientTape()
            as tape2:
                # Get R(st), [b,1]
                v_target = tf.expand_dims(tf.gather(Rs, index,
                axis=0), axis=1)
                # Calculate the predicted value of v(s), which
                is the bias b, we will introduce why it is
                written as v later
                v = self.critic(tf.gather(state, index, axis=0))
                delta = v_target - v # Calculating
                advantage value
                advantage = tf.stop_gradient(delta)
                # Disconnect the gradient
```

```python
# Because TF's gather_nd and pytorch's
gather function are different, it needs to be
constructed
 # Coordinate parameters required by gather_nd
 need to be constructed, indices:[b, 2]
# pi_a = pi.gather(1, a) # pytorch only need
oneline implementation
a = tf.gather(action, index, axis=0) # Take out
the action
# batch's action distribution pi(a|st)
pi = self.actor(tf.gather(state, index, axis=0))
indices = tf.expand_dims(tf.range(
a.shape[0]), axis=1)
indices = tf.concat([indices, a], axis=1)
pi_a = tf.gather_nd(pi, indices)
  # The probability of action, pi(at|st), [b]
pi_a = tf.expand_dims(pi_a, axis=1)
  # [b]=> [b,1]
# Importance sampling
ratio = (pi_a / tf.gather(old_action_log_prob,
index, axis=0))
surr1 = ratio * advantage
surr2 = tf.clip_by_value(ratio, 1 - epsilon,
1 + epsilon) * advantage
# PPO error function
policy_loss = -tf.reduce_mean(
tf.minimum(surr1, surr2))
# For the bias v, it is hoped that the R(st)
estimated by MC is as close as possible
value_loss = losses.MSE(v_target, v)
```

```
# Optimize policy network
grads = tape1.gradient(policy_loss, self.actor.
trainable_variables)
self.actor_optimizer.apply_gradients(zip(grads,
self.actor.trainable_variables))
# Optimize bias network
grads = tape2.gradient(value_loss, self.critic.
trainable_variables)
self.critic_optimizer.apply_gradients(zip(grads,
self.critic.trainable_variables))

self.buffer = []  # Empty trained data
```

Training results. After 500 rounds of training, we draw the total return curve, as shown in Figure 14-11, we can see that for a simple game such as a balance bar, the PPO algorithm appears to be easy to use.

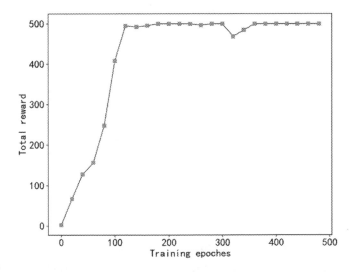

Figure 14-11. *Return curve of PPO algorithm*

14.4 Value Function Method

A better policy model can be obtained using the policy gradient method by directly optimizing the policy network parameters. In the field of reinforcement learning, in addition to the policy gradient method, there is another type of method that indirectly obtains the policy by modeling the value function, which we collectively call the value function method.

Next, we will introduce the definition of common value functions, how to estimate value functions, and how value functions help generate policies.

14.4.1 Value Function

In reinforcement learning, there are two types of value functions: state value function and state-action value function, both of which represent the definition of the starting point of the expected return trajectory is different under the strategy π.

State value function (V function for short), which is defined as the expected return value that can be obtained from the state s_t under the control of the strategy π:

$$V^{\pi}(s_t) = E_{\tau \sim p(\tau)} \left[R(\tau_{t:T}) | \tau_{s_t} = s_t \right]$$

Expand $R(\tau_{t:T})$ as:

$$R((\tau_{t:T})) = r_t + \gamma r_{t+1} + \gamma^2 r_{t+2} + \dots$$

$$= r_t + \gamma \left(r_{t+1} + \gamma^1 r_{t+2} + \dots \right)$$

$$= r_t + \gamma R((\tau_{t+1:T}))$$

So:

$$V^{\pi}\left(s_{t}\right) = E_{\tau \sim p(\tau)}\left[r_{t} + \gamma R\left(\tau_{t+1:T}\right)\right]$$

$$= E_{\tau \sim p(\tau)}\left[r_{t} + \gamma V^{\pi}\left(s_{t+1}\right)\right]$$

This is also called the Bellman equation of the state value function. Among all policies, the optimal policy π^* refers to the policy that can obtain the maximum value of $V^{\pi}(s)$, namely:

$$\pi^{*} = V^{\pi}\left(s\right) \ \forall s \in S$$

At this time, the state value function achieves the maximum value:

$$V^{*}\left(s\right) = V^{\pi}\left(s\right) \forall s \in S$$

For the optimal policy, Bellman's equation is also satisfied:

$$V^{*}\left(s_{t}\right) = E_{\tau \sim p(\tau)}\left[r_{t} + \gamma V^{*}\left(s_{t+1}\right)\right]$$

which is called Bellman optimal equation of the state value function.

Consider the maze problem in Figure 14-12. In the 3 × 4 grid, the grid with coordinates (2,2) is impassable, and the grid with coordinates (4,2) has a reward of -10, and the grid with coordinates (4,3) has a reward of is 10. The agent can start from any position, and the reward is -1 for every additional step. The goal of the game is to maximize the return. For this simple maze, the optimal vector for each position can be drawn directly, that is, at any starting point, the optimal strategy $\pi^*(a|s)$ is a deterministic policy, and the actions are marked in Figure 14-12(b) . Let $\gamma = 0.9$, then:

- Starting from $s_{(4,3)}$, that is, coordinates $(4, 3)$, the optimal policy is $V^*(s_{(4,3)}) = 10$

- Starting from $s_{(3,3)}$, $V^*(s_{(4,3)}) = -1 + 0.9 \cdot 10 = 8$

Starting from $s_{(2,1)}$, $V^*(s_{(2,1)}) = -1 - 0.9 \cdot 1 - 0.9^2 \cdot 1 - 0.9^3 \cdot 1 + 0.9^4 \cdot 10 = 3.122$

It should be noted that the premise of the state value function is that under a certain strategy π, all the preceding calculations are to calculate the state value function under the optimal strategy.

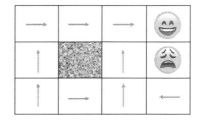

(3.1)	(3.2)	(3.3)	
(2.1)		(2.3)	
(1.1)	(2.1)	(3.1)	(4.1)

(a)Maze and coordinates   (b)Optimal strategy

Figure 14-12. *Maze problem-V function*

The value of the state value function reflects the quality of the state under the current policy. The larger $V^\pi(s_t)$, the greater the total return expectation of the current state. Take the space invader game that is more in line with the actual situation as an example. The agent needs to fire at the flying saucers, squids, crabs, octopuses, and other objects, and score points when it hit them. At the same time, it must avoid being concentrated by these objects. A red shield can protect the agent, but the shield can be gradually destroyed by hits. In Figure 14-13, in the initial state of the game, there are many objects in the figure. Under a good policy π, a larger $V^\pi(s)$ value should be obtained. In Figure 14-14, there are fewer objects. No matter how good the policy is, it is impossible to obtain a larger value of $V^\pi(s)$. The quality of the policy will also affect the value of $V^\pi(s)$. As shown in Figure 14-15, a bad policy (such as moving to the right) will cause the agent to be hit. Therefore, $V^\pi(s)=0$. A good policy can shoot down the objects in the picture and obtain a certain reward.

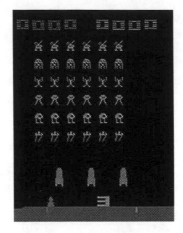

Figure 14-13. *$V^\pi(s)$ may be larger under the policy π*

Figure 14-14. *$V^\pi(s)$ is small under any policy π*

Figure 14-15. *Bad policy (such as to the right) will end the game*
$V^\pi(s) = 0$, *good policy can still get a small return*

State-action value function (Q function for short), which is defined
as the expected return value that can be obtained under the control of
strategy π from the dual setting of state s_t and execution of action a_t:

$$Q^\pi \left(s_t, a_t \right) = E_{\tau \sim p(\tau)} \left[R \left(\tau_{t:T} \right) \middle|, \tau_{a_t} = a_t, \tau_{s_t} = s_t \right]$$

Although both the Q function and the V function are expected return
values, the action a_t of the Q function is a prerequisite, which is different
from the definition of the V function. Expand the Q function to:

$$Q^\pi \left(s_t, a_t \right) = E_{\tau \sim p(\tau)} \left[r \left(s_t, a_t \right) + \gamma r_{t+1} + \gamma^2 r_{t+2} + \ldots \right]$$

$$= E_{\tau \sim p(\tau)} \left[r \left(s_t, a_t \right) + r_t + \gamma \left(r_{t+1} + \gamma^1 r_{t+2} + \ldots \right) \right]$$

So:

$$Q^\pi \left(s_t, a_t \right) = E_{\tau \sim p(\tau)} \left[r \left(s_t, a_t \right) + \gamma V^\pi \left(s_{t+1} \right) \right]$$

645

Because s_t *and* a_t are fixed, $r(s_t, a_t)$ is also fixed.

The Q function and the V function have the following relationship:

$$V^\pi\left(s_t\right) = E_{a_t \sim \pi\left(s_t\right)}\left[Q^\pi\left(s_t, a_t\right)\right]$$

That is, when a_t is sampled from policy $\pi(s_t)$, the expected value of $Q^\pi(s_t, a_t)$ is equal to $V^\pi(s_t)$. Under the optimal policy $\pi^*(a|s)$, there is the following relationship:

$$Q^*\left(s_t, a_t\right) = Q^\pi\left(s_t, a_t\right)$$

$$\pi^* = Q^*\left(s_t, a_t\right)$$

It also means:

$$V^*\left(s_t\right)Q^*\left(s_t, a_t\right)$$

At this time:

$$Q^*\left(s_t, a_t\right) = E_{\tau \sim p(\tau)}\left[r\left(s_t, a_t\right) + \gamma V^*\left(s_{t+1}\right)\right]$$

$$= E_{\tau \sim p(\tau)}\left[r\left(s_t, a_t\right) + \gamma Q^*\left(s_{t+1}, a_{t+1}\right)\right]$$

The preceding formula is called the Bellman optimal equation of the Q function.

We define the difference between $Q^\pi(s_t, a_t)$ and $V^\pi(s)$ as the advantage value function:

$$A^\pi\left(s, a\right) \triangleq Q^\pi\left(s, a\right) - V^\pi\left(s\right)$$

It shows the degree of advantage of taking action a in state s over the average level: $A^\pi(s, a) > 0$ indicates that taking action a is better than the average level; otherwise, it is worse than the average level. In fact, we have already applied the idea of advantage value function in the section of REINFORCE algorithm with bias.

Continuing to consider the example of the maze, let the initial state be $s_{(2,1)}$, a_t can be right or left. For function $Q^*(s_t, a_t)$, $Q^*(s_{(2,1)}, right) = -1 - 0.9 \cdot 1 - 0.9^2 \cdot 1 - 0.9^3 \cdot 1 + 0.9^4 \cdot 10 = 3.122$, $Q^*(s_{(2,1)}, left) = -1 - 0.9 \cdot 1 - 0.9^2 \cdot 1 - 0.9^3 \cdot 1 - 0.9^4 \cdot 1 - 0.9^5 \cdot 1 + 0.9^6 \cdot 10 = 0.629$. We have calculated $V^*(s_{(2,1)}) = 3.122$, and we can intuitively see that they satisfy $V^*(s_t)Q^*(s_t, a_t)$.

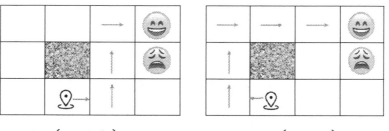

(b)$Q^*(s_{2,1}, right)$ (b) $Q^*(s_{2,1}, left)$

***Figure 14-16.** Maze problem-Q function*

Take the space invader game as an example to intuitively understand the concept of the Q function. In Figure 14-17, the agent in the figure is under the protective cover. If you choose to fire at this time, it is generally considered a bad action. Therefore, under a good policy π, $Q^\pi(s, no\, fire) > Q^\pi(s, fire)$. If you choose to move to the left at this time in Figure 14-18, you may miss the object on the right due to insufficient time, so $Q^\pi(s, left)$ may be small. If the agent moves to the right and fires in Figure 14-19, $Q^\pi(s, right)$ will be larger.

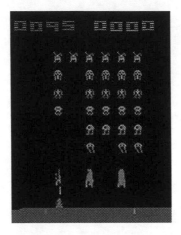

Figure 14-17. *$Q^\pi(s, no\ fire)$ may be larger than $Q^\pi(s, fire)$*

Figure 14-18. *$Q^\pi(s, left)$ may be smaller*

Figure 14-19. *Under a good policy π, $Q^\pi(s, right)$ can still get some rewards*

After introducing the definition of the Q function and the V function, we will mainly answer the following two questions:

- How is the value function estimated?

- How to derive the policy from the value function?

14.4.2 Value Function Estimation

The estimation of value function mainly includes Monte Carlo method and temporal difference method.

Monte Carlo method

The Monte Carlo method is actually to estimate the V function and the Q function through multiple trajectories $\{\tau^{(n)}\}$ generated by the sampling policy $\pi(a|s)$. Consider the definition of the Q function:

$$Q^\pi (s,a) = E_{\tau \sim p(\tau)} \left[R\left(\tau_{s_0 = s, a_0 = a} \right) \right]$$

According to the law of large numbers, it can be estimated by sampling:

$$Q^\pi(s,a) \approx \widehat{Q}^\pi(s,a) = \frac{1}{N} \sum_{n=1}^{N} R\left(\tau^{(n)}_{s_0=s,a_0=a}\right)$$

where $\tau^{(n)}_{s_0=s,a_0=a}$ represents the n-th sampled trajectory, $n \in [1, N]$. The actual state of each sampled trajectory is s, the initial action is a, and N is the total number of trajectories. The V function can be estimated according to the same method:

$$V^\pi(s) \approx \widehat{V}^\pi(s) = \frac{1}{N} \sum_{n=1}^{N} R\left(\tau^{(n)}_{s_0=s}\right)$$

This method of estimating the expected return by sampling the total return of the trajectory is called the Monte Carlo method (MC method for short).

When the Q function or V function is parameterized through a neural network, the output of the network is recorded as $Q^\pi(s, a)$ or $V^\pi(s)$, and its true label is recorded as the Monte Carlo estimate $\widehat{Q}^\pi(s,a)$ or $\widehat{V}^\pi(s)$, the direct error between the network output value and the estimated value can be calculated through an error function such as the mean square error. The gradient descent algorithm is used to optimize the neural network. From this perspective, the estimation of the value function can be understood as a regression problem. The Monte Carlo method is simple and easy to implement, but it needs to obtain the complete trajectory, so the calculation efficiency is low, and there is no clear end state in some environments.

Temporal difference

Temporal difference (TD method for short) utilizes the Bellman equation properties of the value function. In the calculation formula, only one or more steps are required to obtain the error of the value function and optimize the update value function network. The Carlo method is more computationally efficient.

Recall the Bellman equation of the V function:

$$V^\pi \left(s_t \right) = E_{\tau \sim p(\tau)} \left[r_t + \gamma V^\pi \left(s_{t+1} \right) \right]$$

Therefore, the TD error term $\delta = r_t + \gamma V^\pi(s_{t+1}) - V^\pi(s_t)$ is constructed and updated as follows:

$$V^\pi \left(s_t \right) \leftarrow V^\pi \left(s_t \right) + \alpha \left(r_t + \gamma V^\pi \left(s_{t+1} \right) - V^\pi \left(s_t \right) \right)$$

where $\alpha \in [0, 1]$ is the update step.

The Bellman optimal equation of the Q function is:

$$Q^* \left(s_t, a_t \right) = E_{\tau \sim p(\tau)} \left[r \left(s_t, a_t \right) + \gamma Q^* \left(s_{t+1}, a_{t+1} \right) \right]$$

Similarly, construct TD error term $\delta = r(s_t, a_t) + \gamma Q^*(s_{t+1}, a_{t+1}) - Q^*(s_t, a_t)$, and use the following equation to update:

$$Q^* \left(s_t, a_t \right) \leftarrow Q^* \left(s_t, a_t \right) + \alpha \left(r \left(s_t, a_t \right) + \gamma Q^* \left(s_{t+1}, a_{t+1} \right) - Q^* \left(s_t, a_t \right) \right)$$

14.4.3 Policy Improvement

The value function estimation method can obtain a more accurate value function estimation, but the policy model is not directly given. Therefore, the policy model needs to be derived indirectly based on the value function.

First, look at how to derive the policy model from the V function:

$$\pi^* = V^\pi \left(s \right) \ \forall s \in S$$

Considering that the state space S and the action space A are usually huge, this way of traversing to obtain the optimal policy is not feasible. So can the policy model be derived from the Q function? Consider:

$$\pi'(s) = arg \max_a Q^\pi(s,a)$$

In this way, an action can be selected by traversing the discrete action space A in any state s. This strategy $\pi'(s)$ is a deterministic policy. Because:

$$V^\pi(s_t) = E_{a_t \sim \pi(s_t)}\left[Q^\pi(s_t,a_t)\right]$$

So:

$$V^{\pi'}(s_t) \geq V^\pi(s_t)$$

That is, the strategy π' is always better than or equal to the strategy π, thus achieving policy improvement.

The deterministic policy produces the same action in the same state, so the trajectory produced by each interaction may be similar. The policy model always tends to exploitation but lacks exploration, thus making the policy model limited to a local area, lack of understanding of global status and actions. In order to be able to add exploration capabilities to the $\pi'(s)$ deterministic policy, we can make the $\pi'(s)$ policy have a small probability ϵ to adopt a random policy to explore unknown actions and states.

$$\pi^\epsilon(s_t) = \{arg \max_a Q^\pi(s,a), \ probability \ of \ 1-\epsilon \ random \ action, \ probability \ of \ \epsilon$$

This policy is called ϵ-greedy method. It makes a small amount of modification on the basis of the original policy and can balance utilization and exploration by controlling the hyperparameter ϵ, achieving simple and efficient.

The training process of the value function is shown in Figure 14-20.

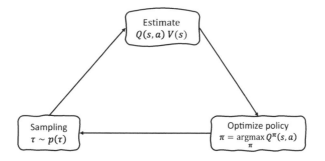

Figure 14-20. *Value function method training process*

14.4.4 SARSA Algorithm

SARSA algorithm [5] uses:

$$Q^{\pi}\left(s_{t},a_{t}\right) \leftarrow Q^{\pi}\left(s_{t},a_{t}\right) + \alpha\left(r\left(s_{t},a_{t}\right) + \gamma Q^{\pi}\left(s_{t+1},a_{t+1}\right) - Q^{\pi}\left(s_{t},a_{t}\right)\right)$$

method to estimate the Q function, at each step of the trajectory, only s_t, a_t, r_t, s_{t+1}, *and* a_{t+1} data can be used to update the Q network once, so it is called SARSA (state action reward state action) algorithm. The s_t, a_t, r_t, s_{t+1}, *and* a_{t+1} of the SARSA algorithm come from the same policy $\pi^c(s_t)$, so they belong to the on-policy algorithm.

14.4.5 DQN Algorithm

In 2015, DeepMind proposed the Q Learning [4] algorithm implemented using deep neural networks, published in Nature [1], and trained and learned on 49 mini games in the Atari game environment, achieving a human level equivalent or even superior. The performance of human level has aroused the strong interest of the industry and the public in the research of reinforcement learning.

Q Learning algorithm uses:

$$Q^*(s_t,a_t) \leftarrow Q^*(s_t,a_t) + \alpha \left(r(s_t,a_t) + \gamma Q^*(s_{t+1},a_{t+1}) - Q^*(s_t,a_t) \right)$$

to estimate the $Q^*(s_t, a_t)$ function and use the $\pi(s_t)$ policy to obtain policy improvement. The Deep Q Network (DQN) uses a deep neural network to parameterize the $Q^*(s_t, a_t)$ function and uses the gradient descent algorithm to update the Q network. The loss function is:

$$L = \left(r_t + \gamma Q_\theta (s_{t+1},a) - Q_\theta (s_t,a_t) \right)^2$$

Since both the training target value $r_t + \gamma Q_\theta(s_{t+1}, a)$ and the predicted value $Q_\theta(s_t, a_t)$ come from the same network, and the training data has a strong correlation, [1] proposed two measures to solve the problem: by adding experience relay buffer to reduce the strong correlation of the data and by freezing target network technology to fix the target estimation network and stabilize the training process.

The replay buffer pool is equivalent to a large data sample buffer pool. During each training, the data pair (s, a, r, s') generated by the latest policy is stored in the replay buffer pool, and then multiple data pairs (s, a, r, s') are randomly sampled from the pool for training. In this way, the strong correlation of the training data can be reduced. It can also be found that the DQN algorithm is an Off-Policy algorithm with high sampling efficiency.

Freezing target network is a training technique. During training, the target network $Q_\theta (s_{t+1},a)$ and the prediction network $Q_\theta(s_t, a_t)$ come from the same network, but the update frequency of $Q_\theta (s_{t+1},a)$ network will be after $Q_\theta(s_t, a_t)$, which is equivalent to being in a frozen state when $Q_\theta (s_{t+1},a)$ is not updated, and then pull latest network parameters from $Q_\theta(s_t, a_t)$ after the freezing is over:

$$L = \left(r_t + \gamma Q_{\underline{\theta}} (s_{t+1},a) - Q_\theta (s_t,a_t) \right)^2$$

In this way, the training process can become more stable.

DQN algorithm is shown in Algorithm 3.

Algorithm 3: DQN algorithm

Randomly initialize θ

repeat

  Reset and get game initial state s

  repeat

    Sample action $a = \pi^\epsilon(s)$

    Interact with environment and get reward r and state s'

    Optimize Q network:

$$\nabla_\theta(r(s_t, a_t) + \gamma Q^*(s_{t+1}, a_{t+1}) - Q^*(s_t, a_t))$$

    Update state $s \leftarrow s'$

  Until game ending

until reach required training times

Output: policy network (s_t)

14.4.6 DQN Variants

Although the DQN algorithm has made a huge breakthrough on the Atari game platform, follow-up studies have found that the Q value in DQN is often overestimated. In view of the defects of the DQN algorithm, some variant algorithms have been proposed.

Double DQN. In [6], the Q network and estimated Q network of target $r_t + \gamma Q\left(s_{t+1}, \max\limits_{a} Q(s_{t+1}, a)\right)$ were separated and updated according to the loss function:

$$L = \left(r_t + \gamma \underline{Q}\left(s_{t+1}, \max\limits_{a} Q(s_{t+1}, a)\right) - Q(s_t, a_t)\right)^2$$

Dueling DQN. [7] separated the network output into $V(s)$ and $A(s, a)$, as shown in Figure 14-21. Then use:

$$Q(s,a) = V(s) + A(s,a)$$

to generate Q function estimate $Q(s, a)$. The rest and DQN remain the same.

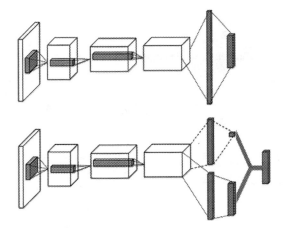

Figure 14-21. *DQN network(upper) and dueling DQN network(lower) [7]*

14.4.7 Hands-On DQN

Here we continue to implement the DQN algorithm based on the balance bar game environment.

Q network. The state of the balance bar game is a vector of length 4. Therefore, the input of the Q network is designed as four nodes. After a 256-256-2 fully connected layer, the distribution of the Q function estimation $Q(s, a)$ with the number of output nodes of 2 is obtained. The implementation of the network is as follows:

```python
class Qnet(keras.Model):
    def __init__(self):
        # Create a Q network, the input is the state vector,
        and the output is the Q value of the action
        super(Qnet, self).__init__()
        self.fc1 = layers.Dense(256, kernel_initializer=
        'he_normal')
        self.fc2 = layers.Dense(256, kernel_initializer=
        'he_normal')
        self.fc3 = layers.Dense(2, kernel_initializer=
        'he_normal')

    def call(self, x, training=None):
        x = tf.nn.relu(self.fc1(x))
        x = tf.nn.relu(self.fc2(x))
        x = self.fc3(x)
        return x
```

Replay buffer pool. The replay buffer pool is used in the DQN algorithm to reduce the strong correlation between data. We use the Deque object in the ReplayBuffer class to implement the buffer pool function. During training, the latest data (s, a, r, s') is stored in the Deque object through the put (transition) method, and n data (s, a, r, s') are randomly sampled from the Deque object using sample(n) method. The implementation of the replay buffer pool is as follows:

```python
class ReplayBuffer():
    # Replay buffer pool
    def __init__(self):
        # Deque
        self.buffer = collections.deque(maxlen=buffer_limit)

    def put(self, transition):
        self.buffer.append(transition)
```

```
def sample(self, n):
    # Sample n samples
    mini_batch = random.sample(self.buffer, n)
    s_lst, a_lst, r_lst, s_prime_lst, done_mask_lst = [],
    [], [], [], []
    # Organize by category
    for transition in mini_batch:
        s, a, r, s_prime, done_mask = transition
        s_lst.append(s)
        a_lst.append([a])
        r_lst.append([r])
        s_prime_lst.append(s_prime)
        done_mask_lst.append([done_mask])
    # Convert to tensor
    return tf.constant(s_lst, dtype=tf.float32),\
                  tf.constant(a_lst, dtype=tf.int32), \
                  tf.constant(r_lst, dtype=tf.float32), \
                  tf.constant(s_prime_lst, dtype=tf.
                  float32), \
                  tf.constant(done_mask_lst, dtype=tf.
                  float32)
```

Policy improvement. The ϵ-greedy method is implemented here. When sampling actions, there a probability of $1 - \epsilon$ to choose *argarg Q^π (s, a)*, and a probability of ϵ to randomly choose an action.

```
def sample_action(self, s, epsilon):
    # Send the state vector to get the strategy: [4]
    s = tf.constant(s, dtype=tf.float32)
    # s: [4] => [1,4]
    s = tf.expand_dims(s, axis=0)
    out = self(s)[0]
```

```
coin = random.random()
# Policy improvement: e-greedy way
if coin < epsilon:
    # epsilon larger
    return random.randint(0, 1)
else:  # Q value is larger
    return int(tf.argmax(out))
```

Network main process. The network trains up to 10,000 rounds. At the beginning of the round, the game is first reset to get the initial state *s*, and an action is sampled from the current Q network to interact with the environment to obtain the data pair (s, a, r, s'), and stored in the replay buffer pool. If the number of samples in the current replay buffer pool is sufficient, sample a batch of data, and optimize the estimation of the Q network according to the TD error until the end of the game.

```
for n_epi in range(10000):  # Training times
        # The epsilon probability will also be attenuated by
        8% to 1%. The more you go, the more you use the action
        with the highest Q value.
        epsilon = max(0.01, 0.08 - 0.01 * (n_epi / 200))
        s = env.reset()  # Reset environment
        for t in range(600):  # Maximum timestamp of a round
            # if n_epi>1000:
            #      env.render()
            # According to the current Q network, extract and
            improve the policy.
            a = q.sample_action(s, epsilon)
            # Use improved strategies to interact with the
            environment
            s_prime, r, done, info = env.step(a)
            done_mask = 0.0 if done else 1.0  # End flag mask
```

```
        # Save
        memory.put((s, a, r / 100.0, s_prime, done_mask))
        s = s_prime  # Update state
        score += r  # Record return
        if done:  # End round
            break
    if memory.size() > 2000:  # train if size is greater
    than 2000
        train(q, q_target, memory, optimizer)
    if n_epi % print_interval == 0 and n_epi != 0:
        for src, dest in zip(q.variables, q_target.
        variables):
            dest.assign(src)  # weights come from Q
```

During training, only the Q_θ network will be updated, while the $Q_{\bar\theta}$ network will be frozen. After the Q_θ network has been updated many times, use the following code to copy the latest parameters from Q_θ to $Q_{\bar\theta}$.

```
for src, dest in zip(q.variables, q_target.variables):
                dest.assign(src)  # weights come from Q
```

Optimize the Q network. When optimizing the Q network, it will train and update ten times at a time. Randomly sample from the replay buffer pool each time, and select the action $Q_{\underline{\theta}}\left(s_{t+1},a\right)$ to construct the TD difference. Here we use the Smooth L1 error to construct the TD error:

$$L = \{0.5*(x-y)^2, |x-y| < 1 | x-y| - 0.5, |x-y| \ge 1$$

In TensorFlow, Smooth L1 error can be implemented using Huber error as follows:

```
def train(q, q_target, memory, optimizer):
    # Construct the error of Bellman equation through Q network
    and shadow network.
```

```
# And only update the Q network, the update of the shadow
network will lag behind the Q network
huber = losses.Huber()
for i in range(10):  # Train 10 times
    # Sample from buffer pool
    s, a, r, s_prime, done_mask = memory.sample(batch_size)
    with tf.GradientTape() as tape:
        # s: [b, 4]
        q_out = q(s)  # Get Q(s,a) distribution
        # Because TF's gather_nd is different from
        pytorch's gather, we need to the coordinates of
        gather_nd, indices:[b, 2]
        # pi_a = pi.gather(1, a) # pytorch only needs
        one line.
        indices = tf.expand_dims(tf.range(
        a.shape[0]), axis=1)
        indices = tf.concat([indices, a], axis=1)
        q_a = tf.gather_nd(q_out, indices) # The
        probability of action, [b]
        q_a = tf.expand_dims(q_a, axis=1) # [b]=> [b,1]
        # Get the maximum value of Q(s',a). It comes from
        the shadow network! [b,4]=>[b,2]=>[b,1]
        max_q_prime = tf.reduce_max(q_target(
        s_prime),axis=1,keepdims=True)
        # Construct the target value of Q(s,a_t)
        target = r + gamma * max_q_prime * done_mask
        # Calcualte error between Q(s,a_t) and target
        loss = huber(q_a, target)
    # Update network
    grads = tape.gradient(loss, q.trainable_variables)
    optimizer.apply_gradients(zip(grads, q.trainable_
    variables))
```

14.5 Actor-Critic Method

When introducing the original policy gradient algorithm, in order to reduce the variance, we introduced the bias b mechanism:

$$\frac{\partial J(\theta)}{\partial \theta} = E_{\tau \sim p_\theta(\tau)}\left[\sum_{t=1}^{T-1}\frac{\partial}{\partial \theta}\log\,log\,\pi_\theta\left(s_t\right)\left(R(\tau)-b\right)\right]$$

where b can be estimated by Monte Carlo method $b = \frac{1}{N}\sum_{n=1}^{N}R\left(\tau^{(n)}\right)$.

If $R(\tau)$ is understood as the estimated value of $Q^\pi(s_t, a_t)$, the bias b is understood as the average level $V^\pi(s_t)$ of state s_t, then $R(\tau) - b$ is (approximately) the advantage value function $A^\pi(s, a)$. Among them, if the bias value function $V^\pi(s_t)$ is estimated using neural networks, it is the Actor-Critic method (AC method for short). The policy network $\pi_\theta(s_t)$ is called Actor, which is used to generate policies and interact with the environment. The $V_\phi^\pi\left(s_t\right)$ value network is called Critic, which is used to evaluate the current state. θ and ϕ are the parameters of the Actor network and the Critic network, respectively.

For the Actor network π_θ, the goal is to maximize the return expectation, and the parameter θ of the policy network is updated through the partial derivative of $\frac{\partial J(\theta)}{\partial \theta}$:

$$\theta' \leftarrow \theta + \eta \cdot \frac{\partial J}{\partial \theta}$$

For the Critic network V_ϕ^π, the goal is to obtain an accurate $V_\phi^\pi\left(s_t\right)$ value function estimate through the MC method or the TD method:

$$\phi = dist\left(V_\phi^\pi\left(s_t\right), V_{target}^\pi\left(s_t\right)\right)$$

where dist(a,b) is the distance measurer of a and b, such as Euclidean distance. $V_{target}^{\pi}\left(s_t\right)$ is the target value of $V_{\phi}^{\pi}\left(s_t\right)$. When estimated by the MC method,

$$V_{target}^{\pi}\left(s_t\right)=R\left(\tau_{t:T}\right)$$

When estimated by the TD method,

$$V_{target}^{\pi}\left(s_t\right)=r_t+\gamma V^{\pi}\left(s_{t+1}\right)$$

14.5.1 Advantage AC Algorithm

The Actor-Critic algorithm using the advantage value function $A^{\pi}(s, a)$ is called the Advantage Actor-Critic algorithm. It is currently one of the mainstream algorithms that use the Actor-Critic idea. In fact, the Actor-Critic series of algorithms do not have to use the advantage value function $A^{\pi}(s, a)$. There are other variants.

When the Advantage Actor-Critic algorithm is trained, the Actor obtains the action a_t according to the current state s_t and the policy π_{θ} sampling, and then interacts with the environment to obtain the next state s_{t+1} and reward r_t. The TD method can estimate the target value $V_{target}^{\pi}\left(s_t\right)$ of each step, thereby updating the Critic network so that the estimation of the value network is closer to the expected return of the real environment. $\hat{A}_t=r_t+\gamma V^{\pi}\left(s_{t+1}\right)-V^{\pi}\left(s_t\right)$ is used to estimate the advantage value of the current action, and the following equation is used to calculate the gradient info of the Actor network. $L^{PG}\left(\theta\right)=\hat{E}_t\left[\log\log\pi_{\theta}\left(s_t\right)\hat{A}_t\right]$

By repeating this process, the Critic network will be more and more accurate, and the Actor network will also adjust its policy to make it better next time.

14.5.2 A3C Algorithm

The full name of the A3C algorithm is the Asynchronous Advantage Actor-Critic algorithm. It is an asynchronous version proposed by DeepMind based on the Advantage Actor-Critic algorithm [8]. The Actor-Critic network is deployed in multiple threads for simultaneous training, and the parameters are synchronized through the global network. . This asynchronous training mode greatly improves the training efficiency; therefore, the training speed is faster and the algorithm performance is better.

As shown in Figure 14-22, the algorithm will create a new global Network and M Worker threads. Global Network contains Actor and Critic networks, and each thread creates a new interactive environment, Actor and Critic networks. In the initialization phase, Global Network initializes parameters θ and ϕ randomly. The Actor-Critic network in Worker pulls parameters synchronously from Global Network to initialize the network. During training, the Actor-Critic network in the Worker first pulls the latest parameters from the Global Network, and then the latest policy $\pi_\theta(s_t)$ will sample actions to interact with the private environment, and calculate the gradients of parameters θ and ϕ according to the Advantage Actor-Critic algorithm. After completing the gradient calculation, each worker submits the gradient information to the Global Network and uses the optimizer of the Global Network to complete the parameter update. In the algorithm testing phase, only Global Network interacts with the environment.

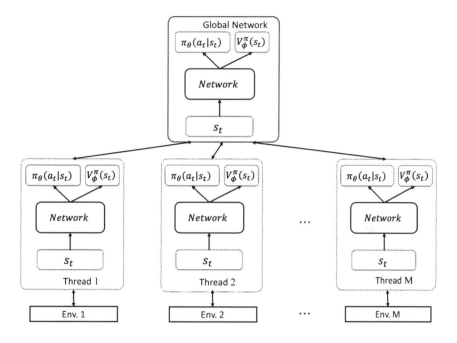

Figure 14-22. *A3C algorithm*

14.5.3 Hands-On A3C

Next we implement the asynchronous A3C algorithm. Like the ordinary Advantage AC algorithm, the Actor-Critic network needs to be created. It contains an Actor sub-network and a Critic sub-network. Sometimes Actor and Critic will share the previous network layers to reduce the amount of network parameters. The balance bar game is relatively simple. We use a two-layer fully connected network to parameterize the Actor network, and another two-layer fully connected network to parameterize the Critic network.

The Actor-Critic network code is as follows:

```python
class ActorCritic(keras.Model):
    # Actor-Critic model
    def __init__(self, state_size, action_size):
        super(ActorCritic, self).__init__()
        self.state_size = state_size # state vector length
        self.action_size = action_size # action size
        # Policy network Actor
        self.dense1 = layers.Dense(128, activation='relu')
        self.policy_logits = layers.Dense(action_size)
        # V network Critic
        self.dense2 = layers.Dense(128, activation='relu')
        self.values = layers.Dense(1)
```

The forward propagation process of Actor-Critic calculates the policy distribution $\pi_\theta(s_t)$ and the V function estimation $V^\pi(s_t)$ separately. The code is as follows:

```python
    def call(self, inputs):
        # Get policy distribution Pi(a|s)
        x = self.dense1(inputs)
        logits = self.policy_logits(x)
        # Get v(s)
        v = self.dense2(inputs)
        values = self.values(v)
        return logits, values
```

Worker thread class. In the Worker thread, the same calculation process as the Advantage AC algorithm is implemented, except that the gradient information of parameters θ and ϕ is not directly used to update the Actor-Critic network of the Worker, instead it is submitted to the Global Network for update. Specifically, in the initialization phase of the Worker class, the server object and the opt object represent the Global Network model and optimizer respectively, and create a private ActorCritic class client and interactive environment env.

```python
class Worker(threading.Thread):
    # The variables created here belong to the class, not to
    the instance, and are shared by all instances
    global_episode = 0 # Round count
    global_avg_return = 0 # Average return
    def __init__(self,  server, opt, result_queue, idx):
        super(Worker, self).__init__()
        self.result_queue = result_queue # Shared queue
        self.server = server # Central model
        self.opt = opt # Central optimizer
        self.client = ActorCritic(4, 2) # Thread private network
        self.worker_idx = idx # Thread id
        self.env = gym.make('CartPole-v0').unwrapped
        self.ep_loss = 0.0
```

In the thread running phase, each thread interacts with the environment for up to 400 rounds. At the beginning of the round, the client network sampling action is used to interact with the environment and saved to the memory object. At the end of the round, train the Actor network and the Critic network to obtain the gradient information of the parameters θ and ϕ, and call the opt optimizer object to update the Global Network.

```python
    def run(self):
        total_step = 1
        mem = Memory() # Each worker maintains a memory
        while Worker.global_episode < 400: # Maximum number of
        frames not reached
            current_state = self.env.reset() # Reset client state
            mem.clear()
```

```
ep_reward = 0.
ep_steps = 0
self.ep_loss = 0
time_count = 0
done = False
while not done:
    # Get Pi(a|s),no softmax
    logits, _ = self.client(tf.constant(
    current_state[None, :],
                            dtype=tf.float32))
    probs = tf.nn.softmax(logits)
    # Random sample action
    action = np.random.choice(2, p=probs.numpy()[0])
    new_state, reward, done, _ = self.env.
    step(action) # Interact
    if done:
        reward = -1
    ep_reward += reward
    mem.store(current_state, action, reward) # Record

    if time_count == 20 or done:
        # Calculate the error of current client
        with tf.GradientTape() as tape:
            total_loss = self.compute_loss(done,
            new_state, mem)
        self.ep_loss += float(total_loss)
        # Calculate error
        grads = tape.gradient(total_loss,
        self.client.trainable_weights)
        # Submit gradient info to server, and
        update gradient
```

```
self.opt.apply_gradients(zip(grads,
                             self.server.
                             trainable_
                             weights))
# Pull latest gradient info from server
self.client.set_weights(self.server.get_
weights())
mem.clear() # Clear Memory
time_count = 0

if done:  # Calcualte return
    Worker.global_avg_return = \
        record(Worker.global_episode,
        ep_reward, self.worker_idx,
            Worker.global_avg_return,
            self.result_queue,
            self.ep_loss, ep_steps)
    Worker.global_episode += 1
ep_steps += 1
time_count += 1
current_state = new_state
total_step += 1
self.result_queue.put(None) # End thread
```

Actor-Critic error calculation. When each Worker class is trained, the error calculation of Actor and Critic network is implemented as follows. Here we use the Monte Carlo method to estimate the target value $V_{target}^{\pi}(s_t)$, and use the distance between $V_{target}^{\pi}(s_t)$ and $V_{\phi}^{\pi}(s_t)$ the two as the error function value_loss of the Critic network. The policy loss function policy_loss of the Actor network comes from $-L^{PG}(\theta) = -\hat{E}_t\left[\log \log \pi_\theta(s_t)\hat{A}_t\right]$ where $-\hat{E}_t\left[\log \log \pi_\theta(s_t)\hat{A}_t\right]$ is implemented by TensorFlow's cross-entropy function. After the various loss functions are aggregated, the total loss function is formed and returned.

```
def compute_loss(self,
                 done,
                 new_state,
                 memory,
                 gamma=0.99):
    if done:
        reward_sum = 0.
    else:
        reward_sum = self.client(tf.constant(new_
        state[None, :],
                                    dtype=tf.float32))[-1].
                                    numpy()[0]
    # Calculate return
    discounted_rewards = []
    for reward in memory.rewards[::-1]:  # reverse buffer r
        reward_sum = reward + gamma * reward_sum
        discounted_rewards.append(reward_sum)
    discounted_rewards.reverse()
    # Get Pi(a|s) and v(s)
    logits, values = self.client(tf.constant(
    np.vstack(memory.states),
                                dtype=tf.float32))
    # Calculate advantage = R() - v(s)
    advantage = tf.constant(np.array(discounted_rewards)
    [:, None],
                                        dtype=tf.
                                        float32) - values

    # Critic network loss
    value_loss = advantage ** 2
    # Policy loss
    policy = tf.nn.softmax(logits)
```

```
policy_loss = tf.nn.sparse_softmax_cross_entropy_
with_logits(
                   labels=memory.actions, logits=logits)
# When calculating the policy network loss, the V
network is not calculated
policy_loss *= tf.stop_gradient(advantage)

entropy = tf.nn.softmax_cross_entropy_with_
logits(labels=policy,
                                        logits=logits)
policy_loss -= 0.01 * entropy
# Aggregate each error
total_loss = tf.reduce_mean((0.5 * value_loss +
policy_loss))
return total_loss
```

Agent. The agent is responsible for the training of the entire A3C algorithm. In the initialization phase, the agent class creates a new Global Network object server and its optimizer object opt.

```
class Agent:
    # Agent, include server
    def __init__(self):
        # server optimizer, no client, pull parameters
        from server
        self.opt = optimizers.Adam(1e-3)
        # Sever model
        self.server = ActorCritic(4, 2) # State vector,
        action size
        self.server(tf.random.normal((2, 4)))
```

At the beginning of training, each Worker thread object is created, and each thread object is started to interact with the environment. When each

Worker object interacts, it will pull the latest network parameters from the Global Network and use the latest policy to interact with the environment and calculate its own loss. Finally, each Worker submits the gradient information to the Global Network, and call the opt object to optimize the Global Network. The training code is as follows:

```
def train(self):
    res_queue = Queue() # Shared queue
    # Create interactive environment
    workers = [Worker(self.server, self.opt, res_queue, i)
               for i in range(multiprocessing.cpu_count())]
    for i, worker in enumerate(workers):
        print("Starting worker {}".format(i))
        worker.start()
    # Plot return curver
    moving_average_rewards = []
    while True:
        reward = res_queue.get()
        if reward is not None:
            moving_average_rewards.append(reward)
        else: # End
            break
    [w.join() for w in workers] # Quit threads
```

14.6 Summary

This chapter introduces the problem setting and basic theory of reinforcement learning and introduces two series of algorithms to solve reinforcement learning problems: policy gradient method and value function method. The policy gradient method directly optimizes the policy model, which is simple and direct, but the sampling efficiency is low. The sampling efficiency of the algorithm can be improved by the importance sampling technique. The value function method has high sampling efficiency and is easy to train, but the policy model needs to be derived indirectly from the value function. Finally, the Actor-Critic method combining the policy gradient method and the value function method is introduced. We also introduced the principles of several typical algorithms, and used the balance bar game environment for algorithm implementation and testing.

14.7 References

[1]   V. Mnih, K. Kavukcuoglu, D. Silver, A. A. Rusu, J. Veness, M. G. Bellemare, A. Graves, M. Riedmiller, A. K. Fidjeland, G. Ostrovski, S. Petersen, C. Beattie, A. Sadik, I. Antonoglou, H. King, D. Kumaran, D. Wierstra, S. Legg and D. Hassabis, "Human-level control through deep reinforcement learning," *Nature,* 518, pp. 529-533, 2 2015.

[2]   D. Silver, A. Huang, C. J. Maddison, A. Guez, L. Sifre, G. Driessche, J. Schrittwieser, I. Antonoglou, V. Panneershelvam, M. Lanctot, S. Dieleman, D. Grewe, J. Nham, N. Kalchbrenner, I. Sutskever, T. Lillicrap, M. Leach, K. Kavukcuoglu, T. Graepel and D. Hassabis, "Mastering the game of Go with deep neural networks and tree search," *Nature,* 529, pp. 484-503, 2016.

[3]   D. Silver, J. Schrittwieser, K. Simonyan, I. Antonoglou, A. Huang, A. Guez, T. Hubert, L. Baker, M. Lai, A. Bolton, Y. Chen, T. Lillicrap, F. Hui, L. Sifre, G. Driessche, T. Graepel and D. Hassabis, "Mastering the game of Go without human knowledge," *Nature,* 550, pp. 354--, 10 2017.

[4].   R. J. Williams, "Simple statistical gradient-following algorithms for connectionist reinforcement learning," *Machine Learning,* 8, pp. 229-256, 01 5 1992.

[5]   G. A. Rummery and M. Niranjan, "On-Line Q-Learning Using Connectionist Systems," 1994.

[6]   H. Hasselt, A. Guez and D. Silver, "Deep Reinforcement Learning with Double Q-learning," *CoRR,* abs/1509.06461, 2015.

[7]   Z. Wang, N. Freitas and M. Lanctot, "Dueling Network Architectures for Deep Reinforcement Learning," *CoRR,* abs/1511.06581, 2015.

[8]   V. Mnih, A. P. Badia, M. Mirza, A. Graves, T. P. Lillicrap, T. Harley, D. Silver and K. Kavukcuoglu, "Asynchronous Methods for Deep Reinforcement Learning," *CoRR,* abs/1602.01783, 2016.

[9]   C. J. C. H. Watkins and P. Dayan, "Q-learning," *Machine Learning,* 1992.

[10]   J. Schulman, S. Levine, P. Abbeel, M. Jordan and P. Moritz, "Trust Region Policy Optimization," *Proceedings of the 32nd International Conference on Machine Learning,* Lille, 2015.

[11]   J. Schulman, F. Wolski, P. Dhariwal, A. Radford and O. Klimov, "Proximal Policy Optimization Algorithms," *CoRR,* abs/1707.06347, 2017.

CHAPTER 15

Customized Dataset

> Spending a year on artificial intelligence is enough to make people believe in the existence of God.
>
> —Alan Paley

Deep learning has been widely used in various industries such as medicine, biology, and finance and has been deployed on various platforms such as the Internet and mobile terminals. When we introduced the algorithm earlier, most of the datasets were commonly used classic datasets. The downloading, loading, and preprocessing of the dataset can be completed with a few lines of TensorFlow code, which greatly improves the research efficiency. In actual applications, the datasets are different for different application scenarios. For customized datasets, using TensorFlow to complete data loading, designing excellent network model training process, and deploying the trained model to platforms such as mobile and the Internet network is an indispensable link for the implementation of deep learning algorithms.

In this chapter, we will take a specific application scenario of image classification as an example to introduce a series of practical technologies such as downloading of customized datasets, data processing, network model design, and transfer learning.

© Liangqu Long and Xiangming Zeng 2022
L. Long and X. Zeng, *Beginning Deep Learning with TensorFlow*,
https://doi.org/10.1007/978-1-4842-7915-1_15

15.1 Pokémon Go Dataset

Pokémon Go is a mobile game that uses augmented reality (AR) technology to capture and train Pokémon elves outdoors, and use them to fight. The game was launched on Android and IOS in July 2016. Once released, it was sought after by players all over the world. At one time, the server was paralyzed due to too many players. As shown in Figure 15-1, a player scanned the real environment with his mobile phone and collected the virtual Pokémon "Pikachu."

Figure 15-1. *Pokémon game screen*

We use the Pokémon dataset crawled from the web to demonstrate how to use customized dataset. The Pokémon dataset collects a total of five elven creatures: Pikachu, Mewtwo, Squirtle, Charmander, and Bulbasaur. The information of each elven is shown in Table 15-1, a total of 1168 pictures. There are incorrectly labeled samples in these pictures, so the wrongly labeled samples were artificially eliminated, and a total of 1,122 valid pictures were obtained.

Table 15-1. *Pokémon dataset information*

Elven	Pikachu	Mewtwo	Squirtle	Charmander	Bulbasaur
Amount	226	239	209	224	224
Sample picture					

Readers can download the provided dataset file by themselves (link: https://drive.google.com/file/d/1Db2O4YID7VDcQ5lKOObnkKy-U1ZZVj7c/view?usp=sharing), and after decompression, we can get the root directory named pokemon, which contains five subfolders, the file name of each subfolder represents the category name of the pictures, and the corresponding category is stored under each subfolder as shown in Figure 15-2.

Name	Date modified	Type	Size
bulbasaur	5/25/2019 10:11 AM	File folder	
charmander	5/25/2019 10:11 AM	File folder	
mewtwo	5/25/2019 10:11 AM	File folder	
pikachu	5/25/2019 10:11 AM	File folder	
squirtle	5/25/2019 10:11 AM	File folder	

Figure 15-2. *Pokémon dataset storage directory*

15.2 Customized Dataset Loading

In practical applications, the storage methods of samples and sample labels may vary. For example, in some occasions, all pictures are stored in the same directory, and the category name can be derived from the picture name, such as a picture with a file name of "pikachu_asxes0132.png". The category information can be extracted from the file name pikachu.

The label information of some data samples is saved in a text file in JSON format, and the label of each sample needs to be queried in JSON format. No matter how the dataset is stored, we can always use logic rules to obtain the path and label information of all samples.

We abstract the loading process of customized data into the following steps.

15.2.1 Create Code Table

The category of the sample is generally marked with the category name of the string type, but for the neural network, the category name needs to be digitally encoded, and then converted into one-hot encoding or other encoding formats when appropriate. Considering a dataset of n categories, we randomly code each category into a number $l \in [0, n-1]$. The mapping relationship between category names and numbers is called a coding table. Once created, it generally cannot be changed.

For the storage format of the Pokémon dataset, we create a coding table in the following way. First, traverse all sub-directories under the pokemon root directory in order. For each sub-target, use the category name as the key of the code table dictionary object name2label, and the number of existing key-value pairs in the code table as the label mapping number of the category, and save it into name2label dictionary object. The implementation is as follows:

```
def load_pokemon(root, mode='train'):
    # Create digital dictionary table
    name2label = {}  # Coding dictionary, "sq...":0
    # Traverse the subfolders under the root directory and sort
    them to ensure that the mapping relationship is fixed
    for name in sorted(os.listdir(os.path.join(root))):
        # Skip non-folder objects
        if not os.path.isdir(os.path.join(root, name)):
```

```
    continue
# Code a number for each category
name2label[name] = len(name2label.keys())
  ...
```

15.2.2 Create Sample and Label Form

After the coding table is determined, we need to obtain the storage path of each sample and its label number according to the actual data storage method, which are represented as two list objects, images and labels, respectively. The images list stores the path string of each sample, and the labels list stores the category number of the sample. The two have the same length, and the elements at the corresponding positions are related to each other.

We store the images and labels information in a csv format file, where the csv file format is a plain text file format with data separated by commas, which can be opened with Notepad or MS Excel software. There are many advantages by storing all sample information in a csv file, such as direct dataset division and batch sampling. The csv file can save the information of all samples in the dataset, or you can create three csv files based on the training set, validation set, and test set. The content of the resulting csv file is shown in Figure 15-3. The first element of each row stores the storage path of the current sample, and the second element stores the category number of the sample.

Figure 15-3. *Path and label saved in CSV file*

The process of creating a csv file is: traverse all pictures in the root directory of pokemon, record the path of the picture, and obtain the code number according to the coding table, and write it into the csv file as a line. The code is as follows:

```
def load_csv(root, filename, name2label):
    # Return images,labels Lists from csv file
    # root: root directory, filename:csv file name,
    name2label:category coding table
    if not os.path.exists(os.path.join(root, filename)):
        # Create csv file if not exist.
        images = []
        for name in name2label.keys(): # Traverse all
        subdirectories to get all pictures
            # Only consider image files with suffix
            png,jpg,jpeg:'pokemon\\mewtwo\\00001.png
            images += glob.glob(os.path.join(root, name, '*.png'))
```

```
        images += glob.glob(os.path.join(root, name,
        '*.jpg'))
        images += glob.glob(os.path.join(root, name,
        '*.jpeg'))
    # Print data info:1167, 'pokemon\\
    bulbasaur\\00000000.png'
    print(len(images), images)
    random.shuffle(images) # Randomly shuffle
    # Create csv file, and store image path and
    corresponding label info
    with open(os.path.join(root, filename), mode='w',
    newline='') as f:
        writer = csv.writer(f)
        for img in images:  # 'pokemon\\bulbasaur
        \\00000000.png'
            name = img.split(os.sep)[-2]
            label = name2label[name]
            # 'pokemon\\bulbasaur\\00000000.png', 0
            writer.writerow([img, label])
        print('written into csv file:', filename)
        ...
```

After creating the csv file, you only need to read the sample path and label information from the csv file next time, instead of generating the csv file every time, which improves the calculation efficiency. The code is as follows:

```
def load_csv(root, filename, name2label):
    ...
    # At this time there is already a csv file on the file
    system, read directly
    images, labels = [], []
```

```
with open(os.path.join(root, filename)) as f:
    reader = csv.reader(f)
    for row in reader:
        # 'pokemon\\bulbasaur\\00000000.png', 0
        img, label = row
        label = int(label)
        images.append(img)
        labels.append(label)
# Return image path list and tag list
return images, labels
```

15.2.3 Dataset Division

The division of the dataset needs to be flexibly adjusted according to the actual situation. When the number of samples in the dataset is large, you can choose a ratio of 80%-10%-10% to allocate to the training set, validation set, and test set; when the number of samples is small, for example, the total number of pictures in the Pokémon dataset here is only 1000; if the ratio of the validation set and test set is only 10%, the number of pictures is about 100, so the validation accuracy and test accuracy may fluctuate greatly. For small datasets, although the sample size is small, it is necessary to appropriately increase the ratio of the validation set and test set to ensure accurate test results. Here we set the ratio of validation set and test set to 20%, that is, there are about 200 pictures for validation and testing.

First, call the load_csv function to load the images and labels list, and load the corresponding pictures and labels according to the current model parameters. Specifically, if the model parameter is train, the first 60% data of images and labels are taken as the training set; if the model parameter is val, the 60% to 80% area data of images and labels are taken as the validation set; if the model parameter is test, the last 20% of images and labels are taken as the test set. The code is implemented as follows:

```
def load_pokemon(root, mode='train'):
    ...
    # Read Label info
    # [file1,file2,], [3,1]
    images, labels = load_csv(root, 'images.csv', name2label)
# Dataset division
    if mode == 'train':  # 60%
        images = images[:int(0.6 * len(images))]
        labels = labels[:int(0.6 * len(labels))]
    elif mode == 'val':  # 20% = 60%->80%
        images = images[int(0.6 * len(images)):int(0.8 *
        len(images))]
        labels = labels[int(0.6 * len(labels)):int(0.8 *
        len(labels))]
    else:  # 20% = 80%->100%
        images = images[int(0.8 * len(images)):]
        labels = labels[int(0.8 * len(labels)):]
    return images, labels, name2label
```

It should be noted that the dataset division scheme for each run needs to be fixed to prevent the use of test set for training, resulting in inaccurate model generalization performance.

15.3 Hands-On Pokémon Dataset

After introducing the loading process of the custom dataset, let's load and train the Pokémon data set.

15.3.1 Create Dataset Object

First, return the images, labels, and coding table information through the load_pokemon function as follows:

```
# Load the pokemon dataset, specify to load the
training set
# Return the sample path list of the training set, the
label number list and the coding table dictionary
images, labels, table = load_pokemon('pokemon', 'train')
print('images:', len(images), images)
print('labels:', len(labels), labels)
print('table:', table)
```

Construct a Dataset object, and complete the random breakup, preprocessing, and batch operation of the dataset. The code is as follows:

```
# images: string path
# labels: number
db = tf.data.Dataset.from_tensor_slices((images, labels))
db = db.shuffle(1000).map(preprocess).batch(32)
```

When we use tf.data.Dataset.from_tensor_slices to construct the dataset, the passed-in parameter is a tuple composed of images and labels, so when the db object is iterated, the tuple object of (X_i, Y_i) is returned, where X_i is the image tensor of the *ith* batch, Y_i is the image label data of the *ith* batch. We can view the image samples of each traversal through TensorBoard visualization as follows:

```
# Create TensorBoard summary object
writter = tf.summary.create_file_writer('logs')
for step, (x,y) in enumerate(db):
    # x: [32, 224, 224, 3]
    # y: [32]
```

```
with writter.as_default():
    x = denormalize(x) # Denormalize
    # Write in image data
    tf.summary.image('img',x,step=step,max_outputs=9)
    time.sleep(5) # Delay 5s
```

15.3.2 Data Preprocessing

We complete the preprocessing of the data by calling the .map(preprocess) function when constructing the data set. Since our images list currently only saves the path information of all images, not the content tensor of the image, it is necessary to complete the image reading and tensor conversion in the preprocessing function.

For the preprocess function (x,y) = preprocess(x,y), its incoming parameters need to be saved in the same format as the parameters given when creating the dataset, and the return parameters need to be saved in the same format as the incoming parameters. In particular, we pass in the (x, y) tuple object when constructing the dataset, where x is the path list of all pictures and y is the label number list of all pictures. Considering that the location of the map function is db = db.shuffle(1000).map(preprocess). batch(32), then the incoming parameters of preprocess are (x_i, y_i), where x_i and y_i are, respectively, the i-th picture path string and label number. If the location of the map function is db = db.shuffle(1000).batch(32). map(preprocess), then the incoming parameters of preprocess are (x_i, y_i), where x_i and y_i are the path and tag list of the i-th batch respectively. The code is as follows:

```
def preprocess(x,y): # preprocess function
    # x: image path, y:image coding number
    x = tf.io.read_file(x) # Read image
    x = tf.image.decode_jpeg(x, channels=3) # Decode image
    x = tf.image.resize(x, [244, 244]) # Resize to 244x244
```

```
# Data augmentation
# x = tf.image.random_flip_up_down(x)
x= tf.image.random_flip_left_right(x) # flip left and right
x = tf.image.random_crop(x, [224, 224, 3]) # Crop
to 224x224
# Convert to tensor and [0, 1] range
# x: [0,255]=> 0~1
x = tf.cast(x, dtype=tf.float32) / 255.
# 0~1 => D(0,1)
x = normalize(x) # Normalize
y = tf.convert_to_tensor(y) # To tensor

return x, y
```

Considering that the scale of our dataset is very small, in order to prevent overfitting, we have done a small amount of data enhancement transformation to obtain more data. Finally, we scale the pixel values in the range of 0~255 to the range of 0~1, and normalize the data, and map the pixels to the distribution around 0, which is beneficial to the optimization of the network. Finally, the data is converted to tensor data and returned. At this time, the data returned will be the tensor data in batch form when iterating over the db object.

The standardized data is suitable for network training and prediction, but when visualizing, the data needs to be mapped back to the range of 0~1. The reverse process of standardization and standardization is as follows:

```
# The mean and std here are calculated based on real data, such
as ImageNet
img_mean = tf.constant([0.485, 0.456, 0.406])
img_std = tf.constant([0.229, 0.224, 0.225])
def normalize(x, mean=img_mean, std=img_std):
```

```
    # Normalization function
    # x: [224, 224, 3]
    # mean: [224, 224, 3], std: [3]
    x = (x - mean)/std
    return x

def denormalize(x, mean=img_mean, std=img_std):
    # Denormalization function
    x = x * std + mean
    return x
```

Using the preceding method, distribute the Dataset objects that create the training set, validation set, and test set. Generally speaking, the validation set and test set do not directly participate in the optimization of network parameters, and there is no need to randomly break the order of samples.

```
batchsz = 128
# Create training dataset
images, labels, table = load_pokemon('pokemon',mode='train')
db_train = tf.data.Dataset.from_tensor_slices((images, labels))
db_train = db_train.shuffle(1000).map(preprocess).
batch(batchsz)
# Create validation dataset
images2, labels2, table = load_pokemon('pokemon',mode='val')
db_val = tf.data.Dataset.from_tensor_slices((images2, labels2))
db_val = db_val.map(preprocess).batch(batchsz)
# Create testing dataset
images3, labels3, table = load_pokemon('pokemon',mode='test')
db_test = tf.data.Dataset.from_tensor_slices((images3,
labels3))
db_test = db_test.map(preprocess).batch(batchsz)
```

15.3.3 Create Model

The mainstream network models such as VGG13 and ResNet18 have been introduced and implemented before, and we will not repeat the specific implementation details of the model here. Commonly used network models are implemented in the keras.applications module, such as VGG series, ResNet series, DenseNet series, and MobileNet series, and these model networks can be created with only one line of code. For example:

```
# Load the DenseNet network model, remove the last fully
connected layer, and set the last pooling layer to max pooling
net = keras.applications.DenseNet121(weights=None, include_
top=False, pooling='max')
# Set trainable to True, i.e. DenseNet's parameters will be
updated.
net.trainable = True
newnet = keras.Sequential([
    net, # Remove last layer of DenseNet121
    layers.Dense(1024, activation='relu'), # Add fully
    connected layer
    layers.BatchNormalization(), # Add BN layer
    layers.Dropout(rate=0.5), # Add Dropout layer
    layers.Dense(5) # Set last layer node to 5 according to
    output categories
])
newnet.build(input_shape=(4,224,224,3))
newnet.summary()
```

The DenseNet121 model is used to create the network. Since the output node of the last layer of DenseNet121 is designed to be 1000, we remove the last layer of DenseNet121 and add a fully connected layer with the number of output nodes of 5 according to the number of categories of the customized dataset. The whole setup is repackaged into a new network model through Sequential containers, where include_top=False indicates that the last fully connected layer is removed, and pooling='max' indicates that the last Pooling layer of DenseNet121 is designed as Max Polling. The network model structure is shown in Figure 15-4.

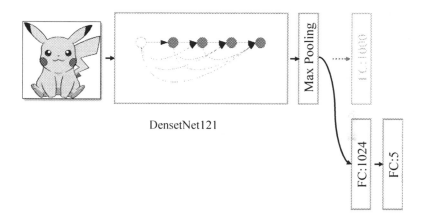

Figure 15-4. *Model structure diagram*

15.3.4 Network Training and Testing

We directly use the Compile&Fit method provided by Keras to compile and train the network. The optimizer uses the most commonly used Adam optimizer, the error function uses the cross-entropy loss function, and sets from_logits=True. The measurement index that we pay attention to during the training process is the accuracy rate. The network model compile code is as follows:

```
# Compile model
newnet.compile(optimizer=optimizers.Adam(lr=5e-4),
               loss=losses.CategoricalCrossentropy(from_
               logits=True),
               metrics=['accuracy'])
```

Use the fit function to train the model on the training set. Each iteration of Epoch tests a validation set. The maximum number of training Epochs is 100. In order to prevent overfitting, we use early stopping technology, and pass early stopping into the callbacks parameter of the fit function as in the following:

```
# Model training, support early stopping
history  = newnet.fit(db_train, validation_data=db_val,
validation_freq=1, epochs=100,
          callbacks=[early_stopping])
```

where early_stopping is the standard EarlyStopping class. The indicator it monitors is the accuracy of the validation set. If the measurement result of the validation set does not increase by 0.001 for three consecutive times, the EarlyStopping condition is triggered and the training ends.

```
# Create Early Stopping class
early_stopping = EarlyStopping(
    monitor='val_accuracy',
    min_delta=0.001,
    patience=3
)
```

We draw the training accuracy rate, validation accuracy rate, and the accuracy rate obtained on the final test set in the training process as a curve, as shown in Figure 15-5. It can be seen that the training accuracy rate has increased rapidly and maintained at a high state, but the validation accuracy rate is relatively lower, and at the same time, it has not been greatly improved. The early stopping condition is triggered, and the training process is quickly terminated. The network has a little bit overfitting problem.

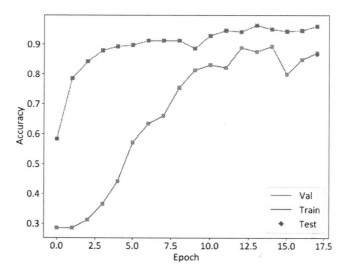

Figure 15-5. *Training DenseNet from random initialization*

So why does overfitting occur? The number of layers of the DensetNet121 model has reached 121, and the number of parameters has reached 7 million, which is a large network model, while our dataset has only about 1,000 samples. According to experience, this is far from enough to train such a large-scale network model, and it is prone to overfitting. In order to reduce overfitting, a network model with a shallower number of layers and fewer parameters can be used, or regularization items can be added, or even the size of the data set can be increased. In addition to these methods, another effective method is transfer learning technology.

15.4 Transfer Learning

15.4.1 Principles of Transfer Learning

Transfer learning is a research direction of machine learning. It mainly studies how to transfer the knowledge learned on task A to task B to improve the generalization performance on task B. For example, task

A is a cat and dog classification problem, and a classifier needs to be trained to better distinguish pictures of cats and dogs, and task B is a cattle and sheep classification problem. It can be found that there is a lot of shared knowledge in task A and task B. For example, these animals can be distinguished from the aspects of hair, body shape, shape, and hair color. Therefore, the classifier obtained in task A has mastered this part of knowledge. When training the classifier of task B, you don't need to start training from scratch, instead you can train or fine-tune the knowledge obtained on task A, which is very similar to the idea of "standing on the shoulders of giants." By transferring the knowledge learned on task A, training the classifier on task B can use fewer samples and lower training costs, and obtain good performance.

We introduce a relatively simple, but very commonly used transfer learning method: network fine-tuning technology. For convolutional neural networks, it is generally believed that it can extract features layer by layer. The abstract feature extraction ability of the network at the end of the layer is stronger. The output layer generally uses the fully connected layer with the same number of output nodes as the classification network as the probability distribution prediction. For similar tasks A and B, if their feature extraction methods are similar, the previous layers of the network can be reused, and the following layers can be trained from scratch according to specific task settings.

As shown in Figure 15-6, the network on the left is trained on task A to learn the knowledge of task A. When migrating to task B, the parameters of the early layers of the network model can be reused, and the later layers can be replaced with new networks and start training from scratch. We call the model trained on task A a pre-trained model. For image classification, the model pre-trained on the ImageNet dataset is a better choice.

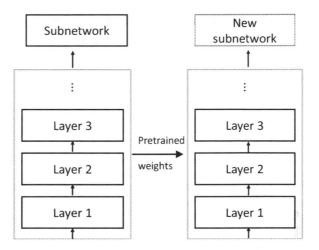

Figure 15-6. *Diagram of neural network transfer learning*

15.4.2 Hands-On Transfer Learning

Based on DenseNet121, we initialize the network with the model parameters pre-trained on the ImageNet dataset, remove the last fully connected layer, add a new classification sub-network, and set the number of output nodes in the last layer to 5.

```
# Load DenseNet model, remove last layer, set last pooling
layer as max pooling
# Initilize with pre-trained parameters
net = keras.applications.DenseNet121(weights='imagenet',
include_top=False, pooling='max')
# Set trainable to False, i.e. fix the DenseNet parameters
net.trainable = False
newnet = keras.Sequential([
    net, #  DenseNet121 with last layer
    layers.Dense(1024, activation='relu'), # Add fully
    connected layer
    layers.BatchNormalization(), # Add BN layer
```

```
    layers.Dropout(rate=0.5), # Add Dropout layer
    layers.Dense(5) # Set the nodes of last layer to 5
])
newnet.build(input_shape=(4,224,224,3))
newnet.summary()
```

When the preceding code creates DenseNet121, the pre-trained DenseNet121 model object can be returned by setting the weights='imagenet' parameter, and the reused network layer and the new sub-classification network are repackaged into a new model newnet through the Sequential container. In the fine-tuning stage, the parameters of the DenseNet121 part can be fixed by setting net.trainable = False, that is, the DenseNet121 part of the network does not need to update the parameters, so only the newly added sub-classification network needs to be trained, which greatly reduces the amount of parameters actually involved in training . Of course, you can also train all parameters like a normal network by setting net.trainable = True. Even so, because the reused part of the network has initialized with a good parameter state, the network can still quickly converge and achieve better performance.

Based on the pre-trained DenseNet121 model, we plot the training accuracy, validation accuracy, and test accuracy in Figure 15-7. Compared with training from scratch approach, with the help of transfer learning, the network learns much faster and only needs a few samples to achieve better performance, and the improvement is very significant.

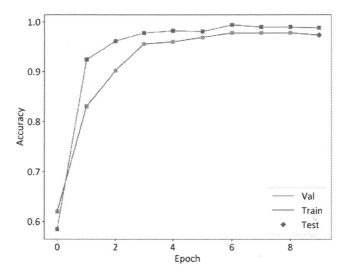

Figure 15-7. *Training DenseNet from pre-trained ImageNet weights*

At this point, you have reached the end of this book. However, your machine learning journey just gets started. Hope this book can help you as one of the reference books during your research or work!

15.5 Summary

Congratulations! You have come a long way to learn both the theories and implementations of deep learning using the popular deep learning framework – TensorFlow 2. Now you should be able to not only understand the fundamental principles of deep learning, but also develop your own deep learning models using TensorFlow 2 to solve real-world problems. For real-world applications, good models are not enough. We need reliable operational systems to consistently produce high-quality model results. This is very challenging given that real-world data changes all the time and often contain noises or errors. Therefore, a reliable machine learning operational system requires a robust data processing pipeline, real-time model performance monitoring, and appropriate mechanisms to retrain or switch models, which leads us to the concept of machine learning operations (MLOps). For readers who are interested in learning more about MLOps and keeping up to date with the latest applications and development of deep learning, deeplearning.ai provides a lot of good resources and courses along with its weekly newsletter – *The Batch*. Hope this book brings you to your own fun journey of deep learning and boosts your career and life!

Index

A

A3C algorithm, 664, 665, 671
Accuracy metric, 306, 307
Activation function
    leaky ReLU, 210, 211
    ReLU, 208–210
    Sigmoid, 206, 207
    Tanh, 212
Actor-Critic method, 628, 662
    A3C, 664, 665
    advantage, 663
    agent, 671, 672
    code, 666, 667
    error calculation, 669
Actor network, 633, 662, 663,
        665, 667
agent.optimizer() function, 636
AlexNet, 5, 10, 11, 14, 15
Amazon's Mechanical Turk
        system, 519
Artificial general intelligence
        (AGI), 2
Artificial intelligence (AI), 3
    definition, 2
    implementation, 2
    stages, 2, 3
Atari game environment, 604, 653

Atari game
        platform, 11, 16, 655
Augmented reality (AR), 676
Autoencoder
    adding noise, 533
    calculation method, 522
    data representation, 522
    denoising diagram, 533
    mapping relationship, 520, 521
    and neural network, 522
    neural network
        parameterization, 521
    optimization goal, 522
    vs. PCA, 522, 523
    training process, 528
    variants
        adversarial autoencoder,
            534, 535
        dropout autoencoder, 534
Autonomous driving, 1, 21

B

Back propagation algorithm
    activation function, derivative
        LeakyReLU function,
            245, 246
        ReLU function, 244, 245

© Liangqu Long and Xiangming Zeng 2022
L. Long and X. Zeng, *Beginning Deep Learning with TensorFlow*,
https://doi.org/10.1007/978-1-4842-7915-1

D

W, X, Y, Z

Printed in the United States
by Baker & Taylor Publisher Services